IET POWER AND ENERGY SERIES 67

Multicore Simulation of Power System Transients

Other volumes in this series:

Volume 1	**Power circuit breaker theory and design** C.H. Flurscheim (Editor)
Volume 4	**Industrial microwave heating** A.C. Metaxas and R.J. Meredith
Volume 7	**Insulators for high voltages** J.S.T. Looms
Volume 8	**Variable frequency AC motor drive systems** D. Finney
Volume 10	**SF6 switchgear** H.M. Ryan and G.R. Jones
Volume 11	**Conduction and induction heating** E.J. Davies
Volume 13	**Statistical techniques for high voltage engineering** W. Hauschild and W. Mosch
Volume 14	**Uninterruptible power supplies** J. Platts and J.D. St Aubyn (Editors)
Volume 15	**Digital protection for power systems** A.T. Johns and S.K. Salman
Volume 16	**Electricity economics and planning** T.W. Berrie
Volume 18	**Vacuum switchgear** A. Greenwood
Volume 19	**Electrical safety: a guide to causes and prevention of hazards** J. Maxwell Adams
Volume 21	**Electricity distribution network design, 2nd edition** E. Lakervi and E.J. Holmes
Volume 22	**Artificial intelligence techniques in power systems** K. Warwick, A.O. Ekwue and R. Aggarwal (Editors)
Volume 24	**Power system commissioning and maintenance practice** K. Harker
Volume 25	**Engineers' handbook of industrial microwave heating** R.J. Meredith
Volume 26	**Small electric motors** H. Moczala *et al.*
Volume 27	**AC-DC power system analysis** J. Arrillaga and B.C. Smith
Volume 29	**High voltage direct current transmission, 2nd edition** J. Arrillaga
Volume 30	**Flexible AC Transmission Systems (FACTS)** Y-H. Song (Editor)
Volume 31	**Embedded generation** N. Jenkins *et al.*
Volume 32	**High voltage engineering and testing, 2nd edition** H.M. Ryan (Editor)
Volume 33	**Overvoltage protection of low-voltage systems, revised edition** P. Hasse
Volume 34	**The lightning flash** V. Cooray
Volume 36	**Voltage quality in electrical power systems** J. Schlabbach *et al.*
Volume 37	**Electrical steels for rotating machines** P. Beckley
Volume 38	**The electric car: development and future of battery, hybrid and fuel-cell cars** M. Westbrook
Volume 39	**Power systems electromagnetic transients simulation** J. Arrillaga and N. Watson
Volume 40	**Advances in high voltage engineering** M. Haddad and D. Warne
Volume 41	**Electrical operation of electrostatic precipitators** K. Parker
Volume 43	**Thermal power plant simulation and control** D. Flynn
Volume 44	**Economic evaluation of projects in the electricity supply industry** H. Khatib
Volume 45	**Propulsion systems for hybrid vehicles** J. Miller
Volume 46	**Distribution switchgear** S. Stewart
Volume 47	**Protection of electricity distribution networks, 2nd edition** J. Gers and E. Holmes
Volume 48	**Wood pole overhead lines** B. Wareing
Volume 49	**Electric fuses, 3rd edition** A. Wright and G. Newbery
Volume 50	**Wind power integration: connection and system operational aspects** B. Fox *et al.*
Volume 51	**Short circuit currents** J. Schlabbach
Volume 52	**Nuclear power** J. Wood
Volume 53	**Condition assessment of high voltage insulation in power system equipment** R.E. James and Q. Su
Volume 55	**Local energy: distributed generation of heat and power** J. Wood
Volume 56	**Condition monitoring of rotating electrical machines** P. Tavner, L. Ran, J. Penman and H. Sedding
Volume 57	**The control techniques drives and controls handbook, 2nd edition** B. Drury
Volume 58	**Lightning protection** V. Cooray (Editor)
Volume 59	**Ultracapacitor applications** J.M. Miller
Volume 62	**Lightning electromagnetics** V. Cooray
Volume 63	**Energy storage for power systems, 2nd edition** A. Ter-Gazarian
Volume 65	**Protection of electricity distribution networks, 3rd edition** J. Gers
Volume 905	**Power system protection**, 4 volumes

Multicore Simulation of Power System Transients

Fabian M. Uriarte

The Institution of Engineering and Technology

Published by The Institution of Engineering and Technology, London, United Kingdom

The Institution of Engineering and Technology is registered as a Charity in England & Wales (no. 211014) and Scotland (no. SC038698).

First published 2013

The Institution of Engineering and Technology
Michael Faraday House
Six Hills Way, Stevenage
Herts, SG1 2AY, United Kingdom

www.theiet.org

British Library Cataloguing in Publication Data
A catalogue record for this product is available from the British Library

ISBN 978-1-84919-572-0 (hardback)
ISBN 978-1-84919-573-7 (PDF)

Typeset in India by MPS Limited
Printed in the UK by CPI Group (UK) Ltd, Croydon

Dedication

To God
and
my wife Veronica, daughter Valentina,
father J. Manuel, mother Angela, and brother Ivan

Contents

List of tables

List of figures

List of snippets

About the author

Fabian Marcel Uriarte holds a PhD in Electrical Engineering from Texas A&M University at College Station in the area of parallel power system simulation, and an MS and BS in Electrical Engineering from Virginia Polytechnic Institute and State University in the area of power systems. He is currently a power systems simulation specialist at the Center for Electromechanics (CEM) of The University of Texas at Austin (UT). He is the technical lead for the development of a parallel power system solver in a project sponsored by the Office of Naval Research and Electric Ship Research and Development Consortium.

At CEM, Dr. Uriarte focuses on power systems, microgrid, and Smart Grid modeling and simulation. He brings these techniques to bear on projects ranging from compute-intense algorithm development and parallelization to modeling and simulation of military and civilian sector power systems. Currently, he leads a research thrust to leverage multicore technology on the supercomputers at UT's Texas Advanced Computing Center to parallelize electromagnetic simulations for a broad range of power systems. His research aims to reduce simulation runtime across a wide range of simulation environments to increase model realm and reduce runtime costs.

Dr. Uriarte is recognized for his expertise in modeling and simulation of Smart Grids at UT. He advises a multi-disciplinary team of faculty and students on computer techniques for assessing the impact of emerging technologies in residential communities. He routinely visits utility companies in Texas and travels throughout the United States and abroad to disseminate his research findings. His research is particularly important in that it contributes to goals set forth by the Departments of Defense and Energy to promote the shift toward emerging technologies on the national grid and in military installations ashore and offshore. This shift imposes larger demands on computing resources than before, which makes Dr. Uriarte's work in parallel simulation especially opportune.

Contacting the author

The author welcomes constructive comments aimed at reducing the boundaries of knowledge or at improving the quality and impact of his work. He also invites intellectual discussions and is interested in scholarly collaboration and coauthoring with specialists in power system simulation.

Due to current proprietary restrictions, the multicore solver developed for this book is not currently available to the public. If the multicore solver and other useful material become available, they will be posted at the author's website at: https://sites.google.com/site/fabianuriarte/

Fabian M. Uriarte
f.uriarte@cem.utexas.edu

Foreword

It is rewarding to be associated with a field for a long enough time to witness the evolution of technology. The field of power system modeling has evolved from reliance on analog simulators to full-scale adoption of digital simulation. This transition has benefited greatly by the continuous improvement that semiconductor manufacturers have made in processor speeds. Early in the use of digital simulation, reduced order models were used to compensate for the relatively slow processing speed.

But viewing the progress from a high level, symbiotic progress was made for decades. As the processing time became faster, increasingly complex power systems were being modeled. Yesterday's reduced order models could be solved exactly in convenient times today. And today's reduced order models would become less necessary as processing speeds increased.

While the processing is changing, there are also changes in the power systems of interest. The transition from a relatively passive analog grid to a highly instrumented, digital, Smart Grid has accelerated over the last decade. The Smart Grid helps make the transmission system more reliable by enhancing situational awareness. But it also opens, in combination with distributed generation, the possibility of microgrids that can be intentionally islanded to enhance local reliability.

The Smart Grid activity worldwide has accelerated the growth of microgrids. Microgrids produce a new demand for different types of modeling. Conventional load flow modeling is still needed to understand overall system performance. But this is augmented by a need for improved transient analysis to characterize the influence of power electronics in a smarter grid and to assess things like reliability and insulation coordination in microgrids. The transient behavior of an isolated microgrid can be different from the transient behavior when the same loads are supplied from the regional grid. A leading example of the growth of microgrids is the increasing number of electric ships. These systems are extreme examples of isolated microgrids and require advanced modeling today.

This change in modern power systems has led to an increasing need for accurate transient analysis. At the same time, in the computer world, the rate of processor speed increases has become too expensive to maintain. To achieve increased speed, desktop and laptop computers now come with multiple processors, just like supercomputers, so that different parts of the problem can be solved in parallel by many processors.

An important challenge, however, is that the preponderance of power system simulation software is not written to be partitioned automatically to take advantage of the number of processors available. This is the change that is occurring now. It is likely to be as significant a change as the transition from analog to digital simulation. If it can be done well and automatically, the decreasing time required for accurate simulation will permit enhanced simulation capability to underpin the augmentation of the current grid with versatile microgrids.

Dr. Uriarte is one of the pioneers in this emerging field. This book captures the important issues in automated partitioning of power systems, which is likely to be the cornerstone on which the next generation of power system simulation will be built. As such, it can be an important resource for those active in power system simulation who want to preview how the field is changing, as well as for those who are leading the change.

Dr. R. E. Hebner, FIEE
IEEE Vice-President of Technical Activities
Director, Center for Electromechanics
The University of Texas at Austin, 2013

Preface

Multicore technology has orchestrated the redesign of simulation approaches throughout the software industry–in particular, the redesign of power system solvers to exploit concurrency. The current adoption of multicore technology in power system simulation is not apparent, but it is growing in importance due to the wide availability of multicore personal computers and in response to the many-core shift. This book is therefore a timely effort to explain how to parallelize long-running power system simulations using existing desktop computer technology.

This book is written for electrical engineers seeking to develop a parallel power system solver using a multicore computer running *Windows* as the operating system. Industry professionals, software developers, scientists, and graduate students will find this writing of particular interest. Building on the numerous good books on power system simulation, this writing addresses the topic of partitioning and evaluates runtime as a power system model is partitioned numerous times. Existing publications on power system simulation often focus on three techniques: (1) the use of node voltages as variables, (2) the formulation of network equations using a single matrix, and (3) simulation based on small models. The first two are in use since the beginning of power system simulation, prior to the arrival of multicore technology. The use of small power system models stands to reason: they are easy to elucidate, understand, and present little room for failure; however, many times they are not realistic. Although these three techniques remain in use today, they are not always the most effective way to simulate power systems—especially in light of the wide availability of multicore computers today.

This book provides a fresh perspective on power system simulation to embrace multicore technology. It differs from existing literature in a few ways: (1) the use of mesh currents *in addition to* node voltages as variables, (2) the formulation and simulation of *partitioned* power system equations, (3) the runtime comparison against a *commercial* tool, and (4) the demonstration of the methods presented on a *large* power system model. This book demonstrates that mesh formulations have comparable sparsity to nodal formulations and that simulating power systems as subsystems results in noticeable performance improvements. The use of a large power system model to demonstrate the parallel simulation methodology reveals that the methodology is not limited to academic examples, and that it has utility for larger, more realistic models.

While the multicore simulation approach presented by this book is not suitable for all simulation scenarios, under certain conditions, the methods presented herein yield simulation speedups near two orders of magnitude. Lastly, this book will describe conditions when using the presented parallel simulation methodology is advantageous and when it is not.

Fabian M. Uriarte
Center for Electromechanics
The University of Texas at Austin, 2013

Acknowledgments

The following people have significantly contributed, guided, or funded the development of this work. The author acknowledges that this work would not have been possible without them. My many thanks to all of them.

- J. Manuel Uriarte (Universidad Nacional Mayor de San Marcos,[1] Lima, Peru)
- Robert E. Hebner (Center for Electromechanics, University of Texas at Austin)
- John Herbst (Center for Electromechanics, University of Texas at Austin)
- Angelo Gattozzi (Center for Electromechanics, University of Texas at Austin)
- Terry Ericsen (Office of Naval Research, Arlington, Virginia)
- Karen L. Butler-Purry (Texas A&M University at College Station)
- Karen L. Miu (Drexel University, Philadelphia)
- Noel Shultz (Kansas State University, Manhattan, Kansas)
- César Malavé (Texas A&M University at College Station)
- Salman Mashayekh (Texas A&M University at College Station)
- Virgilio Centeno (Virginia Tech, Blacksburg, Virginia)
- Neal Kegley (Virginia Tech, Blacksburg, Virginia)
- Benjamin Cormier (British Petroleum, Houston, Texas)
- Carter A. Hunt (Pennsylvania State University, University Park, Pennsylvania)
- Luis Mattos (Dell Services, Fairfax, Virginia)
- Laurence Thomas (Lauren Engineers & Associates, Abilene, Texas)
- Neville Watson (University of Canterbury, New Zealand)

Many others have also contributed in person, through telephone, and electronic communication by providing exemplary advice, acting as role models, giving moral support, encouragement, and guidance to develop and complete this manuscript:

Alessio Cacciatori, University of Brescia, Italy; Ali Abur, Northeastern University; Anne Ocheltree, Department of State, Washington, DC; Ayorinde Akinnikawe, Lauren Engineers & Associates; Blake Langland, University of South Carolina; Chenyan Guo, Texas A&M University at College Station; Cinan Singh, Texas A&M University at College Station; Corey Reed, Virginia Tech; Guillermo Andrés Jiménez Tuto, Universidad de Chile; Hamid Toliyat, Texas A&M University at College Station;

[1] Translated as San Marcos Major National University (UNMSM). San Marcos University is the oldest university of the Americas (1551). Prof. J. M. Uriarte, graduated from The American University, Washington, D.C., authored *Transnational Banks and the Dynamics of the Peruvian Foreign Debt and Inflation*, New York, Praeger Publishers Ltd., 1987.

Hamed Funmilayo, Centerpoint Energy; Hung-ming Chou, Texas A&M University at College Station; Jaime De Laree, Virginia Tech; Jim Glanville, Virginia Tech; Jost Allmeling, Plexim GmbH; Juancarlo Depablos, Virginia Tech; Kai Strunz, Technische Universitat Berlin, Germany; Mario Ríos, Universidad de Los Andes, Bogotá, Colombia; Mirrasoul Mousavi, ABB (ASEA Brown Boveri); Mischa Steurer, Florida State University; Natalia Parada, Universidade Estadual de Campinas, Brazil (UNICAMP); Nour Dib, Virginia Tech, Blacksburg, Virginia; Om Nayak, Nayak Corp.; Peng Li, Texas A&M University at College Station; Trevor Misfeldt, CenterSpace Software, Inc.

Fabian M. Uriarte

Chapter 1
Introduction

Electromagnetic transient simulation[2] is required to assess transient stress in advance of and after the installation of electric power systems. This type of simulation, however, is notoriously slow, limits the number of case studies that can be run in a day, and can consume excessive man-hours, machine hours, and other research resources. There are several causes for such long-running simulations: the simulation timestep, the simulation stop time, the model order and complexity, finite computer resources, programming efficiency, and the sequential (i.e., non-parallel) nature of the solution.

Although parallel solutions are important, there are workarounds to the problem of long-running simulations. Most of these workarounds, however, trade accuracy for performance (e.g., the use of reduced-order models or larger simulation timesteps, among other approaches). These approaches allow users to run more case studies per day, but many times obfuscate transient detail and do not capture system-wide transient phenomena. In fact, they are sought out of necessity.

The inability to obtain timely simulations of large power systems models and the recent ubiquity of multicore technology has prompted the scientific community to re-visit orthodox simulation techniques. This book demonstrates a methodology to reduce the runtime of power system transient simulations by parallelizing the solution on multicore desktop computers.[3]

1.1 Scope and purpose

This book discusses the development of a proprietary multicore power system solver. This development covers discretization of power apparatus models, power system partitioning, subsystem formulations, and relevant details of a multithreaded implementation. Although the scope of the book is demonstrated for one power system model (and its sub-variants), the methodology presented is, generally, model agnostic.

[2] This simulation type is also known as time domain, waveform-level, or transient simulation.

[3] The methodology is presented using a quad-core computer running *Windows* as the operating system, but other multicore environments are suitable as well.

Readers are apprised of the following cautionary detail. This book is not about power system *modeling*. It is about power system *partitioning*. With this proviso in mind, it is underlined that readers will *not* find details on advanced power apparatus models, case studies on disturbances, or what is readily available in existing books on power system simulation. Instead, readers will learn the mathematics necessary to develop their own power system simulator; they will also learn how to formulate power system equations for time domain simulation, and how to partition these equations to solve them on a multicore computer running *Windows* as the operating system.

It is assumed that readers are looking for a well-documented technique to partition and parallelize power system simulations and for a method independent of power apparatus complexity. Consistent with this scope, treatment of complex power apparatus models such as machines, transmission lines, or control networks is elided. However, the exclusion (or inclusion) of such models does *not* change the partitioning method presented in this book. The concepts presented in this book are extensible to a wide range of power system models, and they are not limited to the ones presented herein. This implies that the programs readers may develop after reading this book will be applicable to a wide range of power system models.

There are many partitioning methods[4] that reduce the runtime simulation of power system transients as well. And in many cases, such methods may be better suited for some simulation scenarios than the method presented here. In all cases, however, readers should pay close attention to the demonstration of partitioning methods (including the ones presented herein) on well-behaved cases as not all partitioning methods perform equally well across different simulation scenarios. Additionally, readers are encouraged to always pay close attention to the *size* of the model being partitioned.

The purpose of this book is to present and demonstrate an alternative simulation approach to simulate power system transients. Toward this end, the approach adopted is tailored to multicore computers, and its usefulness is justified by comparing the runtime against a widely-used commercial simulation tool. Readers investing resources in the development of a multicore solver should be aware that such development draws expertise from two (unrelated) engineering areas: power system engineering and software engineering. Bridging these two areas often requires the collaboration of a team of professionals to complete the development in acceptable time.

Moreover, readers implementing the methodology presented by this book will likely observe different speedup results than the ones presented here due to differences in software design, programming efficiency, and the choice of numerical library.[5] The decision to invest resources in the development of a multicore solver as the one presented in this book must be carefully examined. Such development may require several man-years to reach acceptable usability and competency.

[4] Different than the one presented in this book.

[5] These facts are important to performance and should not be overlooked.

1.2 Assumed background

For pedagogical considerations, it is assumed that readers are planning, or are in the development stage of a multicore power system solver for *Windows* machines. To develop a solver requires background in electric power systems, matrix algebra, graph theory, programming, and simulation theory.

The topics on power systems in this book are introduced without major explanations. Instead, they are primarily introduced with a focus on implementation rather than on their physical derivation, justification, or meaning. The topics on matrix algebra assume readers are familiar with transformations, matrix and vector notations, algebraic manipulation, matrix properties, and that readers can formulate nodal and mesh matrices by visual inspection of simple circuits. The topics on partitioning assume readers understand circuit theory and can identify independent variables in voltage and current equations. The topics on graph theory assume readers understand basic graph concepts. The topic on programming assumes readers understand the definition of a thread. The topics in electromagnetic transient simulation theory assume readers have read existing books on simulation theory. Throughout the book, literature references and their available sources are provided for readers seeking additional background on all topics presented.

1.3 Contributions

The contributions of this book include: illustrating a method to formulate sparse mesh equations without using graph theory; partitioning power systems by node and mesh-tearing instead of by diakoptics; showing the runtime reductions of the nodal and mesh methods against each other and against a commercial tool; the treatment of a power system model large enough to demonstrate when partitioning is useful and when it is not; and a demonstration of potential speedups for an ordinary *Windows* desktop machine (instead of dedicated machines running alternative operating systems).

On forming sparse matrices, this book addresses two common misconceptions: that mesh analysis is unsuitable for non-planar networks, and that graph theory is required to form mesh resistance matrices. Both of these misconceptions are refuted here by showing how the mesh resistance matrix for a non-planar power system can be formed without resorting to graph theory.

On partitioning without diakoptics, it will be shown that boundary branches are *not* required to tear a model as required by diakoptics. Removal of this restriction permits partitioning (segregating) power systems as fine-grained as one power apparatus per subsystem. Removal of this limitation increases the number of possible disconnection points, increases the maximum-possible number of partitions, and reduces programming complexity.

On runtime reductions, a performance comparison between the nodal and mesh methods under one book cover is rare—especially, in the context of parallel simulation. An additional frequent misconception is that the nodal equations are always the best formulation candidate for simulation. While in several cases this is true, it is usual to overlook how the nodal method scales in partitioned models.

The current literature on partitioning methods is rich and varied, but it often supplies readers with runtime comparisons against the same authors' unpartitioned runtime. While these comparisons are utterly useful and important, they often do not inform readers how well commercial tools perform in the same scenarios. This book provides, for all models treated, runtime comparisons against a commercial tool to show how speedups vary as the number of partitions increases on models of different sizes.

On model size, it is common to demonstrate partitioning methods on small power system models. While several other partitioning methods are valid for such cases, such methods may not scale well with model order and complexity. This book presents a partitioning approach and demonstrates it on a large-order power system.[6] Additionally, and for completeness, the partitioning method presented here is tested against several smaller variants of the *same* power system model to show how (if) the partitioning approach scales with system order.

1.4 Statement of the problem and hypothesis

The inability to perform *timely* transient simulations of large power system models on *Windows*-based desktop computers provides a rationale for this book. This book is written in an effort to attain an important goal, which primarily centers on the ability to produce electromagnetic transient simulations of large-order power system models in reasonable (run) time by using *existing* resources.[7]

The principal reasons for long-running power system simulation were stated early in the chapter. The hypothesis of this book is that low-order, sparse, and partitioned simulations implemented (properly) on a multicore desktop computer can significantly reduce the runtime of power system simulations without having to acquire specialized hardware and software. Considering that faster simulations of power systems are highly sought across the research community, this works attempts to explain and demonstrate that the stated working proposition holds true and that in this course of action such primary goal can be achieved.

1.5 Organization

This book is organized into ten chapters. Chapter 2 introduces a notional power system (and its variants) treated specifically for the development of this book to demonstrate power system partitioning and parallelization.

[6] *Large* is a subjective perception on a unit of measure. This term is specifically defined in Chapter 2.
[7] Refers to the computing resources likely owned by the reader (e.g., a multicore laptop or desktop computer).

Chapter 3 introduces fundamental concepts (and notations) of time domain simulation and explains, in the general sense, why simulations are and can be lengthy.

Chapter 4 covers discretization techniques, which explains how continuous power system models are prepared for discrete computer simulation.

Chapter 5 introduces the power apparatus and control blocks included in the notional power system model.

Chapter 6 explains how to formulate the coefficient matrices for the electrical network. The formation of these matrices resorts to multi-terminal component theory in order to provide a general approach that is independent of power apparatus type.

Chapter 7 elucidates how to partition a power system model. Diakoptics is presented first to establish the partitioning basis from which node and mesh tearing are subsequently derived.

Chapter 8 discusses basic concepts of multithreaded programming and shows how to implement a program similar to the one developed for this book.

Chapter 9 benchmarks the multicore solver against various performance metrics. The analysis of each model variant encompasses concluding remarks on how the partitioning method performed on models of different sizes.

Chapter 10 summarizes and concludes the book with important general observations; the appendices and references follow it.

As noticed from the Table of Contents, most chapters are accompanied with a relevant summary at the end. These summaries include observations related to the material contained in their chapters. The two appendixes included in the book also have relevance to the text and complement the discussions and figures furnished in their related chapters.

Lastly, it should be pointed out that research in parallel time domain simulation is not finished; it has just started. At the time of this writing, available desktop and laptop computers ship with multicore processors having two or four physical cores.[8] The development of multicore solvers is gaining attraction in anticipation of the many-core shift, where all desktop computers (and net books, laptops, phones, and tablets) are expected to include many cores as their standard configuration in the near future.

[8] Eight-core desktop computers are available today via hyper-threading in processors with four physical cores, or by having two four-core processors on a dual-socket motherboard. Neither of these two options are mainstream, however; thus, they are not considered here.

Chapter 2
The power system model

To understand something, you must be able to hang a number on it.

–*Allan Greenwood*[9]

The simulation methodology presented in this book is demonstrated by parallelizing the simulation of a notional Navy shipboard power system model on a quad-core desktop computer running *Windows* 7. The model presents characteristics of Navy shipboard power systems (AC[10]-radial type), but it does not represent any particular vessel.

Navy shipboard power systems are considerably different from terrestrial power systems, but some technical cross-fertilization has been reported [1]. Shipboard power systems are megawatt-level floating microgrids having strict requirements for redundancy, reliability, and service quality that can exceed those of terrestrial grids. Navy shipboard power systems have been the investigation focus of expert K. L. Butler-Purry of Texas A&M University[11] for nearly 20 years of rigorous research. Her expertise includes power distribution automation, catastrophic failure identification, intelligent reconfiguration, geographic information systems for distribution systems, predictive reconfiguration, automated damage control, distributed power management, battle damage assessment, dynamic balancing, prognosis of aging power apparatus, fault studies, load flow analysis, transient simulation, fast and real time simulation, hardware in the loop control, among other areas. Her important contributions to these areas are available in References 2–25.

The information in the references above and in References 26–30 was used to construct the notional Navy shipboard power system model in *MATLAB/Simulink* [31] using the *SimPowerSystems* blockset.[12] This power system model, subsequently, was imported into the multicore solver program developed for this book. The reasons to

[9] Quote taken from *Electrical Transients in Power Systems*, 2nd ed., New York, Wiley & Sons, 1987, page 11, textbook authored by Allan Greenwood. This quotation motivated ways to quantify the size of the power system models presented in this chapter—more specifically, ways to define the term *large*.
[10] Alternating current. DC is direct current.
[11] Power System Automation Lab: http://psalserver.tamu.edu/.
[12] *Simulink* refers to the *MATLAB/Simulink* environment, whereas *SimPowerSystems* refers specifically to the blockset and its features.

construct the power model in *Simulink* first were to compare its runtime performance against the multicore solver developed for this book and to provide readers with a runtime reference on how the model performs on an established simulation platform readers are likely to have.

This chapter introduces the notional power system model used throughout the book; the chapter also characterizes the model size and complexity, and presents (three) less complex variants of the same model. Using less complex variants of an intended model is common practice in the simulation field to reduce design time, speed up simulations, attain reasonable confidence early in the design phase, and—most importantly—to be able to *revert* to simpler variants when inexplicable results arise from complex models. Additionally, the reduction of design time by using simpler model variants allows its users to *incrementally* introduce model complexities and run tests without incurring significant wait times.[13]

2.1 Power system model

The one-line (or single-line [32]) diagram of the notional shipboard power system model is shown in Fig. 2.1; its base quantities are listed in Table 2.1. It should be noted this power system model represents the ship-service side of a notional electromechanical shipboard power system, and it is *not* related to an integrated power system [33] or all-electric ship [33].

Three-phase power is generated at 450 V (AC, line-to-line, RMS[14]) at 60 Hz and distributed radially from the generators to the ring bus, load centers, and loads served from the mains. Although there are more than two generators on shipboards, only two are considered to be online. The installed generation capacity across both generators is 5 MW, which (similar to terrestrial microgrids [1]) are power systems of finite inertia offering enhanced energy security [34]. To increase the reliability of service, electric power is distributed via a three-wire, delta-ungrounded power distribution network.

The power apparatus types and count for the notional model are listed in Table 2.2 and arranged graphically in a bar chart by count in Fig. 2.2. As noticed, this power system model comprises 369 power apparatus and is significantly larger than those used in academic examples.[15] Due to observed component count, power system simulators experience an added computational burden when simulating this type of model.

Similarly, the bus types and their corresponding counts are listed in Table 2.3 and arranged graphically in a bar chart by count in Fig. 2.3. Table 2.3 shows that this power system model comprises nearly 300 buses, pre-dominantly three-phase AC.

[13] This approach is a "bottom-up" one, which is common practice in the simulation community.

[14] RMS is the abbreviation of root mean square, which is the square root of the arithmetic mean of the squares of a number set. This concept is further explained in Chapter 4.

[15] The models of each power apparatus will be presented in Chapter 5.

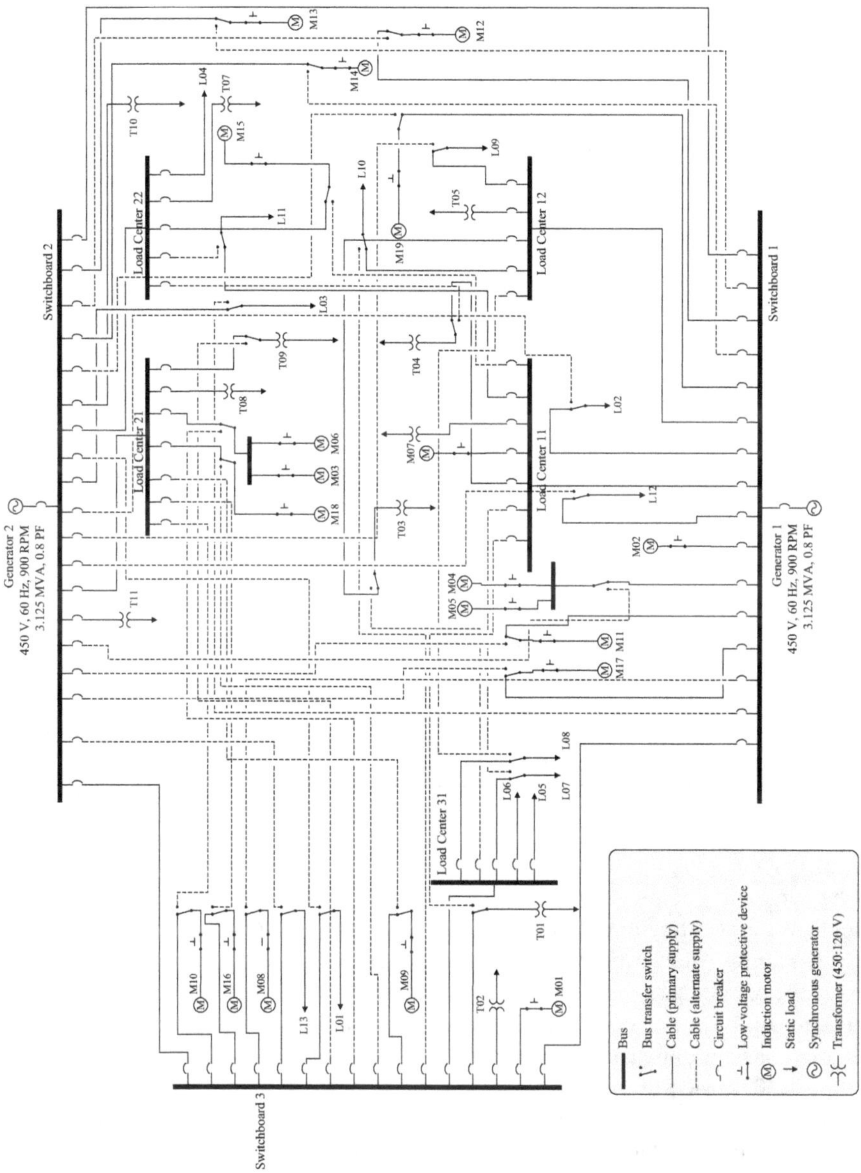

Fig. 2.1 Notional power system model used in this book to demonstrate parallelization[16]

[16]This is the principal model developed for the book, and it will be referred to as *System* 4. Black bars identify buses interconnecting three or more power apparatus. Buses interconnecting two power apparatus are implicit.

Table 2.1 Power system base quantities

Description	Value	Unit
Base power	3.125	MVA
Base line voltage	450	Volts (RMS)
Base line current	4,009	Amps (RMS)
Base impedance	0.1944	Ohms
Base frequency	60	Hz

Table 2.2 Power apparatus: types and counts

Type	Count
Bus transfers	28
Cables	107
Circuit breakers	83
DC filters	19
Induction motors	19
Inverters	19
Low-voltage protective devices	19
Rotor loads	19
Rectifiers	19
Static loads behind transformers	11
Static loads on mains	13
Synchronous generators	2
Transformers	11
Total power apparatus:	369

The salient complexities of the notional power system under analysis are its bus and power apparatus counts, its equation count (shown later in this chapter), and its 19 motor drives amounting to 342 valves.[17] This large number of valves[18] makes the power system model time-variant, computationally expensive, and a model suitable to demonstrate parallelization on multicore computers.

The loads of the power system model are listed in Table 2.4, which are all three-phase static (constant impedance) loads. Since the focus of the simulation is on the three-phase (450 V) side of the model, all single-phase loads, lighting circuits, and cables were lumped as one three-phase static load behind their respective three-phase distribution transformers. This load aggregation is common practice in simulation [35] to reduce computational burden when the details of the lumped portions are not of

[17] The terms valve and switch will be used interchangeably to refer to power electronic switches.

[18] The valve count excludes switches internal to protective devices.

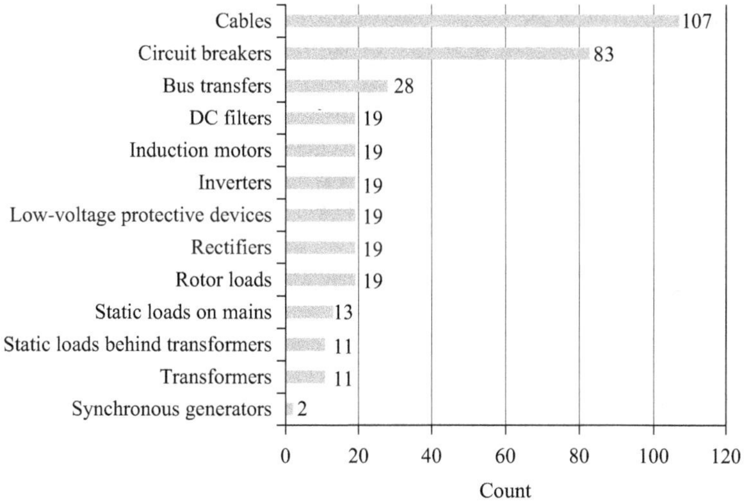

Fig. 2.2 Bar chart of power apparatus types and counts

Table 2.3 Power system bus types

Type	Count
DC	19
AC	277
Total buses:	296

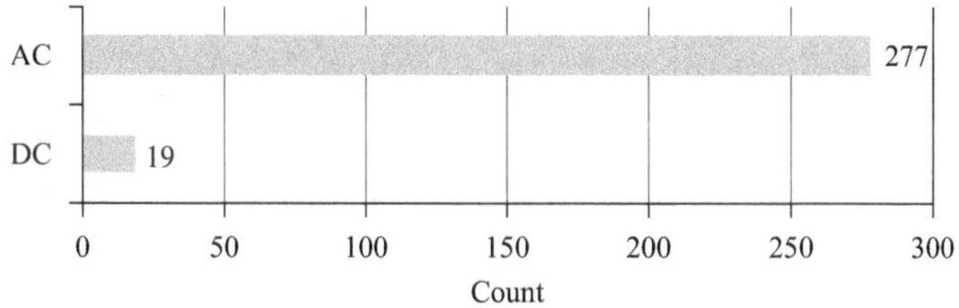

Fig. 2.3 Graphical representation of bus types and their counts

particular interest. The power consumption of these lumped loads was set to 50 kVA,[19] which is the capacity of their upstream transformers (also a common distribution transformer size [36]).

The ratings of the three-phase loads and motors are shown in Tables 2.4 and 2.5, respectively. The bottom row in each table shows the total load: ~1.1 MVA of three-phase loads and ~2 MVA of motor loads. The three-phase static loads are connected directly to the 450 V mains, whereas the motor loads are served from their respective motor drives.

As noticed from Fig. 2.2, there are more cables than any other power apparatus in the notional shipboard power system model. These cables vary in number of conductors, conductor size, insulation, construction, length, and their inclusion is critical to simulation models. Including cables in a simulation model allows assessing voltage drops, power losses, charging currents, and ampacity violations to say the least.

The impedances of most power cables included in the notional power system were taken from Table XIII in References 27, 28 (*LSTSGU* three-conductor, shipboard power cable, 450 V, three-phase, 60 Hz). Most conductor sizes were selected according to the loads they serve. For example, the ampacity of cables upstream of the loads listed in Table 2.4 are approximately 125% [37] of the load's full-load amps (FLA). When ampacity requirements exceed the possible conductor sizes, single-conductor cables of higher ampacity were selected according to Table VI(a)(1) in Reference 28. Cables from switchboards to load centers, between generators and

Table 2.4 Three-phase static loads[20]

Name	kVA	kW	kVar	PF	FLA
L01	6.60	6.60	–	1.00	8.48
L02	6.60	6.60	–	1.00	8.48
L03	6.60	6.60	–	1.00	8.48
L04	11.77	10.00	6.20	0.85	15.10
L05	35.29	30.00	18.59	0.85	45.30
L06	35.29	30.00	18.59	0.85	45.30
L07	38.11	36.20	11.90	0.95	48.90
L08	38.11	36.20	11.90	0.95	48.90
L09	38.11	36.20	11.90	0.95	48.90
L10	70.00	42.00	56.00	0.60	89.90
L11	190.53	181.00	59.49	0.95	245.00
L12	315.79	300.00	98.61	0.95	406.00
L13	333.33	300.00	145.30	0.90	428.00
Total:	1,126.12	1,021.40	438.48		

[19] Each delta phase on the secondary side of the transformers serves a 16.67-kVA load. As a reference of transformer capacity, a 50-kVA single-phase transformer typically serves eight homes in terrestrial residential areas.

[20] The headings mean kVA: total power; kW: real power; kVar: reactive power; PF: power factor; FLA: full-load amps; (–) represents zero.

Table 2.5 Motor loads[21]

Name	HP	PF	Eff. (%)	FLA	RPM	kVA	kW	kVar
M01	5	0.82	88	6.64	3,600	4.55	3.73	2.60
M02	5	0.82	89	6.57	3,600	4.55	3.73	2.60
M03	50	0.87	88	62.58	3,600	42.87	37.30	21.14
M04	60	0.86	88	75.97	3,600	52.05	44.76	26.56
M05	60	0.86	89	75.12	3,600	52.05	44.76	26.56
M06	60	0.86	86	77.74	3,600	52.05	44.76	26.56
M07	60	0.86	95	70.37	3,600	52.05	44.76	26.56
M08	100	0.86	87	128.07	3,600	86.74	74.60	44.27
M09	100	0.86	86	129.56	3,600	86.74	74.60	44.27
M10	150	0.83	89	194.58	3,600	134.82	111.90	75.20
M11	150	0.83	94	184.23	3,600	134.82	111.90	75.20
M12	150	0.83	94	184.23	3,600	134.82	111.90	75.20
M13	150	0.83	89	194.58	3,600	134.82	111.90	75.20
M14	150	0.83	91	190.31	3,600	134.82	111.90	75.20
M15	150	0.83	86	201.37	3,600	134.82	111.90	75.20
M16	250	0.90	92	289.33	3,600	207.22	186.50	90.33
M17	250	0.90	93	286.22	3,600	207.22	186.50	90.33
M18	250	0.90	85	313.15	3,600	207.22	186.50	90.33
M19	250	0.90	90	295.76	3,600	207.22	186.50	90.33
					Totals:	2,071.45	1,790.40	1,033.60

switchboards, and around the ring bus, due to their higher ampacity requirement, were modeled as sufficient single-conductor cables (*LSSSGU* cable type) connected in parallel to meet their required ampacity.

2.2 System size

The power apparatus and bus counts suggest that the power system model under analysis is large. However, the term *large* may have various interpretations. To some, it may mean system capacity (in watts or volt-amperes); to others, it may mean spatial dimensions (in m^2 or m^3). Other interpretations of such term may refer to the number of power apparatus, the number of three-phase buses, the number of single-phase buses, the number of power electronic valves, the number of state-variables, the number of nodal (or mesh) equations, the number of branches, or to the number of non-zeros in the network coefficient matrix or in its factors. Because of the many possible interpretations of the term *large*, different metrics are presented below to characterize the size of the notional power system used in this book.

[21] The headings mean: HP: horsepower; PF: power factor; Eff.: motor efficiency, given in percentage; FLA: full-load amps; RPM: rotor speed in revolutions per minute.

SimPowerSystems includes a netlist[22] utility that can be used to characterize model size. Snippet 2.1 shows *MATLAB* code to create a netlist and display it in *Windows Notepad.*

Snippet 2.1 How to produce a netlist with SimPowerSystems[23]

```
% MATLAB code
clc
disp('Creating netlist...')
power_analyze(gcs, 'net')
system(['notepad ', which([gcs,'.net']), ' &'])
```

Snippet 2.1 was used on the notional shipboard power system model shown in Fig. 2.1, and returned 885 nodes, 1,333 inductors and capacitors, and 702 switches.[24] These values, along with the number of independent state variables, will be represented graphically after introducing the aforementioned power system model variants.

2.3 System variants

The *Simulink* model for the notional power system shown in Fig. 2.1 requires nearly 49 minutes (min) to complete a run when using a stop time of 1 second (s), the backward Euler discrete solver, and a fixed timestep of $\Delta t = 46.296$ μs. This total runtime is acceptable for *one* run, but in practice, it is necessary to run the model hundreds of times during its design and analysis phases.[25] Consequently, waiting 49 min (or more) per run is *not* acceptable.[26]

Due to the inability to obtain timely simulation results for large power system models on ordinary desktop computers, there is merit in creating variants of less order and complexity than the power system model shown in Fig. 2.1. It is common practice to start with simpler models that run fast and—progressively—grow them into larger ones. Consistent with this practice, three simpler variants of the model shown in Fig. 2.1 were created for this book as well. While the purpose of the book is to parallelize the model in Fig. 2.1 (*System* 4), parallelization will also be demonstrated on its three variants (*System* 1, 2, and 3). The description of each variant follows.

System 1

System 1 is the simplest variant of the notional shipboard power system model presented in this book, and it is shown in Fig. 2.4. *System* 1 has the same sources as those indicated for the notional model in Fig. 2.1, but the ring bus has less circuit breakers and there is only one load per bus. The load names are annotated on the schematic and are described in Table 2.4. This load set was chosen to produce

[22] Netlist is a common term that refers to a text file listing branches, parameters, and node information.

[23] After opening a *Simulink* model, typing *gcs* in the *MATLAB* prompt returns the model's name.

[24] This switch count includes valves and protective devices.

[25] Model construction and parameterization requires numerous runs before reaching credible realm.

[26] It is usual for more complex power system models to take several hours to complete a simulation.

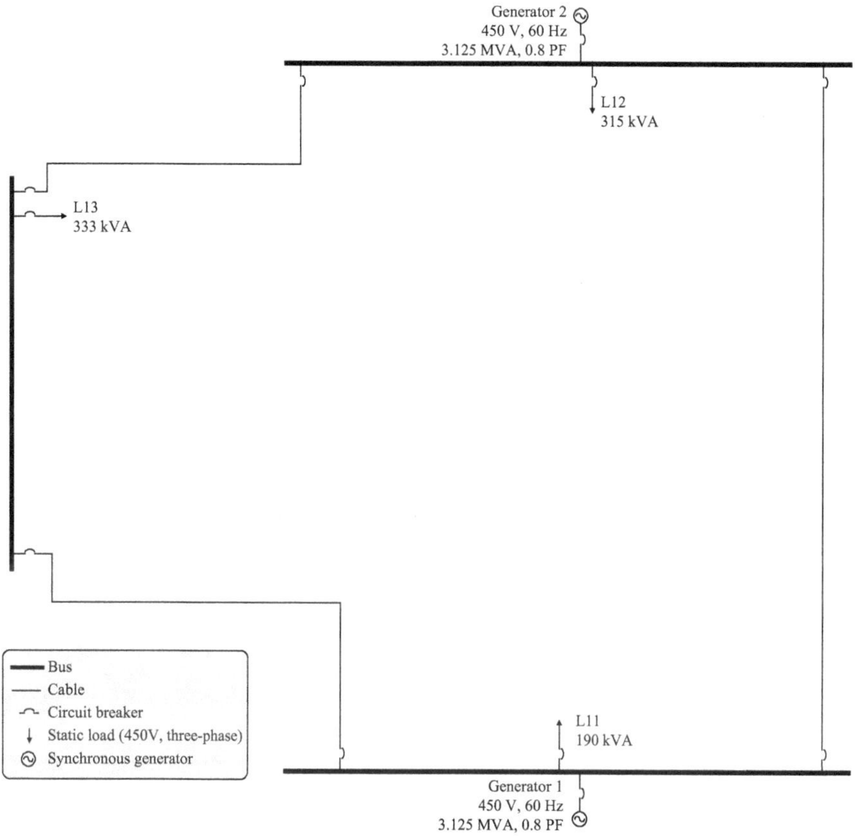

Fig. 2.4 First variant of the notional power system model used in this book (System 1)

noticeable power flows around the ring. This variant exemplifies the first step taken toward progressively building the model shown in Fig. 2.1.

System 2

System 2 builds on *System* 1 and is illustrated in Fig. 2.5. *System* 2 replaces the loads of *System* 1 with motor drives and also increases the load count. The motor names annotated on the schematic are described in Table 2.5. This model also represents a small system, but it increases simulation complexity by including nine three-phase rectifiers and nine three-phase inverters as the motor drives. The rectifier valves self-commutate while the inverter valves receive firing signals from their modulating controllers.

System 3

Following the succession in context, *System* 3 is a leap forward that skips over many elided variants created after *System* 2. The explanations for marginally different

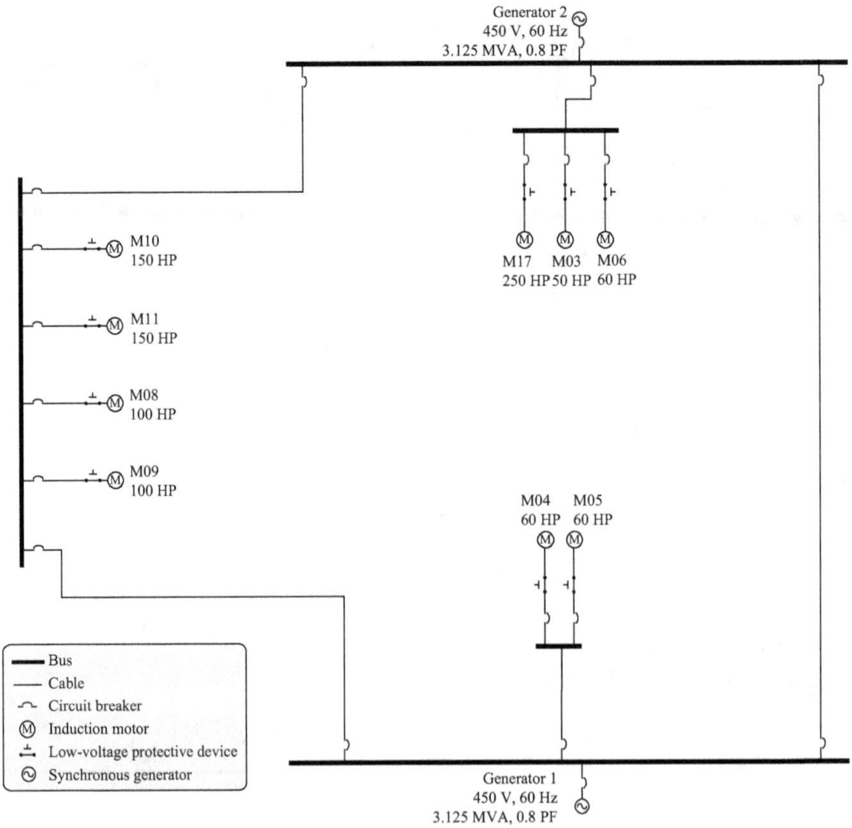

Fig. 2.5 Second variant of the notional power system model used in this book (System 2)

variants (i.e., those that progressively grew from *System* 2 to *System* 3 which are not shown) are redundant and do not present noticeable increases in size or complexity; therefore, *System* 3 exemplifies a model that becomes available toward the end of the progressive model-construction stage. *System* 3 is shown in Fig. 2.6 and is almost identical to the notional model shown in Fig. 2.1. The major difference between *Systems* 3 and 4 is the absence of motor drives. In *System* 3, three-phase static loads are used in place of motor drives. Doing so excludes all power electronic valves from the model.

System 4

System 4 is not a variant model. *System* 4 is the other name given to the notional shipboard power system model used in this book to demonstrate parallelization. (*System* 4 was shown in Fig. 2.1.) This name (*System* 4) follows a sequence of increasing model complexity established in order by *Systems* 1, 2, and 3. That is, *Systems* 1, 2, 3, and 4

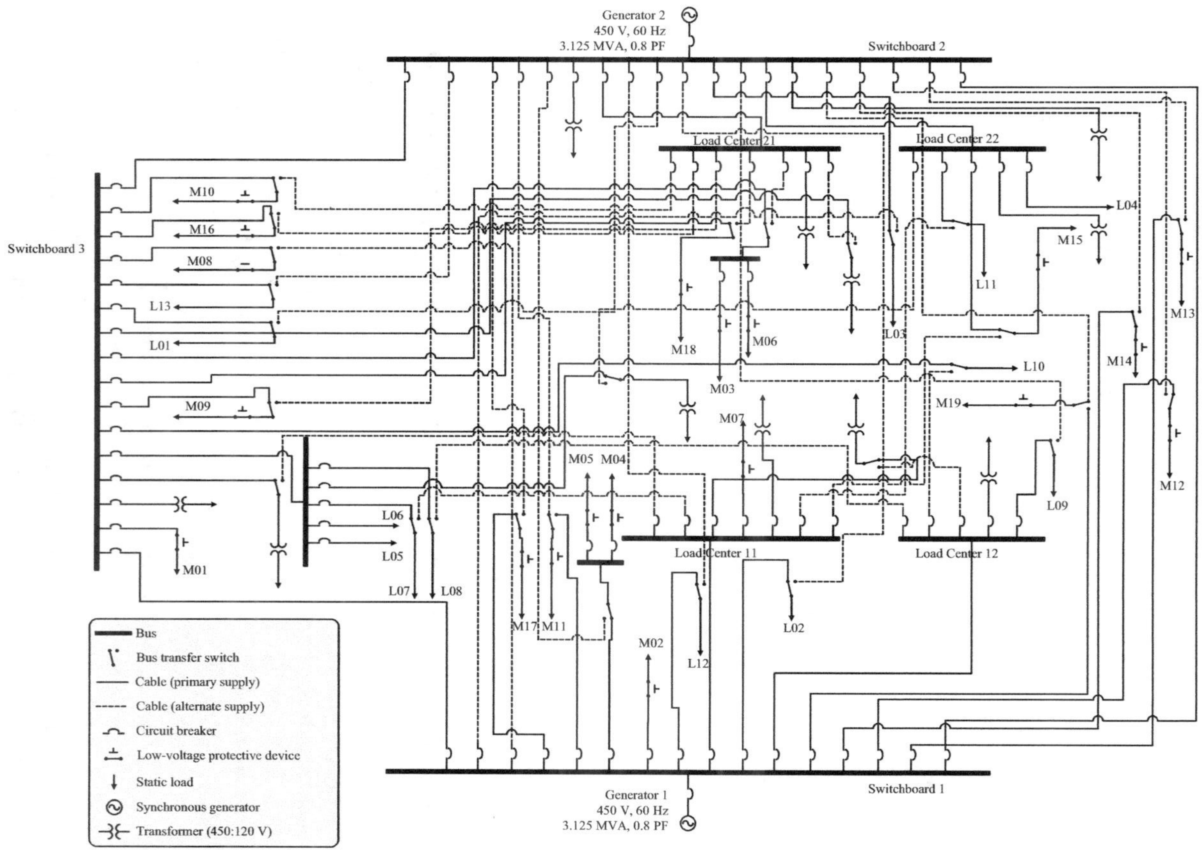

Fig. 2.6 Third variant of the notional power system model used in this book (System 3)

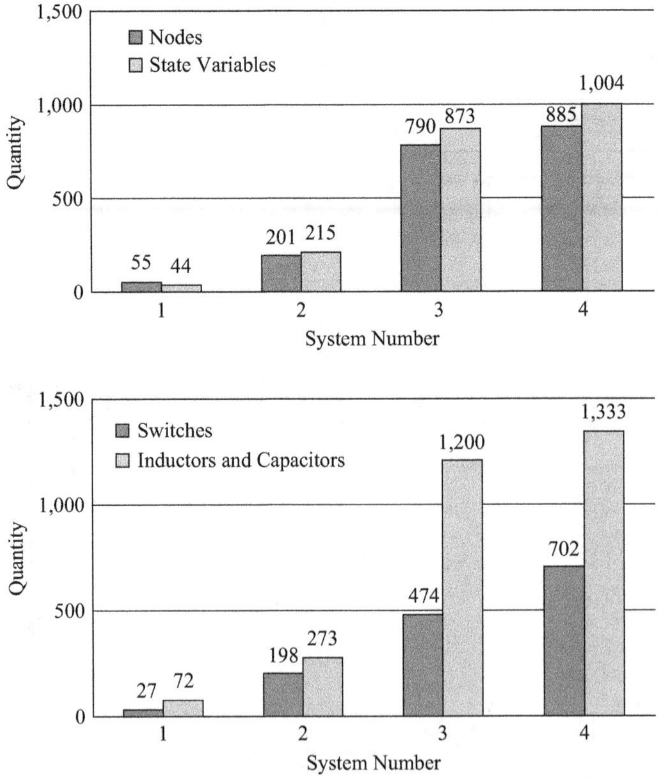

Fig. 2.7 Size assessment for Systems 1, 2, 3, and 4

are numbered sequentially in order of increasing model complexity, system size, and simulation runtime.

To assess the size of each of the four variants, a netlist was produced for each variant using Snippet 2.1. A comparison of the number of nodes and state-variables, and switches and inductors and capacitors for each system is shown in Fig. 2.7. Although state-variable counts are displayed in Fig. 2.7, netlists do not readily provide this information. The state-variable counts were obtained using Snippet 2.2, which–additionally–shows how to programmatically obtain the state matrix and display its non-zero structure in *MATLAB/Simulink*.

Snippet 2.2 was used on *Systems* 1–4 to produce Figs. 2.8 through 2.11. The left-hand side of each figure shows the state matrix (normally termed **A**) of a continuous state-space formulation. The right-hand side of each figure shows the discrete-equivalent state matrix after discretization.[27] The caption under each matrix type shows the square dimensions, the number of non-zeros, and its sparsity[28] as a percent

[27] The concept and function of discretization will be introduced in Chapter 4.

[28] Sparsity is the percentage of zero-entries in a matrix: a highly desirable characteristic of coefficient matrices when solving system equations.

Snippet 2.2 Code to obtain and display a state matrix in MATLAB/Simulink

```
% MATLAB code
clc
disp('Gathering state-space information...');
sps = power_analyze(gcs, 'structure');

% Continuos plot
A = sps.A;
subplot(1,2,1)
spy(A);
title('[Acontinuous]')

% Sparsity of [A]
sA = 100-100*nnz(A)/(size(A,1)*size(A,2));
% Sparsity of [I]
sI = 100*(1-1/size(A,1));
xlabel({['Dimensions: ', num2str(size(A,1)), ' x ', num2str(size(A,2)),
'; Non-zeros: ', num2str(nnz(A)), ';'], ['Sparsity: ', num2str(sA, 4),
' of ', num2str(sI,4), '% (~',num2str(sI-sA,2),'% short)']})

% Discrete plot
A = sps.Adiscrete;
subplot(1,2,2)
spy(A);
title('[Adiscrete]');

% Sparsity of [A]
sA = 100-100*nnz(A)/(size(A,1)*size(A,2));

% Sparsity of [I]
sI = 100*(1-1/size(A,1));
xlabel({['Dimensions: ', num2str(size(A,1)), ' x ', num2str(size(A,2)),
'; Non-zeros: ', num2str(nnz(A)), ';'], ['Sparsity: ', num2str(sA, 4),
' of ', num2str(sI,4), '% (~',num2str(sI-sA,2),'% short)']})
```

(first term) of the sparsity of an identity matrix of equal dimensions (second term). The difference (shortage) between these two sparsity terms is shown in parenthesis, which indicates how close the matrix's sparsity is to the theoretical maximum.

The equation formulation method used by *Simulink* is based on state-variables. This book uses nodal and mesh formulations instead. The choice of a formulation method varies by simulation program. For example, *PSCAD/EMTDC* [38], *VTB* [39], and most other power system simulators [40] (and circuit simulators [41,42]) use the nodal method due to its simple implementation and noticeable sparsity. There are also hybrid methods that combine the state-variable [43] and nodal methods, and also methods that combine the nodal and mesh methods as well [44]. The formulation method is one of the earliest and most important choices that one must make when designing a power system solver.

The initialization, solution, and total runtime experienced when using *Simulink* to simulate all four systems under analysis are listed in Table 2.6 and illustrated with the four charts in Fig. 2.12. The top chart in Fig. 2.12 shows the time it takes for *Simulink* to initialize each power system variant. Initialization time is proportional to model size, but it has a negligible impact on total runtime when the solution time is large (e.g., *System* 4).

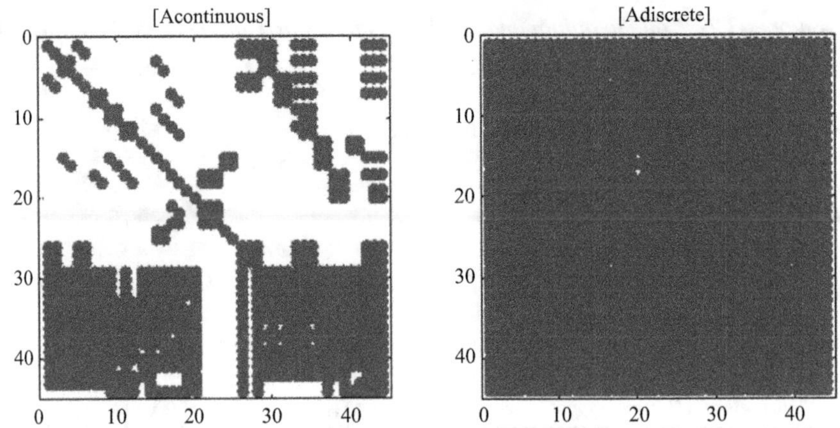

Dimensions: 44 × 44; Non-zeros: 757; Sparsity: 60.9 of 97.73% (~37% short)

Dimensions: 44 × 44; Non-zeros: 1934; Sparsity: 0.1033 of 97.73% (~98% short)

Fig. 2.8 State matrix information for System 1

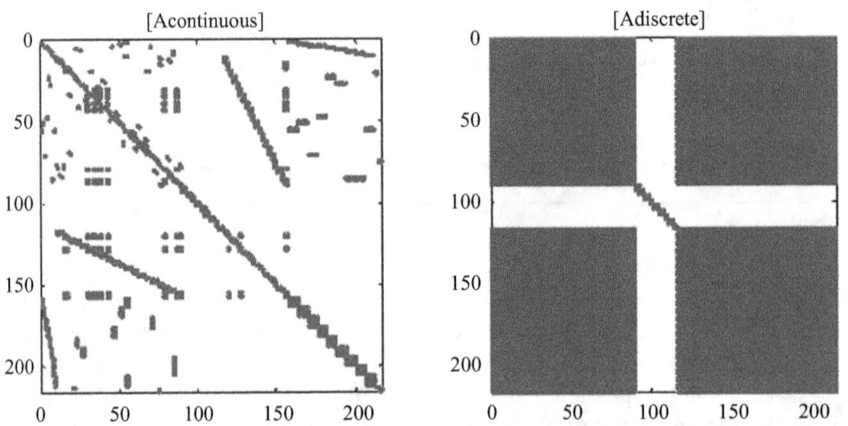

Dimensions: 215 × 215; Non-zeros: 1672; Sparsity: 96.38 of 99.53% (~3.2% short)

Dimensions: 215 × 215; Non-zeros: 35353; Sparsity: 23.52 of 99.53% (~76% short)

Fig. 2.9 State matrix information for System 2

The second chart shows *Simulink's* solution time to each model (i.e., excluding initialization). Note that the solution time in minutes increases when going from *System* 3 to *System* 4. The difference stems from the replacement of static loads with motor drives. (It was mentioned earlier that several variants exist between *Systems* 2 and 3, and, analogously, between *Systems* 3 and 4 variants exist as well. As before, these many variants are not shown in this book in order to keep the explanations clear, short and avoid redundancy.) The solution time (not the total runtime) is the time used to measure speedup in Chapter 9.

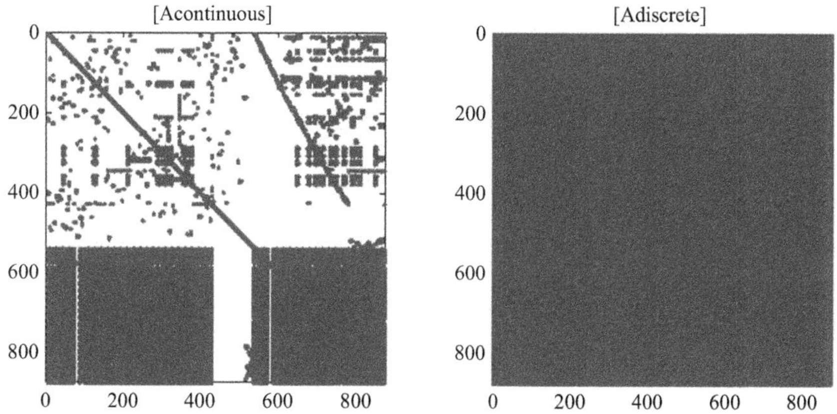

Dimensions: 873 × 873; Non-zeros: 137657;
Sparsity: 81.94 of 99.89% (~18% short)

Dimensions: 873 × 873; Non-zeros: 761847;
Sparsity: 0.037 of 99.89% (~1e+02% short)

Fig. 2.10 State matrix information for System 3

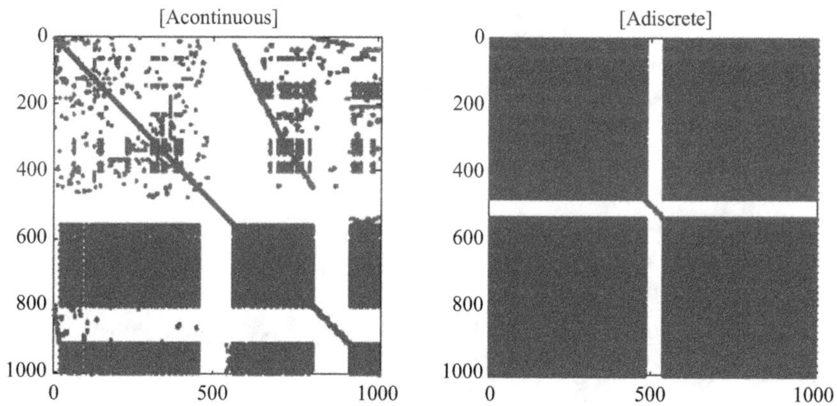

Dimensions: 1004 × 1004; Non-zeros: 129371;
Sparsity: 87.17 of 99.9% (~13% short)

Dimensions: 1004 × 1004; Non-zeros: 896659;
Sparsity: 11.05 of 99.9% (~89% short)

Fig. 2.11 State matrix information for System 4

The third chart in Fig. 2.12 shows the simulation speed in frames per second (fps) and in frame time (milliseconds, ms). Frames per second is the average rate at which the simulation proceeds forward, where *frame* refers to a simulation step. The frame time (ms) is the reciprocal of the frame rate; that is to say, the average time it takes a solver to complete one simulation step.

Finally, the bottom chart in Fig. 2.12 depicts the total runtime for each system under consideration, which represents the "wall clock" wait time[29] users experience

[29] These are approximate times and vary with computer hardware.

Table 2.6 Runtimes obtained with Matlab/Simulink

System number	Initialization time		Solution time*		Total runtime		Simulation speed	
	In seconds	In minutes	In seconds	In minutes	In seconds	In minutes	Frame time (ms)	Frame rate (fps)
1	3.50	0.06	4.80	0.08	8.30	0.14	0.22	4500.00
2	4.11	0.07	88.54	1.48	92.65	1.54	4.10	243.96
3	20.57	0.34	184.77	3.08	205.34	3.42	8.55	116.91
4	30.00	0.50	2925.00	48.75	2955.00	49.25	135.42	7.38

*This is the time benchmarked against the multicore solver discussed in Chapter 9; it is not the *total* runtime.

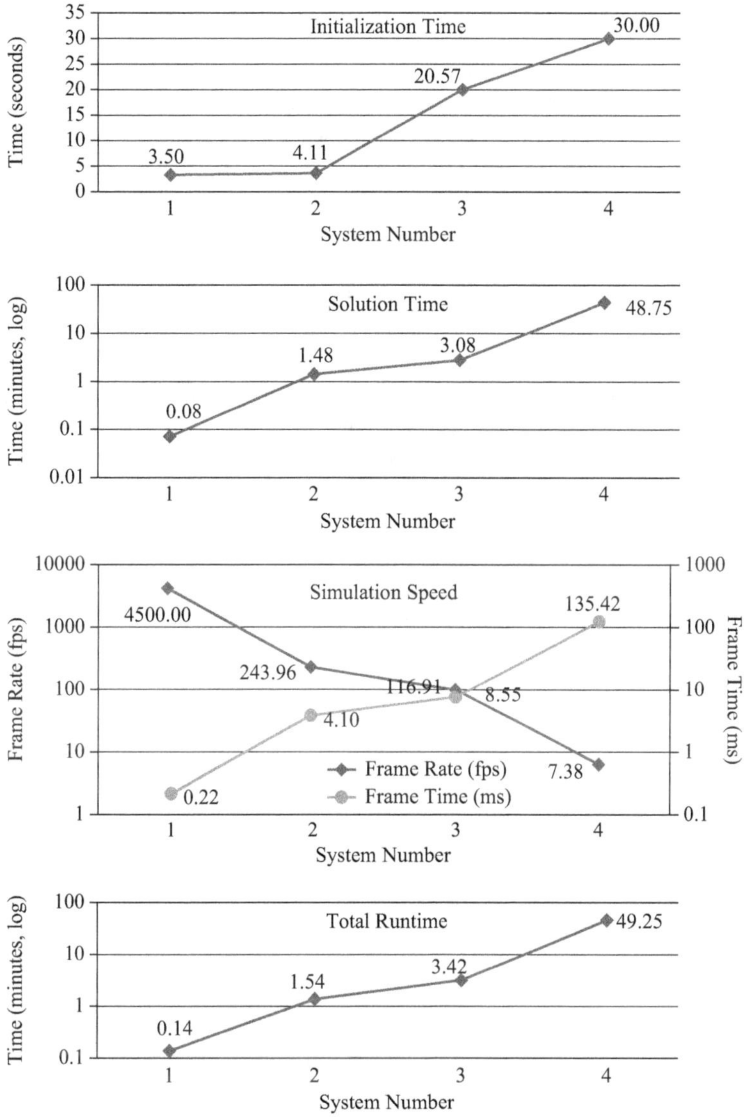

Fig. 2.12 Simulation runtime for Systems 1, 2, 3, and 4 obtained with MATLAB/Simulink

when the models run in *MATLAB/Simulink*. It is noticed though how rapidly the wait time increases with the addition of power electronic converters. It should also be noticed that although the total runtime is the wait time a user experiences when running the models in *Simulink*, the *solution time* dominates the wait time.

2.4 Summary

This chapter introduced the notional shipboard power system model (*System* 4) used to demonstrate parallel simulation on a multicore computer. Along with this model, three smaller and less complex model variants were introduced (*System* 1, 2, and 3). Netlists were used to contrast the sizes of all four power models in terms of the number of their nodes, state-variables switches, inductors and capacitors, number of non-zeros, and sparsity. Readers are encouraged to compare these system sizes against their own models using the code snippets provided in this chapter. These snippets produce "fast numbers" that can be used to compare other power system models against the ones presented here. Finally, it was also shown how long *MATLAB*/*Simulink* 2012a it might take to initialize and run each of these four models. The solution time of each of these models (not their total runtimes) will be used as reference times when measuring the parallelization speedups in Chapter 9. The solution time is the *crux* of a power system simulation and corresponds to the time spent in the time loop (explained in Chapter 3, section 3.3).

Chapter 3
Time domain simulation

The development of this book required writing a multicore solver for *Windows* to demonstrate the performance of the partitioning method presented herein. Such program requires implementing several time domain simulation concepts not always transparent to users.

This chapter introduces such concepts and their notations. In addition to what is presented in this chapter, readers are referred to numerous good books [45–52] and papers [40,53–61] in the area of time domain simulation. Excellent resources to learn power system simulation theory are the books written by H. W. Dommel [47] and N. Watson and J. Arrillaga [51].

Not much theoretical background in time domain simulation is assumed to follow the discussion presented in this chapter. Instead, this chapter was written to provide a foundation of key principles required to develop a time domain solver that can address intermittent high-frequency switching actions. The concepts presented in this chapter will be frequently referred to in subsequent chapters.

3.1 The time grid

Time domain simulation is the simulation of a system over an intended period of time. As the simulation time advances in incremental steps of Δt (in seconds), the computer program produces and stores a new solution at each timestep until the user-specified stop time t_{stop} is reached. The collective data saved through t_{stop} is the result of the simulation.

The time span of a simulation may be effectively abstracted [51] by using a time grid as illustrated in Fig. 3.1. This time grid represents a continuous progression of solutions from left to right, and it is distinguished by several grid divisions (dashed vertical lines). These line divisions indicate the instances at which solutions are produced and results saved. The time increment (or distance) Δt between time grid divisions is defined as the timestep size of the simulation and is commonly a fixed value of 50 μs.

As seen from the time grid shown in Fig. 3.1, there are four related labels: the step number k, the simulation time t, the relative step number, and the relative simulation time. The step number k is an integer counter that tracks where the simulation is along the time grid. The simulation time t represents the simulation time, not the elapsed 'wall-clock' time a user waits for the simulation to complete.

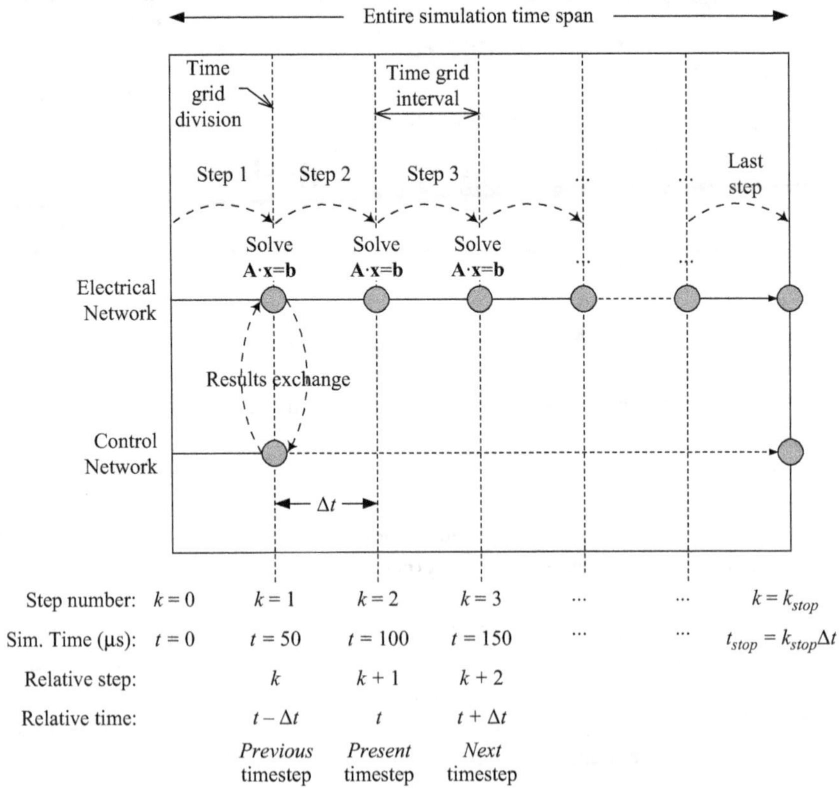

Fig. 3.1 Time domain simulation illustrated with a time grid

The relative step number and relative simulation time are only used for analytical purposes; they do not serve any purpose inside the solver's code. For example, when referring to an arbitrary step number during simulation, $k+1$ represents this arbitrary step, k represents the *previous* step, and $k+2$ represents the next step.[30] Likewise, the same corresponding relations apply to the relative simulation time; that is, $t - \Delta t$, t, and $t + \Delta t$ represent the previous, the current, and the next step's simulation time, respectively. (Some authors prefer to use $k-1$ to refer to the previous timestep, k to refer to the present timestep, and $k+1$ to refer to the next timestep. Both of these conventions are common.)

Referring to the left side of Fig. 3.1, there are two horizontal axes advancing from left to right. One represents the time line for the electrical network and the other the time line for the control network. The solution of these two networks represents the total work that a solver undergoes at each simulation timestep. Although, physically speaking, a power system is *one* network, it is common (and convenient

[30] These superscripts will be elided from most explanations and will be used to clarify ambiguity only.

for analytical and practical purposes) to represent a power system as having an electrical side and a control side. This is a deliberate separation of concerns to facilitate verbal and textual explanations, software design, and to reduce software maintenance and debugging time. This separation of concerns also makes sense, logically, as the electrical network represents the solution to the voltages and currents everywhere, and the control network represents the solution of all actionable commands, measurements, and decision logic.

At each time grid division, a solution is produced for the electrical network first by solving a set of discrete equations in the form of $\mathbf{A}^{k+1}\mathbf{x}^{k+1} = \mathbf{b}^{k+1}$, where $\mathbf{A}^{k+1}$ is the electrical network's coefficient matrix[31] at timestep $k+1$, $\mathbf{x}^{k+1}$ is the vector of unknown variables, and $\mathbf{b}^{k+1}$ is the excitation (or input) vector. The superscripts $k+1$ in $\mathbf{A}^{k+1}\mathbf{x}^{k+1} = \mathbf{b}^{k+1}$ imply that these quantities[32] are time varying and valid at timestep $k+1$ only—that is, at $k+2$ these quantities *may* be different.

Following the solution of the electrical network, the electrical network sends its results to the control network as inputs. At the same time grid division, the control network uses these inputs (and its own exogenous ones) to calculate the outputs of controllers, measurements, Boolean logic, and several other post-processing actions. Although the solution of the control network is available to the electrical network at the same time grid division, this solution is normally not used until the next simulation timestep.

It appears that the solution of these two networks, as shown in Fig. 3.1, always occurs on time grid divisions. While these solutions are always *saved* at these divisions, intermediate solutions are often produced *within* time grid intervals. These intermediate solutions are produced when a solver responds to events occurring between time grid intervals. The intermediate solutions are handled by a special interpolation routine which will be discussed later in this chapter.

Another effective way to represent the sequential solution of the electrical and control networks is through the circular data flow diagram shown in Fig. 3.2 [47]. This circle diagram facilitates the comprehension of the input–output relationships between the electrical and control networks, and it presents enumerated annotations in which the actions take place at each timestep. For instance, after advancing the simulation step from $k = 0$ to $k = 1$, the first action is to produce a solution for the electrical network. These solution results are then passed (via shared memory) to the control network as inputs. The control network, next, uses these inputs, as well other exogenous inputs, to produce its solution. Assuming for the moment that an intermediate solution between time grid divisions solution is not required, the simulation time advances by Δt. An earlier elucidation of this circular process may be seen in Reference 47.

Readers are reminded that there are *many* reasons why the simulation of large power systems is slow. The collective contribution of these reasons can be measured

[31] The network or immittance matrix refers to the nodal conductance matrix in nodal formulations and to the mesh resistance matrix in mesh formulations.

[32] In this book, uppercase bold letters represent matrices and lowercase bold letters represent vectors.

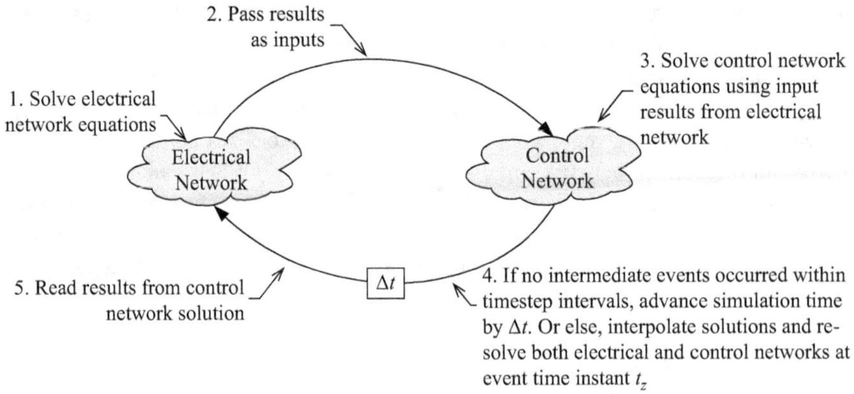

Fig. 3.2 Solution of electrical and control networks

by recording the average time required by a solver to advance one simulation time-step. For example, if a user sets $\Delta t = 50$ μs and $t_{stop} = 20$ s, the solver must produce $k_{stop} = \frac{t_{stop}}{\Delta t} = \frac{20}{50\ \mu s} = 400{,}000$ solutions to complete the simulation. If a solver takes 70 ms (on average) to produce one of these solutions the total simulation runtime will be nearly eight hours.

This illustrative value of 70 ms is called frame time.[33] Readers interested in developing simulation programs should monitor this value and that of its inverse; that is, the frame rate in frames per second (fps). Real time simulators often have frame times of 50 μs, or less, which is required for real time simulation [62–64]. In the realm of personal computers running offline simulations, a frame time of 1 ms is acceptable, but it is difficult to achieve for large and complex models.

The illustration in Fig. 3.2 is an effective diagrammatic device for power system engineers to convey meaning to software engineers. However, an important underlying implication of this diagram is that the simulation is unpartitioned. In partitioned scenarios, the diagram in Fig. 3.2 must be extended to demonstrate what occurs inside each power system partition.

Consider the partitioned power system paradigm shown in Fig. 3.3. This paradigm features four partitions, where each partition[34] contains one electrical subsystem and one control subsystem. These internal electrical and control subsystems are virtually interconnected among themselves and relay information to one another as explained previously in detail.

Despite the partitioned arrangement of a power system model as shown in Fig. 3.3, the partitions remain physically and numerically coupled. These couplings mean that the solution in each partition affects the solutions in the other partitions; additionally,

[33] The term *frame time* is adopted from game programming where game engines aim to maintain a constant frame times regardless of hardware speed.

[34] The terms *partition* and *subsystem* have the same interpretation in this book.

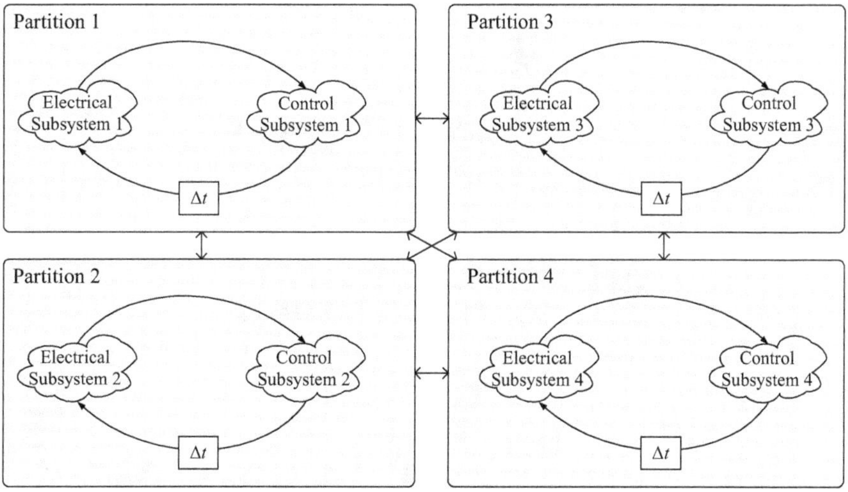

Fig. 3.3 Paradigm of partitioned electrical and control network solutions

an event occurring in one partition requires all partitions to produce intermediate solutions as well. Details on how to form and solve these partitions are given in subsequent chapters.

3.2 Time interpolation

In fixed-timestep simulations, where Δt is constant, the system's state during the interval $\langle t - \Delta t, t\rangle$ is unknown. When an intermediate event occurs *between* time grid divisions, the solver must interpolate (roll back) its electrical and control network solutions (in each partition) to the time instant t_z, where the first event of all partitions occurred. Although determining t_z is not trivial, most approaches resort to linear interpolation [54,55,65] to *estimate* t_z. (A modern approach including extrapolation is given in Reference 66; a comparative review of various interpolation techniques is available in Reference 67.) After detecting and addressing these intermediate events, the network solution (in each partition) is re-synchronized to the time grid and continues to action forward as normal.

To illustrate the concept of time interpolation, consider the time grid excerpt between two divisions showing two events (E1 and E2) between $t - \Delta t$ and t in Fig. 3.4. These events are estimated to have occurred at t_{z1} and t_{z2}, respectively.

The numbered annotations in Fig. 3.4 are described below:

1. The simulation advances from $t - \Delta t$ to t as normal. After producing the electrical and control network solutions (in each partition) at t, it is determined that one or more intermediate events occurred between time grid divisions. (The event(s) could have occurred inside either the electrical or control network

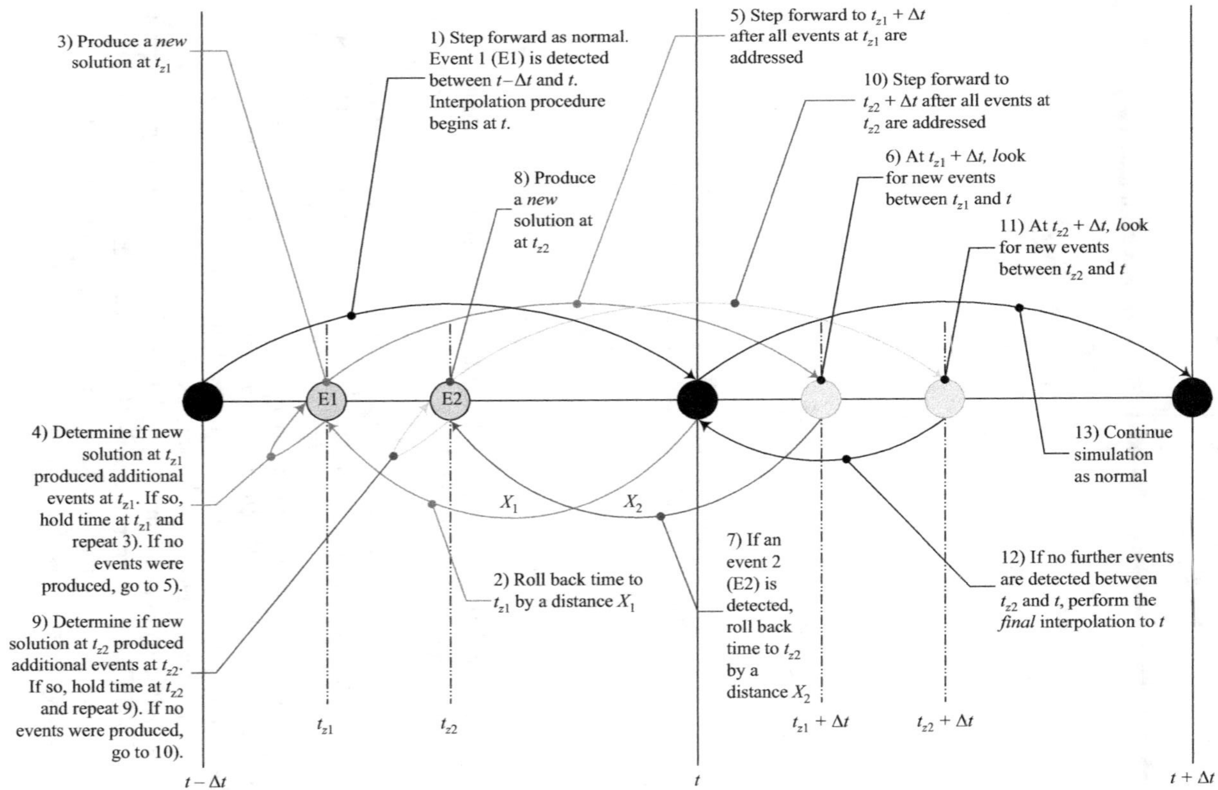

Fig. 3.4 Illustration of time interpolation showing intermediate events between time grid divisions

of any partition.) After comparing all events, it is determined[35] that the *earliest* event occurred at t_{z1}.

2. To address the first event (E1), the simulation rolls back by a distance X_1 to t_{z1}. This distance $0 < X_1 < 1$ is a fraction of Δt computed as $X_1 = (t - t_{z1})/\Delta t$. It should be noticed that the time grid simulation time t is held constant during interpolation. A dummy time variable t_{z1} is created, instead, to hold interim values of time during interpolation.
3. When at t_{z1}, the solver produces new solutions for the electrical and control networks (in each partition). This solution is not aligned with the time grid; therefore, its results are not saved. They are temporarily stored in memory, however.
4. Because one event can instantaneously excite another (e.g., forcing switches *off* can cause free-wheeling diodes to turn *on*), the interpolation routine seeks for additional events that may have been initiated at t_{z1} as a result of the newly produced solution at t_{z1}.
5. If no further events are detected at t_{z1}, the simulation advances as normal from t_{z1} to $t_{z1} + \Delta t$.
6. At $t_{z1} + \Delta t$, it is determined whether additional events occurred between t_{z1} and t (events between t and $t_{z1} + \Delta t$ are ignored). In this example, it is assumed that a second intermediate event occurred at t_{z2}.
7. To address the second event (E2) at t_{z2}, the simulation rolls back to t_{z2} by a distance X_2.
8. A new solution is produced at t_{z2}.
9. After the new solution at t_{z2}, it is determined whether the intermediate event action E2 produced any new switching events at t_{z2}.
10. If there are no more events at t_{z2}, the simulation advances as normal to $t_{z2} + \Delta t$.
11. At $t_{z2} + \Delta t$, it is determined whether further events occurred between t_{z2} and t (any event occurring between t and $t_{z2} + \Delta t$ is ignored). In this example, it is assumed no further events occur.
12. As no more intermediate events are detected, the simulation rolls back to t. A new solution is produced at t only if an event is detected at t; otherwise, the interpolated solution to t is considered to be the correct solution. This last roll back re-aligns the simulation time to the time grid and the data is saved.
13. The simulation continues forward as normal.

Interpolation is a *critical* aspect of power system simulation and is treated further when introducing the power electronic switch model later in the book. If interpolation is not implemented in a solver, spurious harmonics and voltage and current spikes can appear in the simulation results [56,59,60,68]. Readers interested in different interpolation algorithms are referred to References 56 and 67. (A comprehensive family of work in interpolation may be found in References 54–56, 58, 65, 67, 69–71.)

[35] Details on how intermediate events are detected may be seen in section 4.2.3.

3.3 Time loop

Introduction of the time grid and interpolation unfolds another explanation on why time domain simulations can be slow; but this time, by the introduction of the *time loop*. The term *time loop* is analogous to the term *game loop* in game theory [72]. In both, time and game loops, a solver (or program) renders the user graphics (e.g., waveforms or 3D animations) and solves its arithmetic iteratively. The major difference between time and game loops is that the latter uses constant frame rates whereas (offline) the former (as used in power system simulators) does not. Real time power system simulators, however, do require constant frame rates as game loops do, but at faster rates (e.g., real time simulator: 20,000 fps; game loop: 60 fps). Readers seeking to implement efficient time loops are encouraged to search for theory in game loops as—apparently—game theory is better documented.

A time loop expresses the concept of a time grid (illustrated in Fig. 3.1) to pseudocode, which is closer in form to an actual implementation. The structure of an example (simple) time loop is listed in (3.1) [38]. The time parameters of this time loop are typically constrained to a timestep of $\Delta t < 100\ \mu\text{s}$ and a stop time of $t_{stop} < 5$ s as set by the user [73].

$$\left\{\begin{array}{ll} \text{for } t = 0; t < t_{end}; t = t + \Delta t & \\ \quad \text{re-factor } \mathbf{A}(t) \text{ (same as } \mathbf{A}^{k+1}) & \text{(a)} \\ \quad \text{solve } \mathbf{A}(t)\mathbf{x}(t) = \mathbf{b}(t) & \text{(b)} \\ \quad \text{interpolate } \mathbf{x}(t) & \text{(c)} \\ \text{end} & \end{array}\right. \tag{3.1}$$

The time loop in (3.1) only highlights the time consuming subroutines: (a) refactorization of the network matrix **A** as its coefficients vary, (b) solution of a sparse system of equations of the form $\mathbf{A} \cdot \mathbf{x} = \mathbf{b}$, and (c) interpolation when events occur. When events occur, the interpolation routine in (3.1)(c) enters a nested (inner) time loop that does not exit it until *all* intermediate events are resolved. This situation typically requires producing many solutions inside the interval $\langle t - \Delta t, t + \Delta t \rangle$ and before the simulation time can advance. This inner loop reduces the solver's frame rate [55,59] and is exacerbated when there are many (e.g., hundreds) of power electronic switches in a model.

Referring to (3.1), it is not redundant to demonstrate that a simulation can take 24 hours to complete. Consider, for instance, the case when $\Delta t = 5\ \mu\text{s}$ and $t_{stop} = 5$ s. These parameters require executing the time loop (at least) $5/5 \times 10^{-6} = 1 \times 10^6$ times (more if there are interpolations). For a frame time of 86.4 ms, this simulation could take (at least) $10^6 \times 86.4 \times 10^{-3} = 24$ hours to complete. This runtime is experienced by many users carrying out power system simulations today and constitutes a costly technological research bottleneck.

3.4 Timestep selection

The simulation (or integration) timestep defines the separation between time grid divisions (in units of seconds). It is frequently recommended that the choice of Δt

should be as small as possible in order to increase accuracy, but as large as possible to reduce simulation runtime. These two recommendations are contradictory, however, which suggests that a thoughtful balance should exist to determine Δt. This balance is often left for users to decide. In passing, it should be mentioned that there are methods that, instead of using fixed Δt values, adjust Δt during runtime to improve simulation performance. Other approaches use various Δt values simultaneously in the same simulation, which is called multi-rate simulation [40,53,74–76]. Variable Δt and multi-rate approaches are not covered here as they are outside the scope of this book.

Improper Δt selection may cause not showing physical transients [77] in the simulation results, can produce spurious harmonics in moving-window type measurements (e.g., average and RMS), and can even cause misfires in forced-commutated devices due to under-sampling a high-frequency carrier signal.

It is helpful in this discussion to distinguish and comment on some of the various approaches to choose Δt. One approach is to exercise eigenvalue analysis[36] to identify the location of *complex* poles (roots) on the left side of the complex-frequency plane. The locations of these roots predict expected electrical oscillations (ringing) in a power system model (see Fig. 8.6 in Reference 32 or Fig. 9.31 in Reference 78), for which Δt can be selected to capture the imaginary part of the fastest eigenvalue of interest. Another approach to choose Δt is to estimate (guess) the fastest probable transient and choose Δt as a fraction (1/10th) of the time constant corresponding to this transient. Perhaps the most commonly used approach is to choose a Δt that works well for many power system models (i.e., a "silver bullet" value for Δt). The latter approach stems from experience and does not require much thought (e.g., $\Delta t = 10\ \mu s$), but it can affect simulation performance in large models having slow dynamics.

A fourth approach to choose Δt is to find the greatest common denominator (GCD) of the periods of interest. This requires some analysis, but a computer can calculate this effortlessly. This approach is exercised in two steps:

1. Determine all user-specified frequencies of interest if known (e.g., power frequency and carrier frequencies).
2. Compute Δt as the GCD of ¼th of these periods. The ¼ coefficient corresponds to a 4x sampling rate, and it ensures signals are sampled at their peaks and zero-crossings. For power signals (e.g., 60 Hz), the ¼ ensures waveforms show their peaks clearly. For carrier signals (e.g., 5,400 Hz), the ¼ avoids aliasing, which can otherwise cause misfiring of power electronic switches.

This two-step approach to choose Δt ensures that Δt fits $4n$ times in *any* of the periods of interest (where $n = 1, 2, 3, \ldots$), avoids problems of improper timestep selection, and keeps Δt rather large.[37] The following two-step approach exemplifies how Δt was chosen for this book.

[36] Finding complex eigenvalues and setting Δt to be 1/10th of the fastest time constant.

[37] Application of this calculation approach for Δt requires knowing the user-specified frequencies of the system.

Step 1) Determine the frequencies f (and periods T) of interest. These are:

a. Power frequency:

$$\begin{cases} f_1 = 60 \text{ Hz} \\ T_1 = \dfrac{1}{f_1} = 16.667 \text{ ms} \end{cases} \tag{3.2}$$

b. Inverter carrier frequency:

$$\begin{cases} f_2 = 5{,}400 \text{ Hz} \\ T_2 = \dfrac{1}{f_2} = 185.185 \ \mu\text{s}. \end{cases} \tag{3.3}$$

Step 2) Compute Δt as the GCD of $1/4n$ of these periods:

$$\Delta t = \text{GCD}\left(\frac{T_1}{4n}, \frac{T_2}{4n}\right)\bigg|_{n=1} = 46.2963 \ \mu\text{s} \tag{3.4}$$

The value of Δt found in (3.4) fits $T_1 = 16.667$ ms exactly 360 times and fits $T_2 = 185.185$ μs exactly four times. Other frequencies for which $\Delta t = 46.296$ μs and $\Delta t = 50$ μs are exactly compatible are listed in Appendix A. The tables in Appendix A can be used to choose different carrier frequencies while leaving Δt fixed at either 46.296 μs or 50 μs. If different power and carrier frequencies are used, the two-step approach shown above can be re-calculated to obtain a new value for Δt. It should be noticed that as power and carrier frequencies may vary during simulation, remedial action may be necessary to re-calculate Δt during runtime.

3.5 Summary

This chapter presented concepts related to time domain simulation, including the time grid, interpolation, and the time loop concepts. While program developers frequently utilize and understand these concepts well, these concepts are not always known to the end users.

The concepts of interpolation and time loop were explained. Regarding interpolation, intermediate events arise frequently in power system simulation including power converters; not addressing them when they occur can produce dubious results. Interpolation may be ignored if using small values for Δt, but such small timestep values can sacrifice performance when using mundane hardware such as a desktop computer. Finally, a few options to choose Δt were provided. The simulations carried out in the remainder of the book will refer to the concepts presented and discussed in this chapter.

The time loop is the crux of a power system solver. Referring back to Fig. 2.12, the *solution time* was the time spent by *Simulink* in its internal time loop. Explaining how to develop a parallel solver that does not spend too much time in its time loop is the focus of this book.

Chapter 4
Discretization

The previous chapter defined time domain simulation as a succession of network solutions in fixed-step intervals. This chapter examines the mathematical concept of discretization and its application to power system simulation. Discretization is a transformation by which continuous differential equations are transformed into difference algebraic equations. The importance of discretization is that it allows to produce solutions at and between time grid divisions.

Discretization occurs at the *electrical* and *control* network levels. At the electrical network level, power apparatus models are discretized by replacing inductors and capacitors with discrete branches having difference equations. If a power apparatus has switches, these branches are *linearized* instead of discretized. At the control network level, state-variable equations, transfer functions, and other blocks are discretized (or linearized) to obtain algebraic relationships as well. Discretization and linearization of the electrical and control networks is said to *discretize* a power system and prepare it for simulation.

Readers should be aware that the contents of the present chapter and of the following three chapters are arranged in order of required background. This chapter introduces an overview of discretization. Chapter 5 explains how to discretize each power apparatus model. Chapter 6 shows how to formulate discretized network equations. Chapter 7 covers how to partition these system-level equations. It is suggested that readers follow the contents of these chapters in the order they are presented.

4.1 Discretization

Discretization in power system simulation is recognized as an important contribution by the pioneer work of H. W. Dommel [47,79]. Discretization (also known as integration[38]) transforms continuous differential equations into discrete difference ones for their solution at fixed-size timestep intervals. The two discretization (or integration) methods used in this book are *tunable* integration (used for stand-alone electrical branches and control blocks) and *root-matching* (used for electrical branch pairs). The following subsections introduce both of these methods.

[38] These two terms are often used interchangeably.

4.1.1 Tunable integration

Consider the continuous differential equation in (4.1), where $x(t)$ represents the state-variable, $u(t)$ a forcing input function, and $h()$ an arbitrary linear function of $x(t)$ and $u(t)$. Discretization of (4.1) using the trapezoidal rule[39] and the backward Euler integration algorithms [50,51] results in formulations (4.2) and (4.3), respectively. The approximate solution of state-variable $x(t)$ in (4.1) is achieved by solving for x^{k+1} in either (4.2) or (4.3), depending on the preferred discretization method.

$$h(x(t), u(t)) = \frac{d}{dt}x(t), \quad x(0) = 0 \tag{4.1}$$

$$\frac{h(x^{k+1}, u^{k+1}) + h(x^k, u^k)}{2} = \frac{x^{k+1} - x^k}{\Delta t} \quad \begin{pmatrix}\text{trapezoidal}\\ \text{rule}\end{pmatrix} \tag{4.2}$$

$$h(x^{k+1}, u^{k+1}) = \frac{x^{k+1} - x^k}{\Delta t} \quad \begin{pmatrix}\text{backward}\\ \text{Euler}\end{pmatrix} \tag{4.3}$$

The tradeoff between these two integration methods is accuracy vs. stability [42]. Although it is widely accepted that the trapezoidal rule is more accurate than the backward Euler integration, the trapezoidal rule is less stable. The choice of integration method in power system solvers typically depends on which power apparatus are included in the model. For example, the trapezoidal rule is recommended for networks where the voltages and currents are expected to be sinusoidal; backward Euler is recommended for networks where the voltages and currents are expected to be piecewise linear [61,70] such as when power converters are present [50,56]. Because the voltages and currents in power systems can be sinusoidal and/or piecewise linear (depending on which power apparatus are included in a model), an elegant integration approach that combines the trapezoidal rule and the backward Euler integration methods is the so-called *tunable integration* [51,61,68,70].

Comparing the left-hand sides of (4.2) and (4.3), it is observed that the discretization algorithm can be changed during runtime (if necessary) using a tunable parameter γ [51] as shown in (4.4). Changing integration methods during runtime can suppress numerical chatter introduced by the trapezoidal rule, which often occurs when changes in state-variables become zero over two consecutive timesteps [42,61,70,80]. Tunable integration is defined in (4.4) [51], and it is an effective approach that benefits from the accuracy of trapezoidal integration and the stability of backward Euler integration.

$$\begin{matrix}\text{Tunable}\\ \text{integration}\end{matrix} \rightarrow \begin{cases} \gamma \cdot h\left(x^{k+1}, u^{k+1}\right) + (1-\gamma)h\left(x^k, u^k\right) = \dfrac{x^{k+1} - x^k}{\Delta t} \\ \gamma \rightarrow \begin{cases} = \frac{1}{2}, \text{ for trapezoidal rule} \\ = 1, \text{ for backward Euler} \end{cases} \end{cases} \tag{4.4}$$

[39] The trapezoidal rule is also known as the Tustin integration algorithm.

4.1.2 Root-matching

An integration method that is becoming prominent in power system simulation is the root-matching method [51,81,82]. This integration or discretization method (used in this book for electrical branch pairs only) forestalls the loss (or misplace) of poles, zeros, or gains when discretizing time domain equations. Accordingly, the root-matching method is suitable to discretize branch *pairs* and avoid the problem of numerical chatter, but it is also less sensitive to the timestep size selection (Δt). Said branch pairs are defined as any of the following branch combinations:

- *Series RL*: a resistor and an inductor connected in series
- *Series RC*: a resistor and a capacitor connected in series
- *Parallel RC*: a resistor and a capacitor connected in parallel
- *Parallel RL*: a resistor and an inductor connected in parallel

The choice of root-matching over other integration techniques is often subjective. The reasons to adopt root-matching for the multicore solver developed for this book were its stability characteristic, added accuracy, and small sensitivity to the size of Δt, which is independent of Δt [61,83]. Additionally, the power apparatus models presented in this book all contain branch pairs, which suites the root-matching method rather well.

In contrast to traditional discretization (e.g., trapezoidal rule or backward Euler), understanding root-matching requires traversing through the frequency domain. Consider the four-quadrant domain illustration in Fig. 4.1, where discretization starts from the continuous time domain (t) quadrant. The root matching method suggests first obtaining the transfer function of a branch pair, which results in an expression in the continuous frequency s-domain. By paying close attention to the location of the poles and zeros in the s-domain, it is possible to convert the transfer function from the s-domain to the z-domain. Finally, after adjusting the transfer function gain in the z-domain and carrying out an inverse z-transform, the z-domain transfer function is transformed into a discrete time domain (discretized) expression. In sum, this domain traversal yields in more accurate results than going directly from the continuous time domain t to the discrete time domain k [51,81].

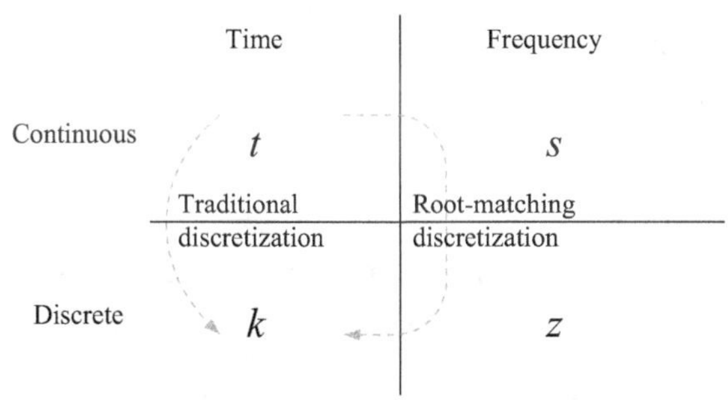

Fig. 4.1 Domain traversal involved in discretization

Additional details of the foregoing domain traversal are given next:

1. Continuous time domain (t)
   - Start with the differential equation of a branch pair.
2. Frequency s-domain (s)
   - Obtain the immittance, continuous transfer function (i.e., s domain) for the desired branch pair.
   - Find the poles and zeros of the transfer function.
   - Using the final value theorem, obtain the final value of the s-domain transfer function under a unit-step input (or ramp input if a final value does not exist).
3. Frequency z-domain (z)
   - Obtain the equivalent immittance transfer function in the z domain.
   - Add an adjustment-gain coefficient to the immittance transfer function.
   - Add poles (and zeros) at the origin of the complex-frequency plane if the z-domain transfer function numerator (or denominator) is constant.
   - Using the final value theorem, find the final value of the z-domain transfer function under a unit-step input (or ramp input if a final value does not exist).
   - Equate the final values of the s- and z-domain transfer functions to find the z-domain gain coefficient.
4. Discrete time domain (k)
   - Take the inverse z-transform of the resulting z-domain transfer function to produce a discrete-time domain equation.
   - Re-arrange the variables of the time domain expression and solve for the branch-pair's total through current (if using nodal formulations) or across voltage (if using mesh formulations).

An overview of discretization was presented by introducing four integration methods: trapezoidal rule, backward Euler, tunable integration, and root-matching. The following sections explain how to apply these transformations to discretize the electrical and control networks of a power system model.

4.2 Electrical network discretization

This section applies tunable and root-matching integration to discretize stand-alone and branch pairs, respectively. It is not redundant to highlight that the objective of introducing discretization is to develop difference, algebraic equations for the power apparatus listed in Table 2.2. These discretized power apparatus equations, subsequently, will yield the electrical network equations of the power system model.

4.2.1 Stand-alone branches

To discretize the power apparatus listed in Table 2.2, stand-alone inductors and capacitors *inside* each power apparatus model are replaced by the equivalent branch shown in Fig. 4.2 [79] (note that the left-hand side is for nodal formulations and the right-hand side is for mesh formulations). On both sides of Fig. 4.2, i^{k+1} represents the total current flowing into the branch, v^{k+1} the voltage drop across the whole branch,

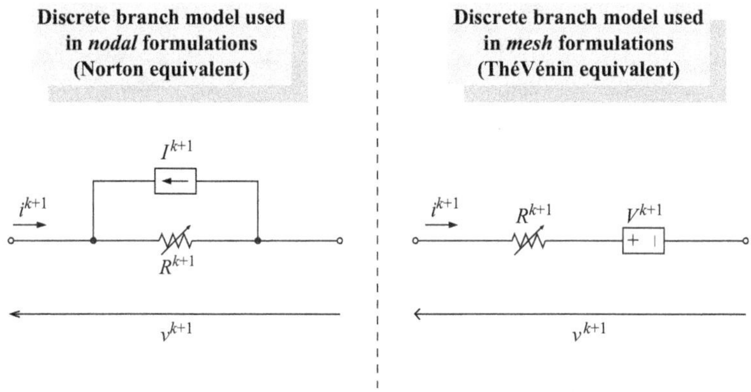

Fig. 4.2 Equivalent branch models (Norton and ThéVénin) of stand-alone or branch pairs

and R^{k+1} the equivalent discrete resistance. Additionally, on the left of Fig. 4.2, I^{k+1} represents a current source injection which value at $k+1$ is a function of the branch voltage and current at k. Equivalently, on the right V^{k+1} represents a voltage source which value at $k+1$ is a function of the branch voltage and current at k. Both I^{k+1} and $V^{k+1}$are termed *historical* sources and are the Norton–ThéVénin duals of each other.

The choice of a discrete branch in Fig. 4.2 depends on the formulation choice. Solvers using nodal equations for the electrical network use the model on the left (also known as resistive-companion form; see References 39, 49). The model on the right is meant for solvers using mesh equations for the electrical network. Readers will find that most work in power system simulation use nodal formulations. This book provides explanations for methods in addition to a pragmatic comparison between nodal analysis and mesh analysis supplied in Appendix B. Throughout this chapter, it will be shown that stand-alone as well as branch pairs conform to the equivalent representations shown in Fig. 4.2.

4.2.1.1 Resistor

Referring to Fig. 4.2, stand-alone resistors are modeled in the discrete time domain by setting $R^{k+1} = R$ (ohms) and the historical term to zero. If a resistance is time varying, R^{k+1} must be updated in the corresponding positions of $\mathbf{A}$ prior to solving $\mathbf{A} \cdot \mathbf{x} = \mathbf{b}$ at the next timestep.

Resistors are simple to model, but are often used excessively. A consequence of using resistors to accomplish a wide-range of tasks is that they may (unintentionally) introduce finite drifts or deviations in simulation results. Examples of resistor overuse can include the modeling of breakers and switches in open positions, using resistors as shunt damping to stabilize numerical oscillations, and using them as voltmeters or ammeters, to name a few. While it often makes sense to use resistors for these cases, readers should be aware that such uses could alter the simulation results.

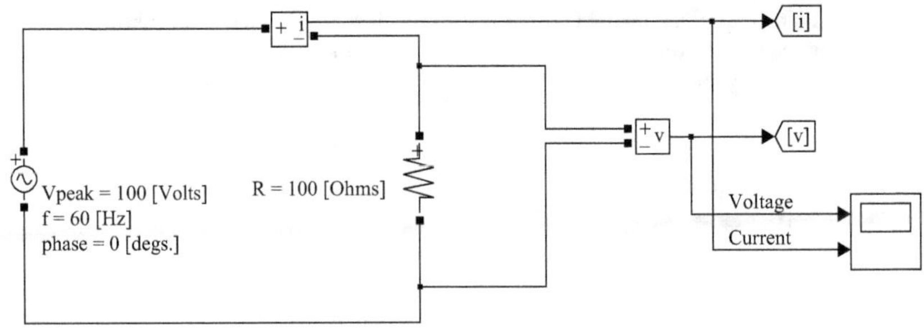

Fig. 4.3 Resistive circuit to exemplify differences in modeling using Simulink

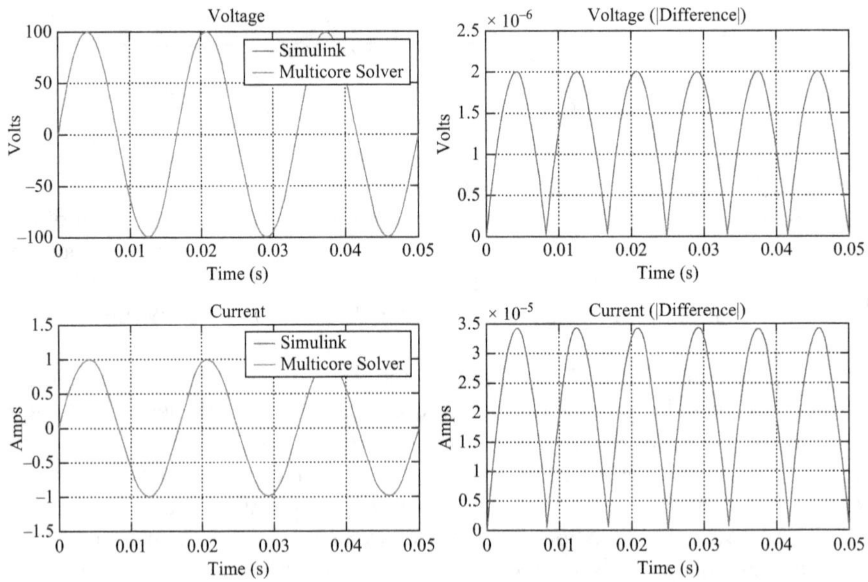

Fig. 4.4 Overlay of resistor voltage and current (left) and the absolute value of their arithmetic difference (right)

To exemplify how a simple resistive circuit may produce unwanted differences, consider the circuit shown in Fig. 4.3 modeled using *Simulink*. The results of solving this circuit using *Simulink* and the multicore solver developed for this book (discussed later) are overlaid on the left of Fig. 4.4. Visual inspection of the left side suggests that there is no difference in the voltage result. However, carrying out a point-by-point comparison[40] of such results on the right-hand side of the figure shows that finite differences do exist in both voltage and current waveforms. The waveforms on the right reveal primarily that the differences are perfectly in step with the source

[40] The difference is calculated as the absolute value of the arithmetic difference: $|x_i - x_j|$.

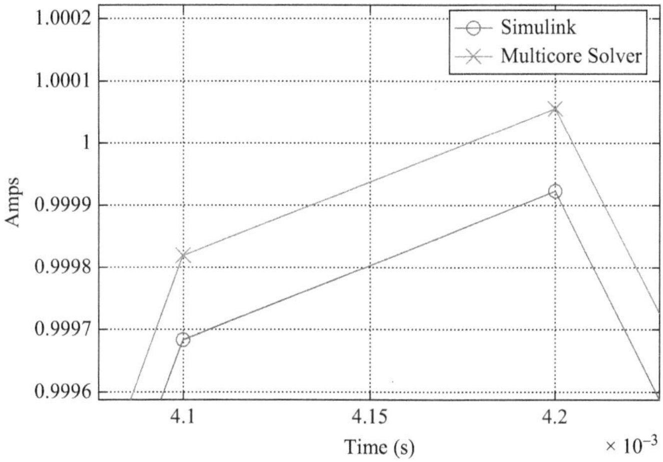

Fig. 4.5 Difference in peak current due to a reduction in effective load resistance

voltage. This equivalency is a clear indication that the resistive load in both solvers is different.

To illustrate the difference in one of the current peaks, consider the close-up in Fig. 4.5. The trace for the multicore solver shows a larger peak than the peak on the trace produced with *Simulink*.[41] The discrepancy is an example of what happens when the voltmeter is modeled as a 1 MΩ resistor in the multicore solver. This internal programming "optimization" effectively reduces the load seen by the voltage source to less than 100 Ω. Unless users are aware of this internal detail, it can be challenging to explain why the peak currents do not match. Examples such as these often arise in practice where, at times, one would like to be able to explain the differences observed in the results produced by different solvers. When *possible*, it is recommended that the results among different solvers be validated against analytical expressions instead of against each other.

4.2.1.2 Inductor

The branch model for a stand-alone inductor is obtained by discretizing the differential equation of an inductor. Using tunable integration, this discretization results in (4.5) for mesh formulations and in (4.6) for nodal formulations. In (4.5) and (4.6), v_L^{k+1} is the voltage across the inductor in volts, L is the inductance in henries, Δt is the discretization timestep in seconds, and i_L^{k+1} is the total current through the inductor in amps. The historical terms in (4.5) and (4.6) are V_{histL}^{k+1} and I_{histL}^{k+1}: both are a function of the inductor's voltage and current at k. The equivalent branches of (4.5) and (4.6) comply with the generic branch model of Fig. 4.2 as shown on the left side of Fig. 4.6.

[41] In this example, both waveforms miss the expected 1 A peak due to an improper timestep selection of $\Delta t = 100$ μs (see section 3.4 for treatment of this topic).

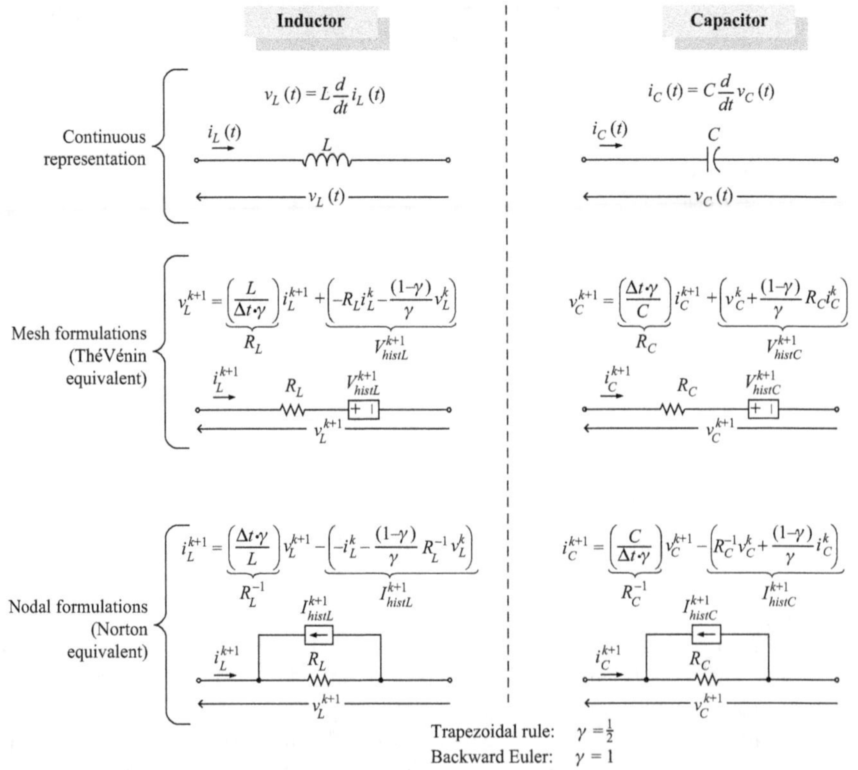

Fig. 4.6 Branch models for stand-alone inductor and capacitor

Mesh

$$v_L(t) = L\frac{d}{dt}i_L(t) \Rightarrow v_L^{k+1} = \underbrace{\left(\frac{L}{\Delta t \cdot \gamma}\right)}_{R_L} i_L^{k+1} + \underbrace{\left(-R_L i_L^k - \frac{(1-\gamma)}{\gamma}v_L^k\right)}_{V_{histL}^{k+1}} \tag{4.5}$$

Nodal

$$v_L(t) = L\frac{d}{dt}i_L(t) \Rightarrow i_L^{k+1} = \underbrace{\left(\frac{\Delta t \cdot \gamma}{L}\right)}_{R_L^{-1}} v_L^{k+1} - \underbrace{\left(-i_L^k - \frac{(1-\gamma)}{\gamma}R_L^{-1}v_L^k\right)}_{I_{histL}^{k+1}} \tag{4.6}$$

4.2.1.3 Capacitor

Like in the case of the inductor, the branch model for the stand-alone capacitor is obtained from the voltage and current relationship expressions in (4.7) and (4.8). In both (4.7) and (4.8), v_C^{k+1} is the voltage across the capacitor in volts, C is the capacitance in farads, i_C^{k+1} is the current through the capacitor in amps, and V_{histC}^{k+1}

and I_{histC}^{k+1} are the historical functions of the capacitor voltage and current at k. The equivalent branches of (4.7) and (4.8) are in accordance with the generic branch model of Fig. 4.2 as shown on the right side of Fig. 4.6.

Mesh

$$v_C(t) = \frac{1}{C}\int i_C(t)dt \Rightarrow v_C^{k+1} = \underbrace{\left(\frac{\Delta t \cdot \gamma}{C}\right)}_{R_C} i_C^{k+1} + \underbrace{\left(v_C^k + \frac{(1-\gamma)}{\gamma} R_C i_C^k\right)}_{V_{histC}^{k+1}} \tag{4.7}$$

Nodal

$$v_C(t) = \frac{1}{C}\int i_C(t)dt \Rightarrow i_C^{k+1} = \underbrace{\left(\frac{C}{\Delta t \cdot \gamma}\right)}_{R_C^{-1}} v_C^{k+1} - \underbrace{\left(R_C^{-1} v_C^k + \frac{(1-\gamma)}{\gamma} i_C^k\right)}_{I_{histC}^{k+1}} \tag{4.8}$$

4.2.1.4 Voltage source

Voltage sources are also modeled using the generic branch model displayed on the right of Fig. 4.2. Referring to the left column of Fig. 4.7, the continuous representation of a voltage source shows the instantaneous voltage $v_E(t)$ as an arbitrary function $f(t)$. In the case of DC sources, $f(t)$ is constant, while for AC sources $f(t)$ is updated using the current simulation time.

Referring to the mesh case of the voltage source shown in Fig. 4.7, the internal resistance R_E conforms with the model illustrated in Fig. 4.2. This resistance can be used to either model internal voltage drops or prevent zero-impedance meshes. The function $f(t)$ is represented by the historical source V_E^{k+1}, which is the electromotive force (EMF) source of the branch. The current through the branch is denoted as i_E^{k+1}, and the total voltage drop is denoted as v_E^{k+1}. When modeling DC or sinusoidal voltage sources, it is common to use V_E^{k+1} as defined in (4.9).

$$V_E^{k+1} \rightarrow V_{dc} + \sqrt{2} \cdot V_{\text{rms}} \sin(\omega \cdot k \cdot \Delta t + \varphi) \tag{4.9}$$

where:

V_{rms} = AC voltage (RMS volts)
V_{dc} = DC offset (V)
ω = frequency (rad/s)
φ = phase angle (rad).

Referring to the nodal case of the voltage source in Fig. 4.7 (left column, last row), the Norton equivalent is obtained from a Norton–ThéVénin transformation. This transformation is recommended (though not necessary) to maintain conformity with the excitation vector **b**, which expects current injections instead of voltage impressions. (Modified nodal analysis [84–87] presents an approach to use voltage sources in nodal analysis, but such approach is not covered here.)

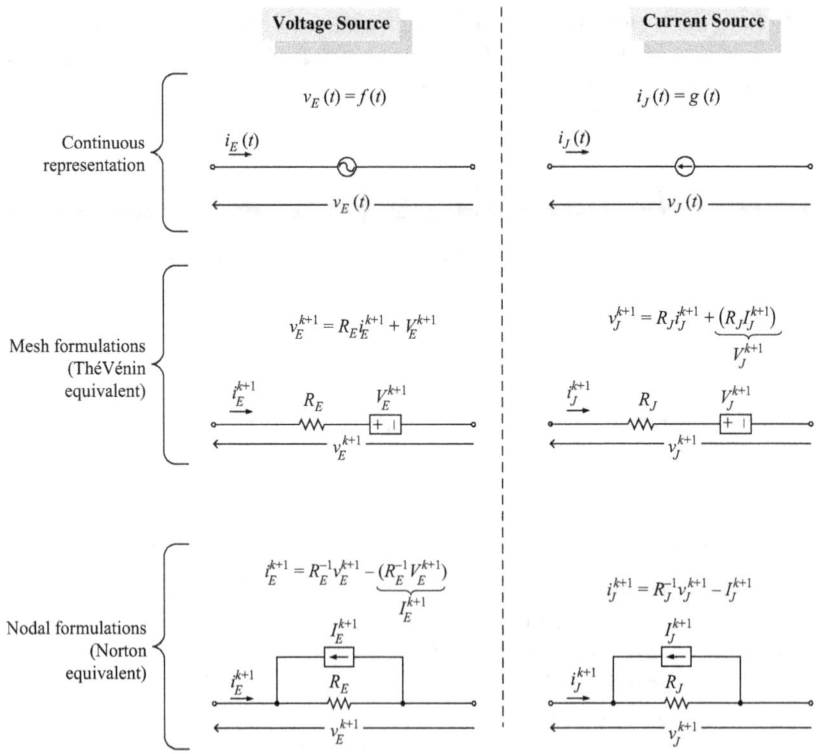

Fig. 4.7 Branch models for voltage and current sources

4.2.1.5 Current source

Current sources are also modeled as the generic branch model shown in Fig. 4.2. Referring to the right column of Fig. 4.7, the continuous representation of a current source also shows its value as an arbitrary function $g(t)$ (in amps).

In the nodal case of the current source (bottom right of Fig. 4.7), an internal resistance R_J of *large* value is included in the model to conform with the model illustrated on the left of Fig. 4.2. This resistance, although it adds fictitious damping to a circuit, is useful to prevent nodes of zero conductance and singularities when partitioning a system (discussed ahead). Similar to voltage sources, a common expression for sinusoidal current injections is given in (4.10).

$$I_J^{k+1} \rightarrow I_{dc} + \sqrt{2} \cdot I_{\text{rms}} \sin(\omega \cdot k \cdot \Delta t + \varphi) \tag{4.10}$$

where:

I_{rms} = AC current RMS value (A)
I_{dc} = DC offset (A)
ω = frequency (rad/s)
φ = phase angle (rad).

The function $g(t)$ is represented by the historical source I_J^{k+1}, which is the current injection of the branch. The total current through the branch is denoted as i_J^{k+1} and its voltage is denoted as v_J^{k+1}.

Referring next to the mesh case of the current source (right column, middle row in Fig. 4.7), the resistance R_J is also *large*—but so is the resulting Norton–ThéVénin transformed historical voltage source V_J^{k+1}. This transformation is recommended (albeit not necessary) to preserve conformity with the right-hand side vector **b**, which expects voltage impressions in each mesh instead of current injection at each node.

4.2.2 Branch pairs

The previous subsection introduced the discretization and representation of *stand-alone* branches as the resistor, inductor, capacitor, and voltage and current sources. This subsection introduces the discretization of branch *pairs* using the root-matching technique. A brief outline of the steps required to perform root-matching discretization of branch pairs was presented in section 4.1.2. To avoid redundancy the aforementioned root-matching discretization steps are re-visited for *series* branch-pairs only as analogous explanations to apply to *parallel* branch pairs. A summary table is provided at the end of this section to show side-by-side comparisons of discretizing series and parallel branch pairs using the root-matching method. These tables are shown in Figs. 4.18 and 4.19, respectively.

4.2.2.1 Series RL

Consider the series resistive-inductive (RL) branch in Fig. 4.8. The following explanations pertain to the explanations presented earlier in section 4.1.2 regarding this branch pair.

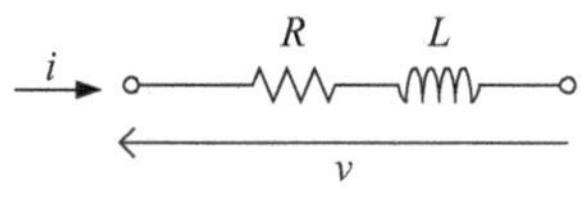

Fig. 4.8 Series RL branch

The s-domain admittance transfer function for this branch is given in (4.11).

$$Y(s) = \frac{I(s)}{V(s)} = \frac{1}{R + sL} \tag{4.11}$$

The pole of this transfer function is given in (4.12).

$$\omega_{pole} = -R/L \tag{4.12}$$

Using the final value theorem under a step input to (4.11) leads to (4.13).

$$\lim_{s \to 0} (sV(s)H(s)) = \lim_{s \to 0} \left(\frac{1}{s} Y(s)s \right) = 1/R \tag{4.13}$$

The equivalent z-domain transfer function to (4.11) is given on the left side of (4.14). After adding the adjustment-gain coefficient k and a zero to the numerator (since it is constant), the transfer function is modified as shown on the right of (4.14).

$$Y[z] = \frac{I[z]}{V[z]} = \frac{1}{z - e^{\omega_{pole}\Delta t}} \Rightarrow \frac{kz}{z - e^{\omega_{pole}\Delta t}} \tag{4.14}$$

Using the final value theorem under a step input to (4.14) results in (4.15).

$$\lim_{z\to 1}\left(\frac{(z-1)}{z}V(z)Y(z)\right) = \lim_{z\to 1}\left(\frac{z}{z-1}Y(z)\frac{z-1}{z}\right) = k/(1-e^{-\Delta t\omega_{pole}}) \tag{4.15}$$

Equating (4.13) and (4.15) results in the adjustment-gain coefficient in (4.16).

$$k = (1 - e^{\omega_{pole}\Delta t})/R \tag{4.16}$$

Substituting (4.16) into (4.14) and solving for the branch voltage and current gives (4.17) and (4.18), respectively.

$$V[z] = I[z]/k + e^{-\Delta t\omega_{pole}}z^{-1}I[z]/k \tag{4.17}$$

$$I[z] = kV[z] + e^{-\Delta t\omega_{pole}}z^{-1}I[z] \tag{4.18}$$

Taking the inverse z-transform of these expressions results in the discrete-time (difference) expressions for branch voltage and current given in (4.19) and (4.20), respectively.

$$v^{k+1} = \frac{1}{k}i^{k+1} + \frac{e^{-\Delta t\omega_{pole}}}{k}i^{k} \tag{4.19}$$

$$i^{k+1} = kv^{k+1} + e^{-\Delta t\omega_{pole}}i^{k} \tag{4.20}$$

From (4.19) and (4.20), the discrete equivalent series RL branch used for the mesh and nodal formulations are shown in Figs. 4.9 and 4.10, respectively. As noticed, these branch models also conform to the generic branch shown in Fig. 4.2.

It should be noticed that the trapezoidal rule, backward Euler, and root-matching integration methods analyzed above can give slightly different results even when simulating a simple series RL circuit. Consider, for example the simple RL circuit shown in Fig. 4.11, where the circuit parameters are indicated as annotations. The voltage and current waveforms of the series RL load are overlaid on the left of Fig. 4.12. These overlays, which appear to be identical, compare the results of using backward Euler integration (in *Simulink*) and the root-matching integration method (multicore solver). A point-by-point comparison[42] between them, shown on the right side,

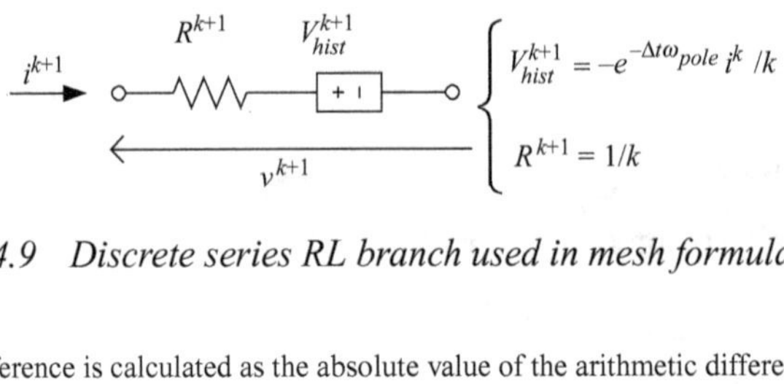

Fig. 4.9 Discrete series RL branch used in mesh formulations

[42] The resulting difference is calculated as the absolute value of the arithmetic difference $|x_i - x_j|$.

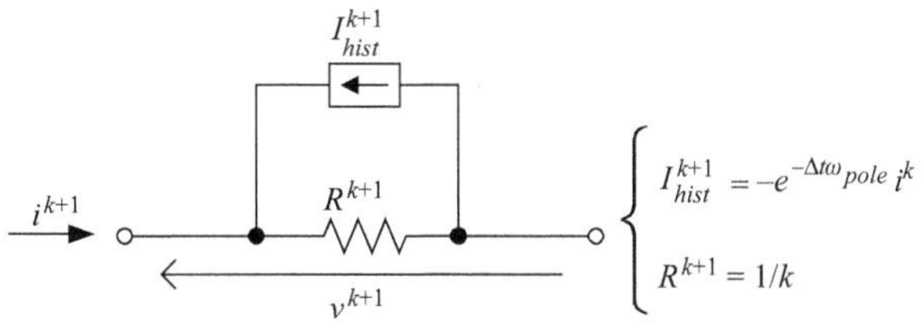

Fig. 4.10 Discrete series RL branch used in nodal formulations

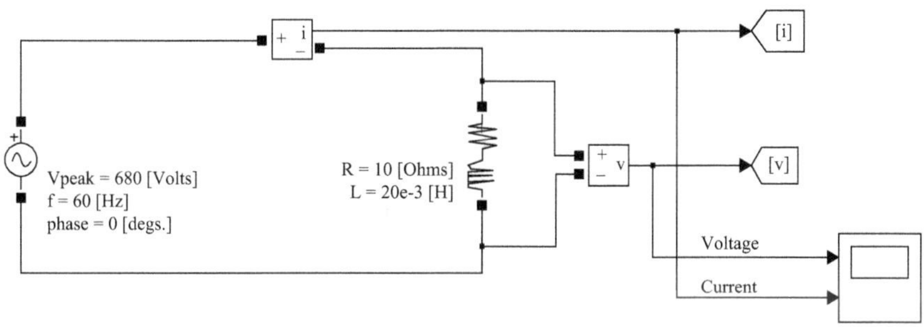

Fig. 4.11 RL circuit to demonstrate finite differences in results

confirms that finite differences exist[43] in both the voltage and current. It is noted that overlays, as shown on the left of Fig. 4.12 appear to be exact, but they are not. Overlaying waveforms as such is often deceiving. Readers are referred to [88,89] for additional examples on comparing root-matching results.

4.2.2.2 Series RC

The second example of root matching is given for the series resistive-capacitive (RC) branch. This series RC branch differs from the series RL branch in its admittance transfer function: the latter has a zero in the numerator and requires a *ramp* input instead of a *step* input to apply the final value theorem.

The s-domain admittance transfer function for the series RC branch is given in (4.21).

$$Y(s) = \frac{I(s)}{V(s)} = \frac{sC}{1 + sRC} \tag{4.21}$$

The zero and pole of this transfer function are as shown in (4.22):

$$\omega_{zero} = 0; \quad \omega_{pole} = -1/RC \tag{4.22}$$

Using the final value theorem with a ramp input yields (4.23):

$$\lim_{s\to 0}\left(\frac{1}{s^2}Y(s)s\right) = C \tag{4.23}$$

[43] The word *error* is not used in this book because the results are not validated against analytical expressions or against experimental data.

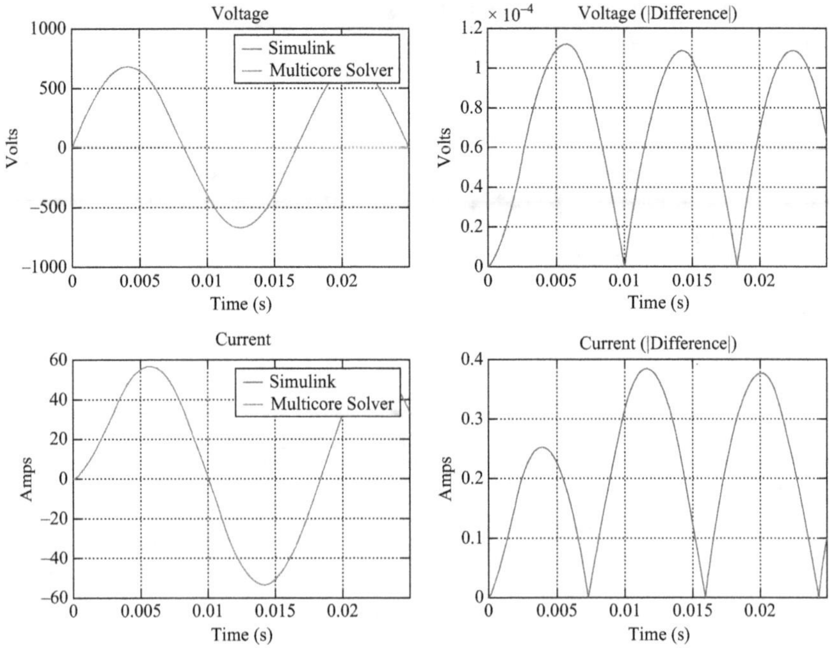

Fig. 4.12 Overlay of series RL voltage and current (left); the absolute value of their arithmetic difference (|difference|) is also shown (right)

Next, transferring the zero and pole to the z-domain, and adding the adjustment-gain coefficient k results in the equivalent z-domain transfer function in (4.24).

$$Y[z] = \frac{I[z]}{V[z]} = \frac{k(z - e^{-\Delta t\omega_{zero}})}{z - e^{-\Delta t\omega_{pole}}} = \frac{k(z-1)}{z - e^{-\Delta t\omega_{pole}}} \tag{4.24}$$

Using the final value theorem with a ramp input applied to (4.24) produces (4.25). Next, by equating (4.23) and (4.25) gives k in (4.26).

$$\lim_{z\to 1}\left(\frac{z\Delta t}{(z-1)^2}Y(z)\frac{z-1}{z}\right) = \Delta tk/(1 - e^{-\Delta t\omega_{pole}}) \tag{4.25}$$

$$k = C(1 - e^{-\Delta t\omega_{pole}})/\Delta t \tag{4.26}$$

Solving for the current in (4.24) results in (4.27). Taking the inverse z-transform of (4.27) leads to the discrete-time current expression in (4.28), which gives the Norton equivalent branch model shown in Fig. 4.13.

$$I[z] = kV[z] - z^{-1}(kV[z] - e^{-\Delta t\omega_{pole}}I[z]) \tag{4.27}$$

$$i^{k+1} = kv^{k+1} - (kv^k - e^{-\Delta t\omega_{pole}}i^k) \tag{4.28}$$

Similar to the derivation of current, solving for the voltage in (4.24) results in (4.29). Taking the inverse z-transform of (4.29) results in the discrete-time current expression given in (4.30), which gives the ThéVénin equivalent branch model illustrated in Fig. 4.14.

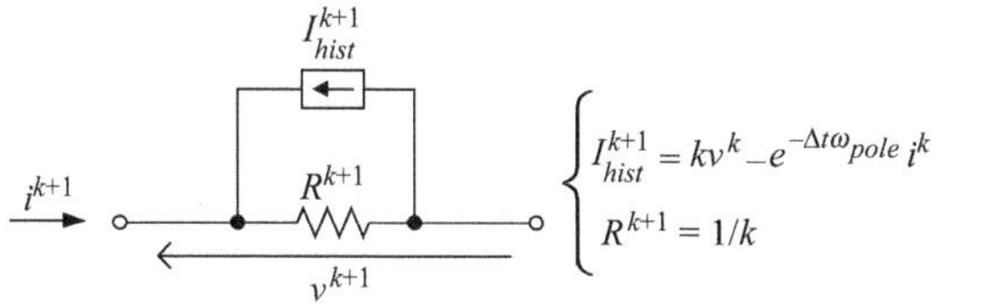

Fig. 4.13 Discrete equivalent series RC branch used in nodal formulations

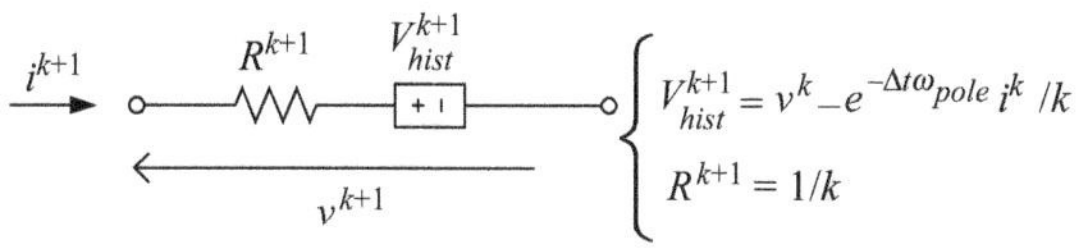

Fig. 4.14 Discrete equivalent series RC branch used in mesh formulations

$$V[z] = I[z]/k + z^{-1}(V[z] - e^{-\Delta t\omega_{pole}} I[z]/k) \tag{4.29}$$

$$v^{k+1} = \frac{1}{k} i^{k+1} + \left(v^k - \frac{e^{-\Delta t\omega_{pole}}}{k} i^k \right) \tag{4.30}$$

Similar to the RL circuit in Fig. 4.11, a contrast of the results produced by the backward Euler and root-matching integration methods is made for the series RC circuit in Fig. 4.15. The voltage and current waveforms of the series RC are shown on the left side of Fig. 4.16. As seen from the right side, finite differences also exist. A close-up of the current's initial rise (where the maximum difference occurs) is shown in Fig. 4.17. As seen, this large difference lasts for one timestep and disappears as the simulation advances. It is important to re-state that overlays such as those on the left of Fig. 4.16 appear to be exact, but discrepancies exist [88,89].

The foregoing root-matching explanations for the *series* RL and RC branch pairs are summarized in Fig. 4.18. Since the derivations for the *parallel* RL and RC branch pairs follow the derivations of their series counterparts, they are expressly

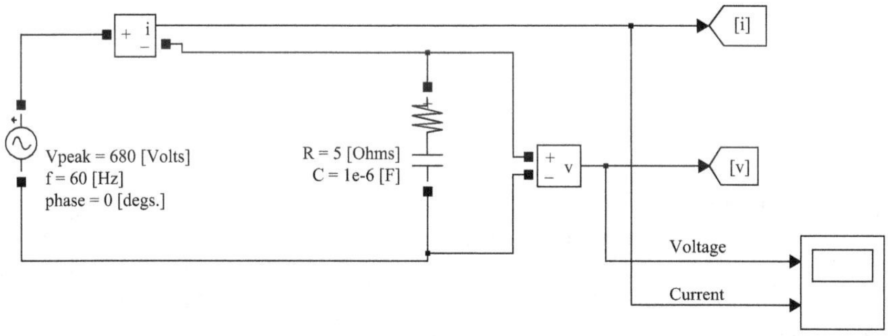

Fig. 4.15 RC circuit to demonstrate finite differences in results

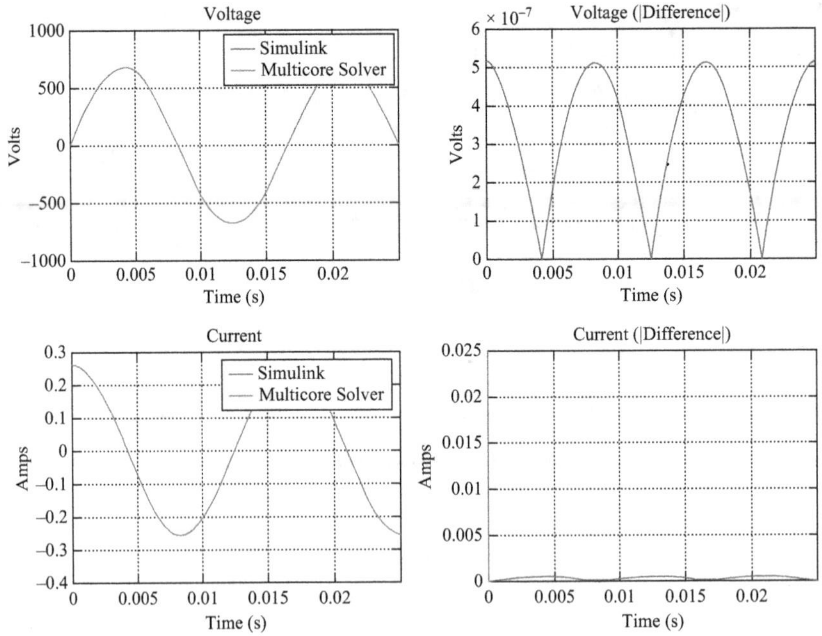

Fig. 4.16 Overlay of series RC voltage and current (left); the absolute value of their arithmetic difference (|difference|) is also shown (right)

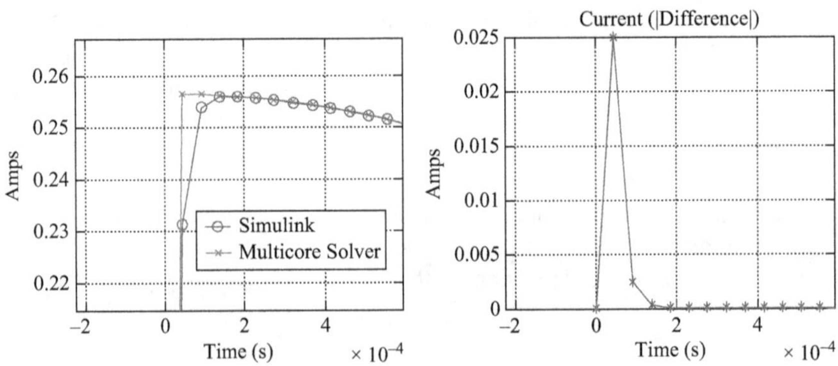

Fig. 4.17 Close-up of initial charging currents. Largest difference lasts about one timestep

omitted from the text. Instead, they are included in Fig. 4.19, which summarizes the application of root-matching to parallel branch pairs.

What is important to note from both summaries in Figs. 4.18 and 4.19 is the form of the transfer functions in the s-domain (admittance vs. impedance); also, whether or not the transfer functions have zeros on the numerators, and whether or not a step or ramp input was used when applying the final value theorem. A close comparison of

		Series RL	Series RC
Time domain (t)	Original branch	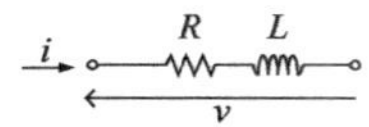	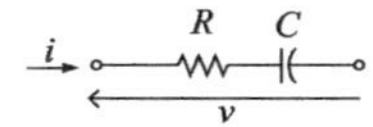
s domain	Transfer function	$Y(s)=\dfrac{I(s)}{V(s)}=\dfrac{1}{R+sL}$	$Y(s)=\dfrac{I(s)}{V(s)}=\dfrac{sC}{1+sRC}$
	Poles and zeros	$\omega_{pole}=-R/L$	$\omega_{zero}=0;\quad \omega_{pole}=-1/RC$
	Final value	$\lim\limits_{s\to 0}\left(\frac{1}{s}Y(s)s\right)=1/R$ (*step* input)	$\lim\limits_{s\to 0}\left(\frac{1}{s^2}Y(s)s\right)=C$ (*ramp* input)
z domain	Transfer function	$Y[z]=\dfrac{I[z]}{V[z]}=\dfrac{kz}{z-e^{\omega_{pole}\Delta t}}$	$Y[z]=\dfrac{I[z]}{V[z]}=\dfrac{k(z-e^{-\Delta t\omega_{zero}})}{z-e^{-\Delta t\omega_{pole}}}=\dfrac{k(z-1)}{z-e^{-\Delta t\omega_{pole}}}$
	Final value	$\lim\limits_{z\to 1}\left(\frac{z}{z-1}Y(z)\frac{z-1}{z}\right)=k\Big/\left(1-e^{-\Delta t\omega_{pole}}\right)$ (*step* input)	$\lim\limits_{z\to 1}\left(\frac{z\Delta t}{(z-1)^2}Y(z)\frac{z-1}{z}\right)=\Delta tk\Big/\left(1-e^{-\Delta t\omega_{pole}}\right)$ (*ramp* input)
	Match final values	$k=\left(1-e^{-\Delta t\omega_{pole}}\right)\big/R$	$k=C\left(1-e^{-\Delta t\omega_{pole}}\right)\big/\Delta t$
	Current	$I[z]=kV[z]+e^{-\Delta t\omega_{pole}}z^{-1}I[z]$	$I[z]=kV[z]-z^{-1}\left(kV[z]-e^{-\Delta t\omega_{pole}}I[z]\right)$
	Voltage	$V[z]=I[z]/k-e^{-\Delta t\omega_{pole}}z^{-1}I[z]/k$	$V[z]=I[z]/k+z^{-1}\left(V[z]-e^{-\Delta t\omega_{pole}}I[z]/k\right)$
Discrete time domain (k)	Current	$i^{k+1}=kv^{k+1}+e^{-\Delta t\omega_{pole}}i^k$	$i^{k+1}=kv^{k+1}-\left(kv^k-e^{-\Delta t\omega_{pole}}i^k\right)$
	Equivalent branch	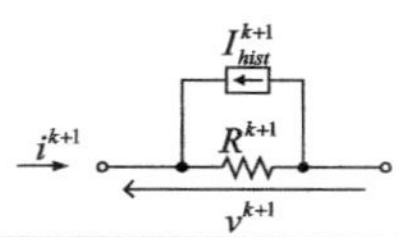 $\begin{cases} I_{hist}^{k+1}=-e^{-\Delta t\omega_{pole}}i^k \\ R^{k+1}=1/k \end{cases}$	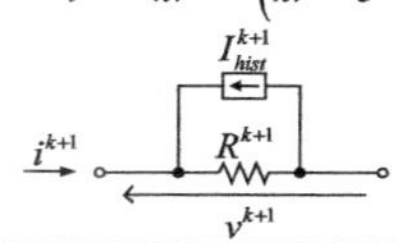 $\begin{cases} I_{hist}^{k+1}=kv^k-e^{-\Delta t\omega_{pole}}i^k \\ R^{k+1}=1/k \end{cases}$
	Voltage	$v^{k+1}=\dfrac{1}{k}i^{k+1}-\dfrac{e^{-\Delta t\omega_{pole}}}{k}i^k$	$v^{k+1}=\dfrac{1}{k}i^{k+1}+\left(v^k-\dfrac{e^{-\Delta t\omega_{pole}}}{k}i^k\right)$
	Equivalent branch	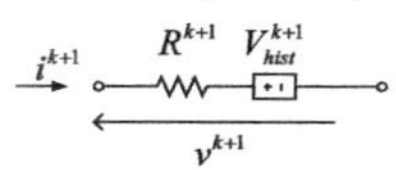 $\begin{cases} V_{hist}^{k+1}=-e^{-\Delta t\omega_{pole}}i^k/k \\ R^{k+1}=1/k \end{cases}$	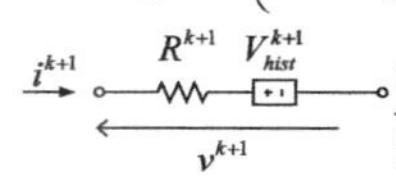 $\begin{cases} V_{hist}^{k+1}=v^k-e^{-\Delta t\omega_{pole}}i^k/k \\ R^{k+1}=1/k \end{cases}$

Fig. 4.18 Root-matching for series branch pairs

		Parallel RC	Parallel RL
Time domain (t)	Original branch	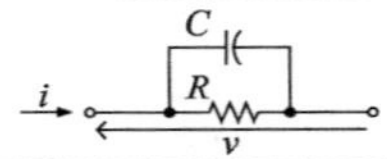	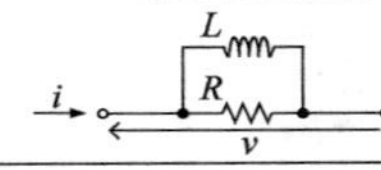
s domain	Admittance transfer function	$Z(s)=\dfrac{V(s)}{I(s)}=\dfrac{R}{1+sRC}$	$Z(s)=\dfrac{V(s)}{I(s)}=\dfrac{sRL}{R+sL}$
	Roots	$\omega_{pole}=-1/RC$	$\omega_{zero}=0;\quad \omega_{pole}=-R/L$
	Final value	R (for a *step* input)	L (for a *ramp* output)
z domain	Admittance transfer function	$Z[z]=\dfrac{V[z]}{I[z]}=\dfrac{kz}{z-e^{-\Delta t\omega_{pole}}}$	$Z[z]=\dfrac{V[z]}{I[z]}=\dfrac{k(z-e^{-\Delta t\omega_{zero}})}{z-e^{-\Delta t\omega_{pole}}}=\dfrac{k(z-1)}{z-e^{-\Delta t\omega_{pole}}}$
	Final value	$k/\left(1-e^{-\Delta t\omega_{pole}}\right)$	$\Delta tk/\left(1-e^{-\Delta t\omega_{pole}}\right)$
	Match final values	$k=R\left(1-e^{-\Delta t\omega_{pole}}\right)$	$k=L\left(1-e^{-\Delta t\omega_{pole}}\right)/\Delta t$
	Branch current	$I[z]=V[z]/k-e^{-\Delta t\omega_{pole}}z^{-1}V[z]/k$	$I[z]=V[z]/k-z^{-1}\left(e^{-\Delta t\omega_{pole}}V[z]/k-I[z]\right)$
	Branch voltage	$V[z]=kI[z]+e^{-\Delta t\omega_{pole}}z^{-1}V[z]$	$V[z]=kI[z]+z^{-1}\left(e^{-\Delta t\omega_{pole}}V[z]-kI[z]\right)$
Discrete time domain (k)	Branch current	$i^{k+1}=\dfrac{1}{k}v^{k+1}-\dfrac{e^{-\Delta t\omega_{pole}}}{k}v^k$	$i^{k+1}=\dfrac{1}{k}v^{k+1}-\left(\dfrac{e^{-\Delta t\omega_{pole}}}{k}v^k-i^k\right)$
	Equivalent branch	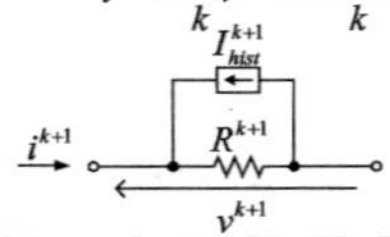 $\begin{cases}I_{hist}^{k+1}=e^{-\Delta t\omega_{pole}}v^k/k\\ R^{k+1}=k\end{cases}$	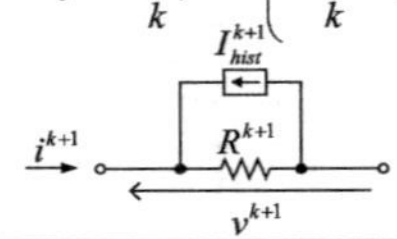 $\begin{cases}I_{hist}^{k+1}=e^{-\Delta t\omega_{pole}}v^k/k-i^k\\ R^{k+1}=k\end{cases}$
	Branch voltage	$v^{k+1}=ki^{k+1}+e^{-\Delta t\omega_{pole}}v^k$	$v^{k+1}=ki^{k+1}+\left(e^{-\Delta t\omega_{pole}}v^k-ki^k\right)$
	Equivalent branch	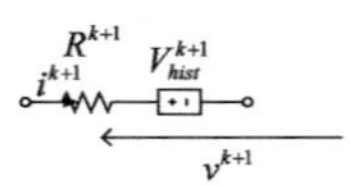 $\begin{cases}V_{hist}^{k+1}=e^{-\Delta t\omega_{pole}}v^k\\ R^{k+1}=k\end{cases}$	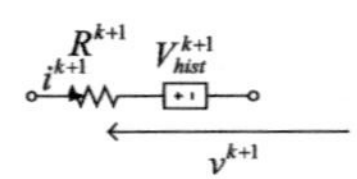 $\begin{cases}V_{hist}^{k+1}=e^{-\Delta t\omega_{pole}}v^k-ki^k\\ R^{k+1}=k\end{cases}$

Fig. 4.19 Root-matching for parallel branch pairs

the columns in Figs. 4.18 and 4.19 suggests that the series RL and series RC branch pairs are congruent with the parallel RC and parallel RL branch pairs, respectively.

4.2.3 Switches

Similar to all branches introduced so far, switches also conform to the branch model of Fig. 4.2. Because the subject of switch modeling is extensive, this book limits the discussion of switches to a few types only. For additional information on modeling switches, readers are referred to References 50, 54, 55, 57, 58, 69, 90–92.

This sub-section introduces six switch variants (referred to as types), the derivations to show they conform to the generic branch models of Fig. 4.2, and presents further details on the interpolation routine mentioned in section 3.2.

4.2.3.1 Switch types

In practice, power electronic switches are categorized by their ratings: power and switching speed [93,94]. Inclusion or absence of these physical traits in a switch model depends on the level of needed fidelity. To reduce development time, it is often convenient to model switches using a one-model-fit-all approach.

Herein, switches are categorized by their commutation type (natural or forced) and by their snubber branch (present or not present). Snubber branches are branches connected in parallel with a switching device to reduce stress levels to safe values. The following list shows the switch types considered in the book:

- Without snubbers
  1. Diode (self-commutated)
  2. IGBT (forced-commutation)
  3. IGBT with diode (self- and forced-commutation)
- With snubbers
  1. Diode with snubber (self-commutated)
  2. IGBT with snubber (forced-commutation)
  3. IGBT with diode and snubber (self- and forced-commutation)

Switches without snubbers can be either stand-alone diodes or insulated-gate bipolar transistors (IGBTs)—or both if connected in anti-parallel. Switches with snubbers, on the other hand, include diodes and/or IGBT combinations with a series-RC branch connected in parallel. These six types of switches are shown in Fig. 4.20.

The top row in Fig. 4.20 shows the switches without snubbers (types 1, 2, and 3). The bottom row shows switches with snubbers (types 4, 5, and 6). In all types, the switches are connected from node p to node q, the total current i_{pq} direction follows the diode convention, and the voltage drop v_{pq} is measured from p to q. Because diodes and IGBT are connected in anti-parallel, $v_{qp} = -v_{pq}$ is defined as the voltage drop in the direction of IGBT, and $i_{qp} = -i_{pq}$ as the current entering the switching device from node q. The measurement direction from p to q or, vice versa, from q to p depends on the switch being analyzed. For example, in rectifier circuits it is convenient to observe switch voltage and currents measured from node p to node q. In inverter circuits, it is more convenient to make measurements from node q to node p.

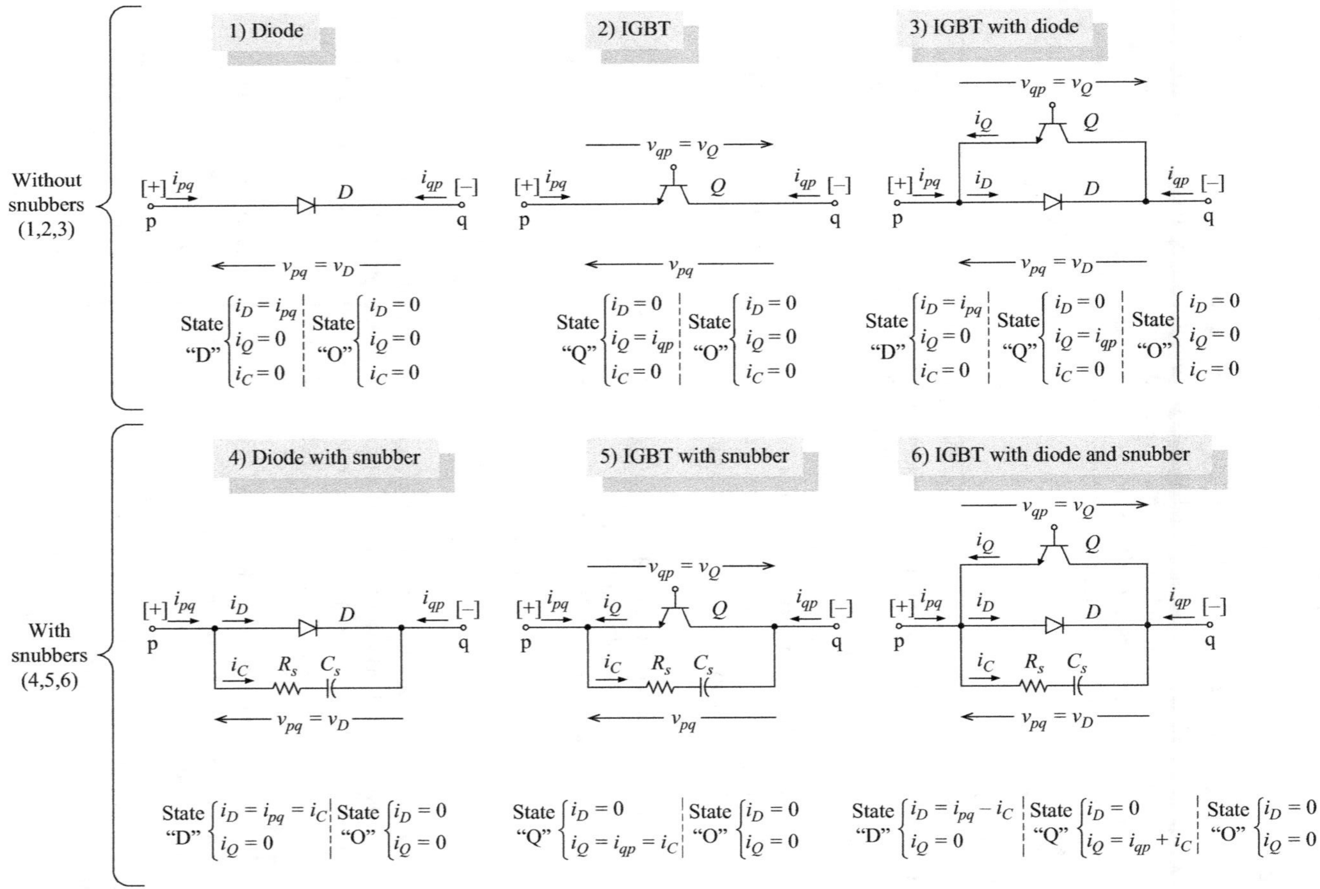

Fig. 4.20 Types of switches and their possible conduction states

Each switch type has a conduction state. These states mean:

- State "D": when only the diode part conducts
- State "Q": when only the IGBT part conducts
- State "O": when neither the diode nor the IGBT parts conducts

The naming conventions for the currents in each state are also annotated in Fig. 4.20. In switch types containing a diode (types 1, 3, 5, and 6), the current through the diode part is i_D. In switches with IGBTs (types 2, 3, 5, and 6), the current through the IGBT part is i_Q. If snubbers are present (switch types 4, 5, and 6), the current through the snubber part is i_C.

Switch type 6 is the most complex switch type as it includes two switching elements (diode and IGBT) and a snubber branch. The diode self-commutates based on its voltage and current, and the IGBT commutates based on the firing signals received from the control network. Regardless of the switch type, the switch variations also conform to the generic branch model shown earlier in Fig. 4.2 and are explained next.

4.2.3.2 Switch branch models

The parallel combination of a variable resistor[44] (to represent a diode or an IGBT), and of a series RC branch to represent a snubber can be reduced to an equivalent resistance R_{pq} and its historical source. Such reductions complies with the generic branch model shown earlier in Fig. 4.2.

To illustrate the reduction of a switch branch to an equivalent resistance R_{pq} and a historical source, which is how all power electronic switches are treated herein, consider the explanatory table in a three-column arrangement with eight boxes ((a)–(h)) in Fig. 4.21. The left column of this table groups the switch types (namely, 1, 2, 3 or 4, 5, 6) and shows the possible values for R_{pq}. (These values of R_{pq} are the same for a nodal or mesh formulation.) The center and right columns show the historical values according to the formulation type.

As seen from the top row of Fig. 4.21, all switch types (1 through 6) conform to the generic branch model shown earlier in Fig. 4.2. For instance, Fig. 4.21(a) shows the equivalent switch model for use in mesh formulations; Fig. 4.21(b) shows the equivalent switch model for use in nodal formulations. (Note that V_{pq} and I_{pq} in Fig. 4.21 correspond to V_{hist} and I_{hist} in Fig. 4.2, respectively.)

Fig. 4.21(c) shows the value of the equivalent (time-varying) resistance R_{pq} across nodes p and q for switch types 1, 2, and 3 (without snubbers). The value $R_{pq} = R_{on}$ is used when *either* the diode or IGBT part conducts; the value $R_{pq} = R_{off}$ is used when *both* the diode and IGBT are off. These two values of R_{pq} are common to both mesh and nodal formulations.

Fig. 4.21(d) shows the voltage impression value V_{pq} between nodes p and q for a *mesh* formulation for switch types (1–3) with no snubbers. This voltage value changes with the state of the switch as indicated by V_{Don} and V_{Qon}, which represent the forward conduction voltage drop of the diode and IGBT, respectively. Similarly, Fig. 4.21(e)

[44] It is common practice to use $R_{on} = 1\ \text{m}\Omega$ and $R_{off} = 1\ \text{M}\Omega$ to represent *on*- and *off*-resistance, respectively, in power electronic valves and protective devices.

Formulation type → / Switch type ↓	Mesh Formulation	Nodal Formulation
	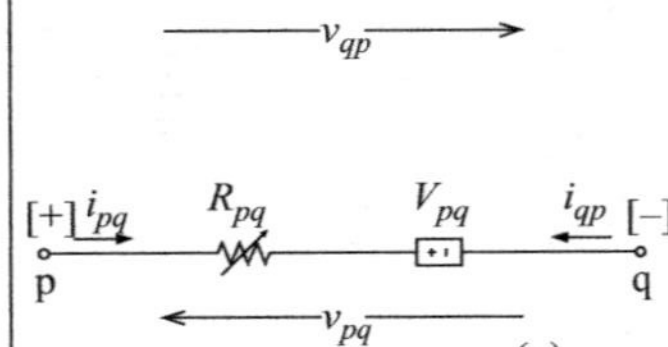 (*a*)	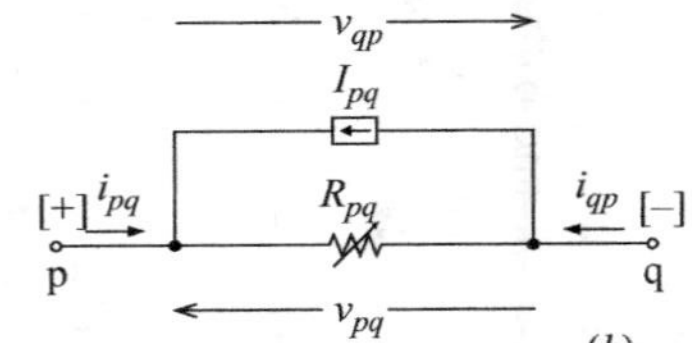 (*b*)
No snubbers (1,2,3)	$R_{pq} \rightarrow$ $= R_{on}$ when diode *or* IGBT is *on* $= R_{off}$ when diode *and* IGBT are *off* $R_{on} = 1\ \text{m}\Omega;\ R_{off} = 1\ \text{M}\Omega$ (*c*)	$V_{pq} \rightarrow$ $= V_{Don}$ when diode is *on* $= -V_{Qon}$ when IGBT is *on* $= 0$ when diode *and* IGBT are *off* (*d*)	$I_{pq} \rightarrow$ $= \dfrac{V_{Don}}{R_{on}}$ when diode is *on* $= \dfrac{-V_{Qon}}{R_{on}}$ when IGBT is *on* $= 0$ when diode *and* IGBT are *off* (*e*)
With snubbers (4,5,6)	$R_{pq} \rightarrow$ $= \dfrac{R_{on}R_{RC}}{R_{on} + R_{RC}}$ when diode *or* IGBT is *on* $= \dfrac{R_{off}R_{RC}}{R_{off} + R_{RC}}$ when diode *and* IGBT are *off* $R_{on} = 1\ \text{m}\Omega;\ R_{off} = 1\ \text{M}\Omega$ (*f*)	$V_{pq} \rightarrow$ $= R_{pq}\left(\dfrac{V_{Don}}{R_{on}} + \dfrac{V_{histRC}}{R_{RC}}\right)$ when diode is *on* $= R_{pq}\left(\dfrac{-V_{Qon}}{R_{on}} + \dfrac{V_{histRC}}{R_{RC}}\right)$ when IGBT is *on* $= V_{histRC}$ when diode *and* IGBT are *off* (*g*)	$I_{pq} \rightarrow$ $= \dfrac{V_{Don}}{R_{on}} + I_{histRC}$ when diode is *on* $= \dfrac{-V_{Qon}}{R_{on}} + I_{histRC}$ when IGBT is *on* $= V_{histRC}$ when diode *and* IGBT are *off* (*h*)

Fig. 4.21 Discrete switch equivalents

shows the values of I_{pq} (the Norton-equivalent of V_{pq}) for use with nodal formulations, which also varies with the state of the switch.

Fig. 4.21(f) shows the value of R_{pq} for switches types 4, 5, and 6 with snubbers. As may be noticed from these expressions, the switch resistance (R_{on} or R_{off}) is paralleled with the snubber resistance denoted R_{RC} (derived earlier for the series RC branch in Fig. 4.18). The parallel combination of resistances into one resistance value R_{pq} is an effective way to model switches with snubber branches. This combination also applies to the switch's historical sources as explained next.

Fig. 4.21(g) shows the values of V_{pq} for switches with snubbers (types 4–6), where V_{histRC} is the historical voltage source of the series RC branch shown in Fig. 4.18 (*supra*). The superposition of terms appearing for V_{pq} results from a series of Norton–ThéVénin transformations (not shown). This transformation combines the historical sources of the diode, IGBT, and snubber branch into one equivalent value V_{pq}, which conforms with the generic branch shown earlier in Fig. 4.2. (The idea of reducing switches to simpler branches through Norton–ThéVénin transformations is suggested in Reference 50.) Finally, Fig. 4.21(h) shows the current injection expressions for switch types 4, 5, and 6 for use in nodal formulations.

4.2.3.3 Interpolation

The concept and importance of interpolation in power system simulation was discussed earlier in section 3.2. It was mentioned that if rollback (or some other compensation method) was not implemented in a solver to account for intermediate events occurring between time grid divisions, the simulation results could show non-physical transient spikes [50,56,58] and, consequently, produce non-physical switching losses and spurious harmonics. Additionally, and referring to Fig. 3.4, it was assumed that the exact time-instant of the events was detected by some means. This was an important assumption to keep in mind. This subsection illustrates how one approach approximate the time-instants of such events by using linear interpolation.

Consider the threshold-crossing situations depicted in Fig. 4.22 that determine whether a diode should toggle *on/off* state. (When compared to Figs. 3.4, 4.22 only shows one event occurring between time grid intervals.) Step (1) in the top diagram of Fig. 4.22 calls for a turn-off action as the diode voltage v_D falls below the threshold voltage V_{Don}. This scenario requires the simulation (Step 2) to roll back from t to the estimated time instant t_z, where $v_D = V_{Don}$. (Steps 3 and 4 were discussed in detail in Fig. 3.4.)

Similar to the top diagram in Fig. 4.22, the bottom diagram illustrates the situation of a diode wanting to turn *on* by its voltage v_D surpassing the threshold V_{Don}. This scenario also requires rolling back the simulation (Step 2) to the estimated time instant t_z where the event occurred.

In sum, the event-detection routine for both turn-*on* and turn-*off* cases is similar. The following explanations summarize the steps annotated in Fig. 4.22:

Step 1) The program begins its search routine and detects that a diode should toggle [68]. Said diode returns to the solver a fractional distance X_i (length) by which the simulation time exceeded the diode's estimated zero-crossing instant t_z.

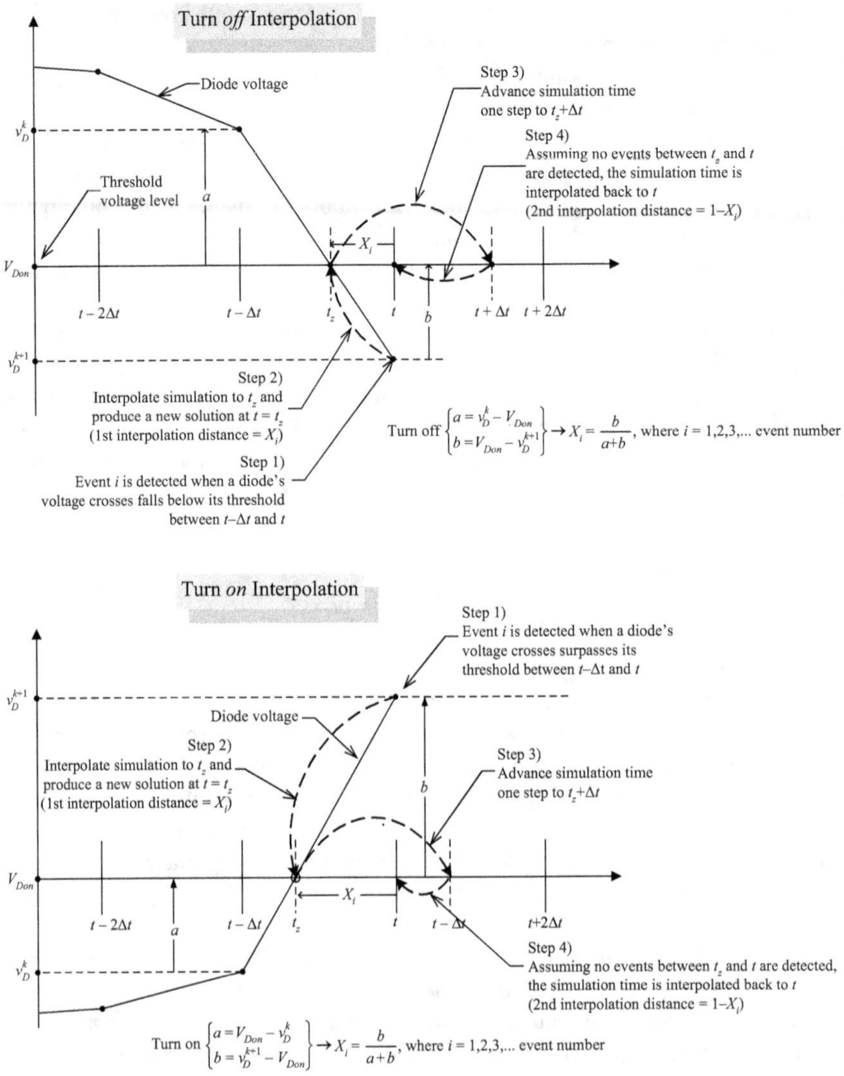

Fig. 4.22 Double interpolation procedure for a diode during turn-off (top) and turn-on (bottom) situations

Step 2) The solver interpolates the simulation time back to t_z (1st interpolation), the diode toggles, and the network is solved again at t_z.

Step 3) The simulation advances as normal and a new solution is produced at $t_z + \Delta t$.

Step 4) If no more intermediate events are detected between t_z and t, the network solution at $t_z + \Delta t$ is interpolated back to t (2nd interpolation).

The fractional distance X_i by which the simulation rolls back to the event's time-instant t_z is computed using (4.31), where a and b, respectively, are the left (historical) and right (present) vertical distances from the threshold shown in Fig. 4.22. The event time instant t_z (in seconds) is computed using (4.32). To produce a new solution at t_z, the solver must update the electrical network's right-hand side vector **b** for a solution at t_z. This is accomplished by, first, interpolating the historical sources of all branch pairs to t_z using (4.33) and, then, re-evaluating all independent (e.g., voltage and current) sources at t_z (except for constant or DC sources).

$$X_i = \frac{b}{a+b} = \frac{h_i}{\Delta t} = \frac{t - t_z}{\Delta t} \tag{4.31}$$

$$t_z = t - \Delta t \cdot X_i \tag{4.32}$$

$$hist_{t_z} = hist_{t-\Delta t} + (1 - X_i)(hist_t - hist_{t-\Delta t}) \quad \text{(1st interpolation)} \tag{4.33}$$

After the solution at t_z the simulation advances from t_z to $t_z + \Delta t$. At this time, a second interpolation rolls back the simulation to t (or to between t_z and t if additional events between t_z and t are detected). During the second (2nd) interpolation, assuming no additional events between t_z and t occur, the simulation re-synchronizes with the time grid by rolling back the excitation vector **b** and solution vector **x** from $t_z + \Delta t$ to t. During the second interpolation, the historical sources (and each value in vector **x**) are interpolated using (4.34).[45]

$$hist_t = hist_{t_z} + X_i(hist_{t_z+\Delta t} - hist_{t_z}) \quad \text{(2nd interpolation)} \tag{4.34}$$

It was underscored earlier that different solvers can produce difference results for the same model [88,89]. To exacerbate this fact, if one considers the uncertainty of using different integration methods, different interpolation techniques [67], and different internal optimizations then one, in all likelihood, should expect further differences in the results of switching networks.

To exemplify these possible differences in results, consider the diode circuit in Fig. 4.23, where the diode's voltage and current are overlaid on the left side

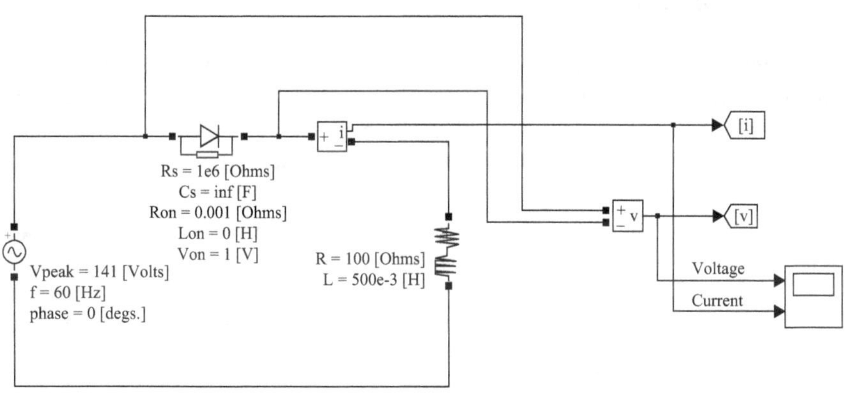

Fig. 4.23 Diode circuit to show difference in discretization

[45] The calculations of the turn-*on* interpolations follow those of the turn-*off* interpolations.

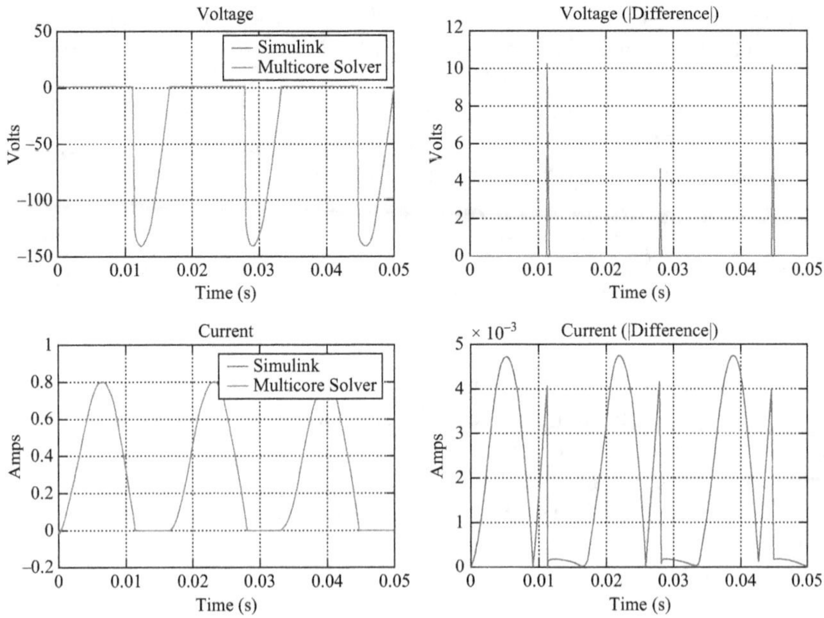

Fig. 4.24 Overlay of diode voltage and current. Results are in reasonable agreement, but they are not exact

of Fig. 4.24. The arithmetic differences of the results produced with *Simulink* and with the multicore solver developed for this book are shown on the right. These differences show that finite differences are possible even in a simple circuit.

4.3 Control network

This section presents the discretization and solution of the power system's control network. Since it is not possible to cover every possible control block, only those blocks used with the power system models of Chapter 2 are covered here. The blocks covered in this section include a state-variable equation set, a transfer function block, moving RMS and moving average blocks, a proportional–integral–derivative (PID) controller block, and a pulse-width modulation (PWM) generator block.

4.3.1 State-variable equations

State-variable equations are commonly used to model dynamic behavior in arbitrary systems (e.g., machine controllers, mechanical loads, among others). Consider the following state-variable equation set given in (4.35)[46]

[46] The *state* matrix **A** is different than the electrical network's immittance matrix **A**. This distinction is made clear in the book where appropriate.

$$\frac{d}{dt}\mathbf{x}(t) = \mathbf{A}\cdot\mathbf{x} + \mathbf{B}\cdot\mathbf{u}(t) \tag{4.35}$$

$$\mathbf{y}(t) = \mathbf{C}\cdot\mathbf{x} + \mathbf{D}\cdot\mathbf{u}(t) \tag{4.36}$$

where:

$\mathbf{x}(t)$ = state vector
$\mathbf{u}(t)$ = input vector
$\mathbf{y}(t)$ = output vector
$\mathbf{A}$ = state matrix
$\mathbf{B}$ = input matrix
$\mathbf{C}$ = state output matrix
$\mathbf{D}$ = input-to-output matrix.

Discretization of (4.35) using tunable integration (sub-section 4.1.1) leads to (4.37). Solving for the state vector $\mathbf{x}^{k+1}$ results in (4.38) [93], where after aggregating terms, results in the discrete state-variable formulation given in (4.39).

$$\frac{1}{\Delta t}(\mathbf{x}^{k+1} - \mathbf{x}^k) = \mathbf{A}\big(\gamma\cdot\mathbf{x}^{k+1} + (1-\gamma)\mathbf{x}^k\big) + \mathbf{B}\big(\gamma\cdot\mathbf{u}^{k+1} + (1-\gamma)\mathbf{u}^k\big) \tag{4.37}$$

$$\mathbf{x}^{k+1} = \underbrace{\overbrace{\left(\frac{1}{\Delta t}\mathbf{I} - \gamma\cdot\mathbf{A}\right)^{-1}}^{\mathbf{Q}}\left(\frac{1}{\Delta t}\mathbf{I} + (1-\gamma)\cdot\mathbf{A}\right)}_{\mathbf{A}_d}\mathbf{x}^k + \underbrace{\mathbf{Q}\cdot\mathbf{B}}_{\mathbf{B}_d}(\gamma\cdot\mathbf{u}^{k+1} + (1-\gamma)\mathbf{u}^k) \tag{4.38}$$

The non-zero structure of the state matrix $\mathbf{A}$ in (4.35) and its discrete counterpart $\mathbf{A}_d$ in (4.39) are *not* the same. A comparison of these two matrices was shown earlier in Figs. 2.8–2.11.

Finally, it should be noted that an automatic (programmatic) formation of state-variable equations is not trivial beyond a few control blocks. For models with a large number of interconnected control blocks, the automated tree approaches to form such equations in References 95–97 are recommended.

$$\text{Discrete state-variable equations} \rightarrow \begin{cases} \mathbf{x}^{k+1} = \mathbf{A}_d\cdot\mathbf{x}^k + \mathbf{B}_d(\gamma\cdot\mathbf{u}^{k+1} + (1-\gamma)\mathbf{u}^k) \\ \mathbf{y}^{k+1} = \mathbf{C}\cdot\mathbf{x}^{k+1} + \mathbf{D}(\gamma\cdot\mathbf{u}^{k+1} + (1-\gamma)\mathbf{u}^k) \\ \mathbf{Q} = \left(\frac{1}{\Delta t}\mathbf{I} - \gamma\cdot\mathbf{A}\right)^{-1} \\ \mathbf{A}_d^{k+1} = \mathbf{Q}\left(\frac{1}{\Delta t}\mathbf{I} + (1-\gamma)\mathbf{A}\right) \\ \mathbf{B}_d^{k+1} = \mathbf{Q}\cdot\mathbf{B} \\ \gamma \rightarrow \begin{cases} = 1/2, & \text{trapezoidal rule} \\ = 1, & \text{backward Euler} \end{cases} \end{cases} \tag{4.39}$$

where:

$\mathbf{x}^{k+1}$ = state vector at $k+1$
$\mathbf{x}^{k}$ = state vector at k
$\mathbf{u}^{k+1}$ = input vector at $k+1$
$\mathbf{u}^{k}$ = input vector at k
$\mathbf{A}$ = continuous state matrix
$\mathbf{A}_d^{k+1}$ = discrete state matrix at $k+1$ ($\neq\mathbf{A}$)
$\mathbf{B}_d^{k+1}$ = discrete input matrix at $k+1$ ($\neq\mathbf{B}$).

4.3.2 First-order transfer functions

Transfer functions are building blocks common to many control systems. These functions describe the input-output relations of linear, time-invariant differential equations [98] using the Laplace[47] operator s.

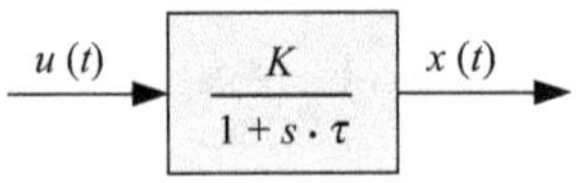

Fig. 4.25 First-order transfer function block

Consider the transfer function block shown in Fig. 4.25 and its corresponding differential equation given in (4.40):

$$\frac{d}{dt}x(t) = -\frac{1}{\tau}x(t) + \frac{K}{\tau}u(t) \tag{4.40}$$

where:

K = transfer function gain
τ = time constant (s)
$u(t)$ = exogeneous input
$x(t)$ = state-variable.

Discretization of (4.40) using the tunable integration method (which allows changing between the trapezoidal rule and backward Euler methods as discussed earlier in the chapter) results in (4.41), where solving for the state-variable x^{k+1} results in (4.42).

$$\frac{x^{k+1} - x^k}{\Delta t} = -\frac{1}{\tau}\left(\gamma \cdot x^{k+1} + (1-\gamma)x^k\right) + \frac{K}{\tau}\left(\gamma \cdot u^{k+1} + (1-\gamma)u^k\right) \tag{4.41}$$

$$x^{k+1} = \frac{\tau + (\gamma - 1)\Delta t}{\tau + \gamma \cdot \Delta t}x^k + \frac{K \cdot \Delta t}{\tau + \gamma \cdot \Delta t}\left(\gamma \cdot u^{k+1} + (1-\gamma)u^k\right) \tag{4.42}$$

[47] Pierre Simon Laplace (1749–1827) was a French applied mathematician and theoretical physicist. His treatise *Mécanique céleste* (1799–1825) is his best-known work.

where:

x^{k+1} = state-variable at $k+1$
x^k = state-variable at k
u^{k+1} = input at $k+1$
u^k = input at k
τ = time constant of differential equation
γ = tunable integration parameter
Δt = simulation time step.

4.3.3 Moving RMS

Root-mean-square (RMS)[48] measurements are of central importance to power system analysts. Its most common use is to assess the effective voltage and current levels throughout a power system. The RMS value of a periodic signal (e.g., voltage or current) of period T_0 is defined by the continuous expression in (4.43)

$$X_{RMS}(t) = \sqrt{\frac{1}{T_0}\int_{T_0}(x(t))^2 dt} \tag{4.43}$$

where:

T_0 = fundamental period in seconds
$x(t)$ = instantaneous measurement (e.g., voltage or current)
$X_{RMS}(t)$ = RMS value of $x(t)$.

The discrete equivalent of the continuous expression in (4.43) is given in (4.44).

$$X_{RMS}^k = \sqrt{\frac{1}{N}\sum_{k=1}^{N}(x^k)^2} \tag{4.44}$$

where:

N = number of samples in period T_0
x^k = data sample at time step k
X_{rms}^k = RMS value at time step k.

The calculation of RMS voltages with (4.44) in each phase (i.e., a, b, c), at every bus, and at each timestep is highly desirable in power system simulation to ensure a system operates within its tolerable voltage range. However, depending on how the RMS calculation is *implemented*, too many RMS measurements may reduce simulation performance. A more efficient way to compute (4.44) is by re-using the RMS result from the previous timestep as follows.

An RMS value can be calculated using a moving window as shown in Fig. 4.26, where the moving window (or array) holds N sample values at any given time. As the simulation time advances, new data enters the array from the right while old data

[48] RMS is defined as an average equal to the square root of the sum of the squares of the values divided by the number values at hand. For example, 3.674 is the RMS of 2, 3, 4, 5.

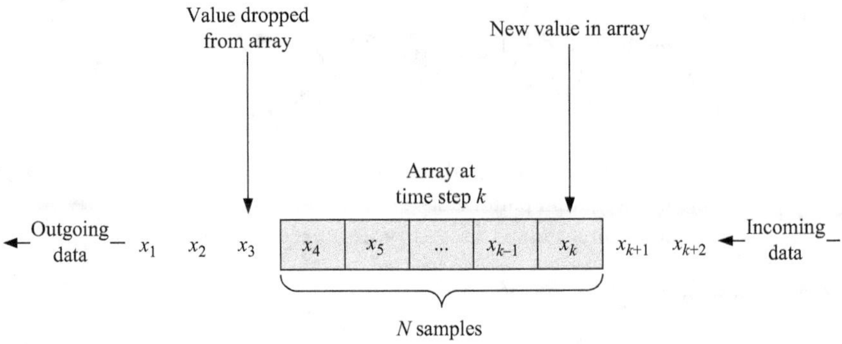

Fig. 4.26 RMS calculation using a moving window at timestep k

is dropped from the left. This type of data array, which appears to move from left to right, is known as moving window, buffer, queue, or first-in-first-out (FIFO)[49] array.

Referring to Fig. 4.26, the RMS value at timestep k is found by expanding (4.44) as:

$$X_{\rm rms}^{k} = \sqrt{\frac{x_4^2 + x_5^2 + \cdots + x_{k-1}^2 + x_k^2}{N}} \tag{4.45}$$

At the following timestep $(k + 1)$, the samples contained in the array are left-shifted by one position—that is, x_4 is dropped from the left, all other samples move left one position, and x_{k+1} is appended to the right as shown in Fig. 4.27. The RMS value at $k + 1$ for the moving window in Fig. 4.27 is computed with (4.46). Subtracting (4.45) from (4.46) results in (4.47).

$$X_{\rm rms}^{k+1} = \sqrt{\frac{x_5^2 + x_6^2 + \cdots + x_k^2 + x_{k+1}^2}{N}} \tag{4.46}$$

$$X_{\rm rms}^{k+1} = \sqrt{\frac{-x_4^2}{N} + \left(X_{\rm rms}^{k}\right)^2 + \frac{x_{k+1}^2}{N}} \tag{4.47}$$

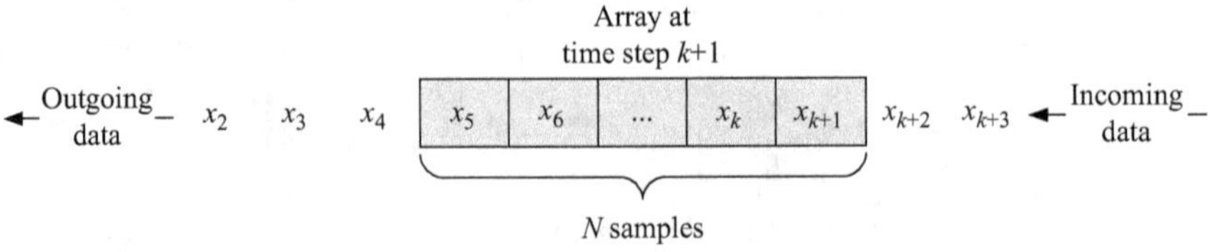

Fig. 4.27 RMS calculation at timestep k

[49] FIFO is an acronym borrowed from the field of accounting to denote one of the inventory valuation methods commonly applied in industry. It is also used in computer science to describe the order in which array elements are retrieved.

Generalizing (4.47) results in formulation (4.48), which is more computationally efficient than calculating an RMS value using (4.44)

$$X_{\text{rms}}^{k+1} = \sqrt{\frac{x_{k+1}^2 - x_{k+1-N}^2}{N} + (X_{\text{rms}}^k)^2} \ \Bigg| \ \begin{pmatrix} x^{k+1-N} = 0 \\ \text{when } (k+1) \leq N \end{pmatrix} \tag{4.48}$$

where:

N = number of samples in the moving array (e.g., $N = 360$ at $\Delta t = 46.296$ μs for a 60 Hz signal)
X_{rms}^{k+1} = RMS value at $k+1$
X_{rms}^{k} = RMS value at k
x_{k+1-N} = last sample value dropped from array (i.e., N steps ago)
x_{k+1} = last sample value to enter array.

While not required, it is recommended that the window size N span exactly one period of the incoming periodic signal. For example, for a 60-Hz signal simulated at $\Delta t = 50$ μs, $N = \frac{1/60 \text{ Hz}}{\Delta t = 50 \ \mu\text{s}} = 333.3\widehat{3}$. If the moving window size N is rounded-up from $N = 333.3\widehat{3}$ to $N = 334$, the resulting RMS value shows a spurious 2nd harmonic (120 Hz) superimposed onto the actual RMS value. To avoid these artificial oscillations, Δt can be reduced such that $N = \frac{1/60 \text{ Hz}}{\Delta t = 46.296 \ \mu\text{s}} = 360$ fits the 60 Hz period exactly 360 times.

To show the effect of the 2nd harmonic due to rounding N, (4.48) was implemented in *Simulink* using both $\Delta t = 46.296$ μs ($N = 360$) and $\Delta t = 50$ μs ($N = 334$). The relevant block diagrams are shown in Figs. 4.28 and 4.29, respectively, and the ensuing results are illustrated in Fig. 4.30. Although the curves on the top two charts appear identical, the close-up on the lower right reveals the 2nd harmonic problem when N is rounded off to 334.

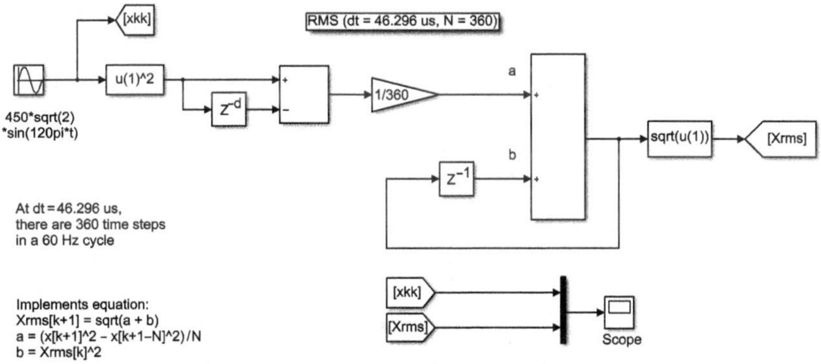

Fig. 4.28 Simulink implementation of a moving RMS block ($\Delta t = 46.296$ μs, $N = 360$)

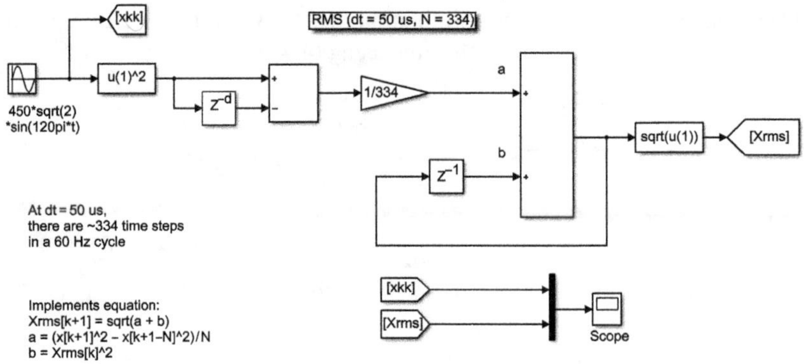

Fig. 4.29 Simulink implementation of a moving RMS block ($\Delta t = 50\ \mu s$, $N = 334$)

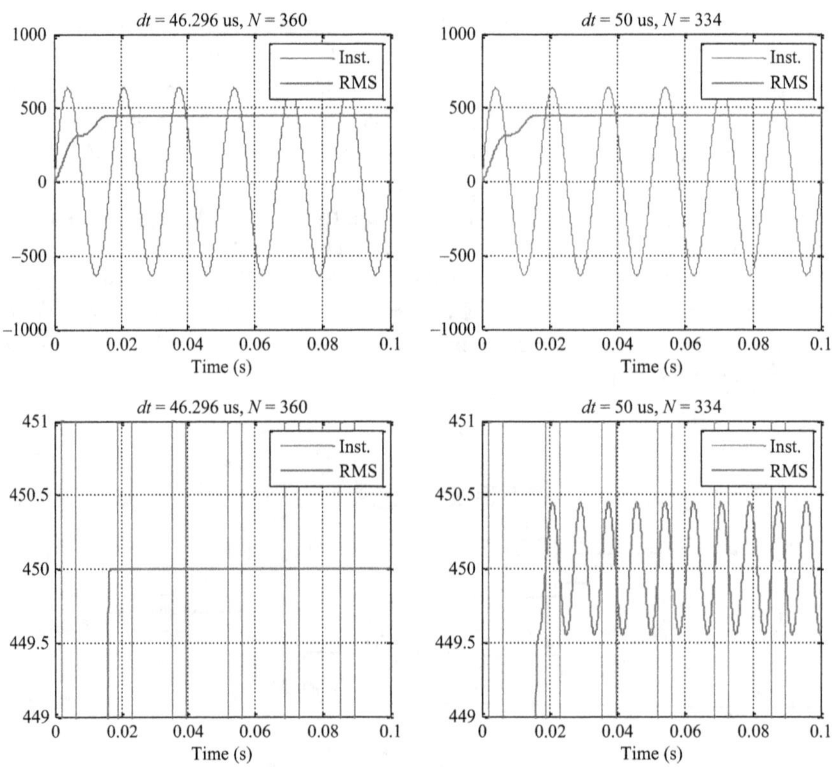

Fig. 4.30 Simulation of a moving RMS block for $\Delta t = 46.296\ \mu s$ (left) and $\Delta t = 50\ \mu s$ (right). Top charts: zoomed-out view, bottom charts: close-up view ("Inst." stands for instantaneous)

Common frequencies in AC power systems are 50 and 60 Hz. Depending on this value, Δt may or may not be naturally compatible with moving-window-based RMS measurements; this incompatibility suggests that the selection of Δt should take into consideration the frequencies of interest to the user as discussed in section 3.4. A *MATLAB* script is included in Appendix A to show frequencies compatible with timesteps $\Delta t = 50\ \mu s$ and $\Delta t = 46.296\ \mu s$.

4.3.4 Moving average

Similar to the foregoing moving RMS measurement, a moving average measurement returns the average of a signal over the last N samples. The continuous expression for a moving average is given in (4.49).

$$X_{avg}(t) = \frac{1}{T_0}\int_{T_0} x(t)dt \tag{4.49}$$

where:

T_0 = fundamental period of x(t) (secs)
$x(t)$ = input signal
$X_{avg}(t)$ = moving average value.

Following the derivation of the RMS measurement presented above, the discretized expression of (4.49) is given in (4.50)

$$X_{avg}^{k+1} = \frac{x^{k+1} - x^{k+1-N}}{N} + X_{avg}^{k} \;\Bigg|\; \begin{pmatrix} x^{k+1-N} = 0 \\ \text{when } (k+1) \leq N \end{pmatrix} \tag{4.50}$$

where:

N = number of sample values in a measurement window (e.g., $N = 360$ at $\Delta t = 46.296\ \mu s$ for a 60 Hz signal)
X_{avg}^{k+1} = average value at time step $k+1$
X_{avg}^{k} = average value computed at time step k
x_{k+1-N} = last sample dropped from array (i.e., N steps ago)
x_{k+1} = latest sample to enter array.

Like in the preceding subsection on moving RMS measurements, (4.50) was also implemented in *Simulink* as shown by the block diagram in Fig. 4.31. The average of the input signal $5 + \sin(120\pi t)$ is shown in Fig. 4.32, where the result returns the DC offset of value 5 as expected.

4.3.5 Power flow

Similar to RMS measurements, power measurements are also of central importance to power system analysts. One way to compute power flows is to use the two-wattmeter method [99]. The two-wattmeter method uses two moving-average blocks W_1 and W_2 to compute real and reactive power, respectively.

Assuming a three-wire three-phase system, W_1 in (4.51) represents the moving average of the signal produced by the product $v_{ab}i_a$, where v_{ab} is the instantaneous line-to-line voltage across phases a and b and i_a is the line current in phase a. Similarly,

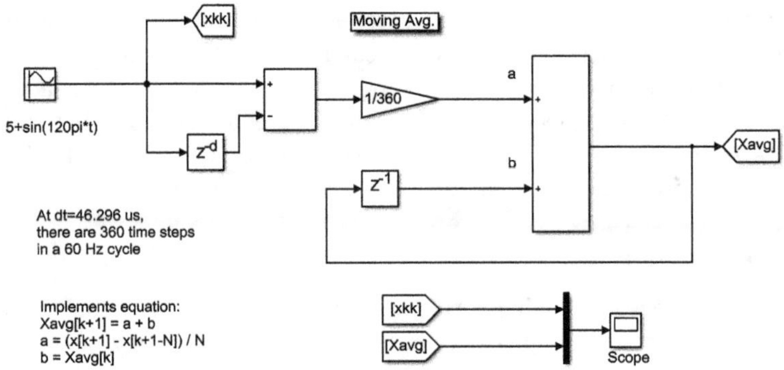

Fig. 4.31 Simulink implementation of a moving average window ($\Delta t = 46.296\ \mu s$, $N = 360$)

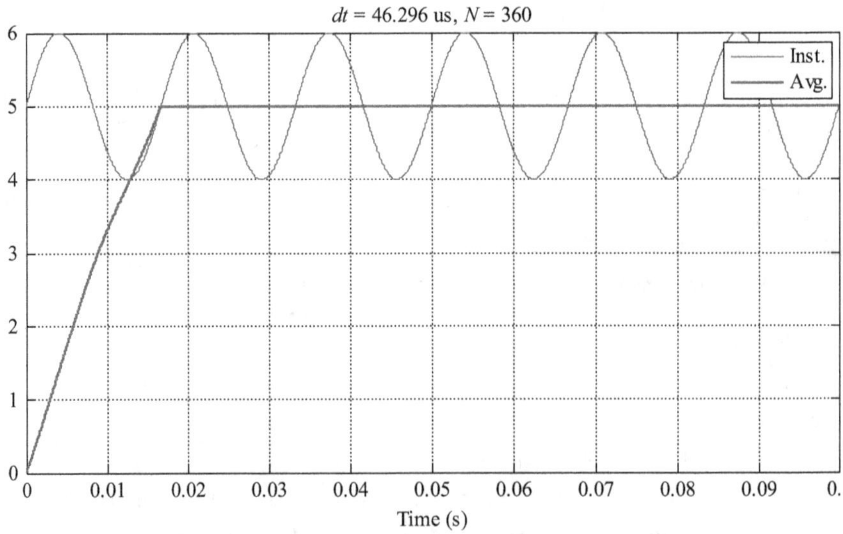

Fig. 4.32 Simulation of a moving average block for $\Delta t = 46.296\ \mu s$ and $N = 360$ ("Inst." stands for instantaneous)

W_2 computes the moving average of the product of voltage $v_{cb}.i_c$. At each simulation timestep, the moving averages W_1 and W_2 are used to compute real and reactive power flow as depicted in Fig. 4.33.

Two-wattmeter measurements are useful when there is no access to a *datum* node connection, or when a *datum* node is not defined (e.g., delta or floating networks). Moreover, the two-wattmeter method is more computationally efficient than using

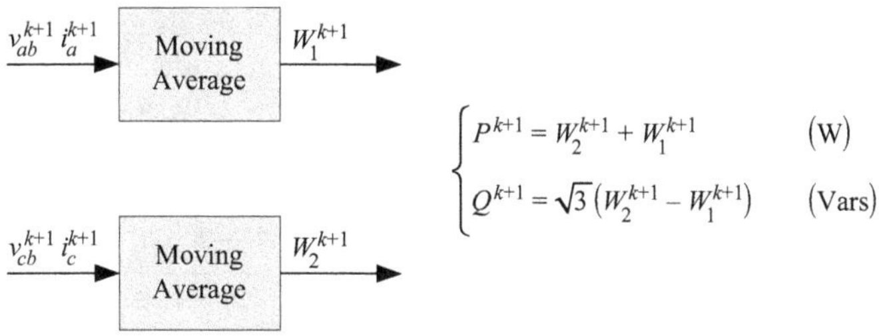

Fig. 4.33 Computation of real and reactive power flow using moving averages

three-phase voltages and currents to calculate power flows. The disadvantage of the two-wattmeter method is that it does not provide per-phase power information.

$$\begin{cases} W_1 = \text{avg}(v_{ab}i_a) \\ W_2 = \text{avg}(v_{cb}i_c) \end{cases} \tag{4.51}$$

4.3.6 PID controller

Proportional–integral–derivative (PID) controllers are commonly used to produce corrective action in power apparatus (e.g., voltage or speed regulation). An illustration of a PID controller is shown in Fig. 4.34, where its governing equations are given in (4.52).

$$\begin{aligned} y(t) &= y_p(t) + y_i(t) + y_d(t) \\ &= K_p e(t) + K_i \int e(t)dt + K_d \frac{d}{dt} e(t) \end{aligned} \tag{4.52}$$

where:

K_p = proportional gain
K_i = integrative gain
K_d = derivate gain
$e(t)$ = error signal
$y(t)$ = output signal.

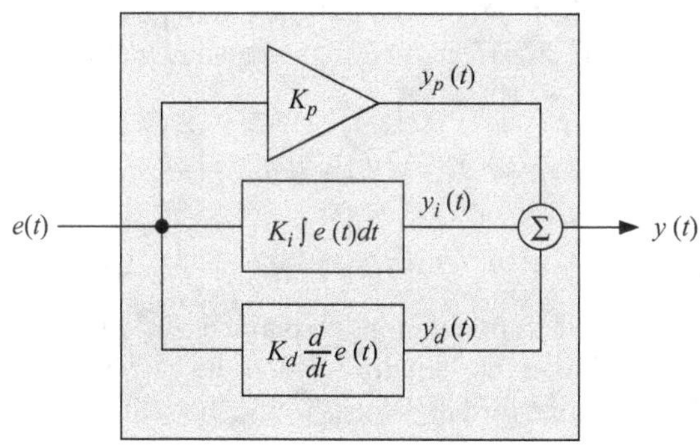

Fig. 4.34 PID controller block

To discretize (4.52), the integral and derivative terms are discretized individually. Discretization of $y_i(t)$ and $y_d(t)$ using tunable integration results in (4.53) and (4.54), respectively. The output of the discrete PID controller at timestep $k+1$ is given in (4.55).

$$y_i(t) = K_i \int e(t)\,dt \Rightarrow y_i^{k+1} = K_i \Delta t \cdot \gamma \cdot e^{k+1} + \underbrace{\left(y_i^k + K_i \Delta t(1-\gamma)e^k\right)}_{\text{historical term}} \tag{4.53}$$

$$y_d(t) = K_d \frac{d}{dt} e(t) \Rightarrow y_d^{k+1} = \left(\frac{K_d}{\Delta t \cdot \gamma}\right) e^{k+1} + \underbrace{\left(\frac{\gamma - 1}{\gamma} y_d^k - \frac{K_d}{\Delta t \cdot \gamma} e^k\right)}_{\text{historical term}} \tag{4.54}$$

$$y^{k+1} = y_p^{k+1} + y_i^{k+1} + y_d^{k+1} \tag{4.55}$$

4.3.7 PWM generator

Pulse-width-modulated (PWM) signal generators produce firing signals to control IGBTs. In sinusoidal PWM-modulation, the firing signals are produced by comparing sinusoidal references (one per phase) against a triangular carrier signal.

Analytical functions for these reference and carrier signals are given in (4.56) [57], where f_{refa} is the sinusoidal reference signal for phase a, $f_r = 60$ Hz is the reference frequency, A_r is the reference amplitude, $f_{carr}(t)$ is the carrier signal, f_c is the carrier frequency in Hz, and A_c is the carrier signal's amplitude. The functions in (4.56) are plotted in Fig. 4.35.

$$\begin{array}{l} \text{Reference signals} \\ \text{(sinusoidal)} \end{array} \rightarrow \left\{ \begin{array}{l} f_{refa}(t) = A_r \sin(2\pi f_r t) \\ f_{refb}(t) = A_r \sin(2\pi f_r t - 120°) \\ f_{refc}(t) = A_r \sin(2\pi f_r t + 120°) \end{array} \right.$$
$$\begin{array}{l} \text{Carrier signal} \\ \text{(triangular)} \end{array} \rightarrow \left\{ f_{carr}(t) = A_c \frac{2}{\pi} \text{asin}()(\sin(2\pi f_c t + 90°)) \right. \tag{4.56}$$

When the inst. value of any reference signal is larger than the carrier signal, the PWM generator produces a *high* output for the corresponding IGBT (a *low* output otherwise). This logic can be programmed into the PWM generator for each reference signal as given in (4.57). (IGBT numberings are shown later in section 5.4.2 in Chapter 5.)

$$\begin{array}{r} \text{PWM} \\ \text{firing} \\ \text{signals} \end{array} \left\{ \begin{array}{llll} \text{IF } (f_{refa}(t) > f_{carr}(t)) & Q_1 = on, Q_4 = off & \text{ELSE} & Q_1 = off, Q_4 = on \\ \text{IF } (f_{refb}(t) > f_{carr}(t)) & Q_3 = on, Q_6 = off & \text{ELSE} & Q_3 = off, Q_6 = on \\ \text{IF } (f_{refc}(t) > f_{carr}(t)) & Q_5 = on, Q_2 = off & \text{ELSE} & Q_5 = off, Q_2 = on \end{array} \right. \tag{4.57}$$

Similar to the natural commutations explained for the diode in Fig. 4.22, PWM generators require time interpolations when the carrier and reference signals cross each other *between* time grid divisions. To illustrate this situation, consider the "scissor" crossing shown in Fig. 4.36, where the sinusoidal reference signal is represented by line 1 and the triangular carrier signal by line 2.

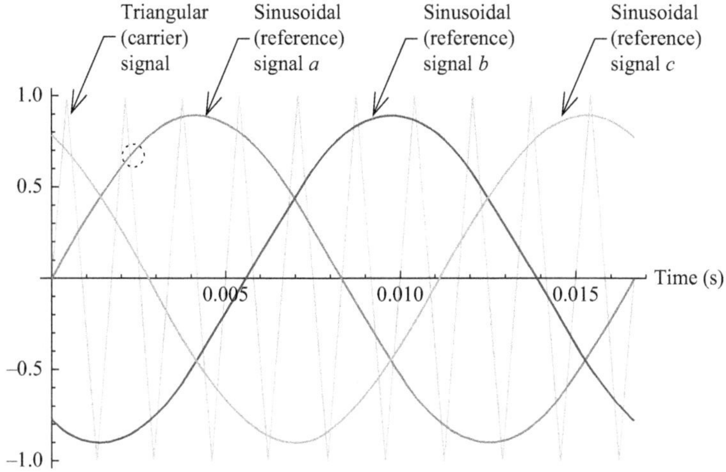

Fig. 4.35 Three reference signals (a, b, c) and one carrier signal. Dashed circle showing a crossing event is discussed in Fig. 4.36

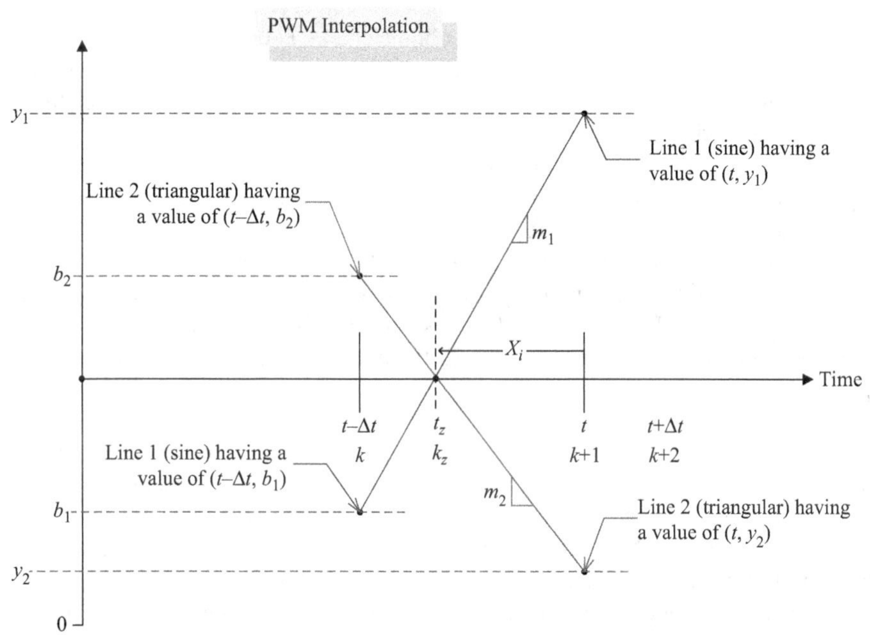

Fig. 4.36 Interpolation due to PWM event represented by encircled area in Fig. 4.35

$$\left.\begin{aligned}\text{Line 1:}\quad m_1 t + b_1 &= \left(\frac{y_1 - b_1}{\Delta t}\right) t + b_1\\ \text{Line 2:}\quad m_2 t + b_2 &= \left(\frac{y_2 - b_2}{\Delta t}\right) t + b_2\end{aligned}\right\}$$

$$\rightarrow t = t_z = \frac{b_2 - b_1}{m_1 - m_2};\quad X_i = \frac{t - t_z}{\Delta t} \tag{4.58}$$

At simulation timestep $k+1$, the line segments for lines 1 and 2 are calculated using (4.58), where m represents the slope of the line and b the line's offset from zero. The time instant where lines 1 and 2 cross each other is labeled t_z; this instant represents the estimated time at which the event (crossing) occurred. Like in switch events, the interpolation of such distance is described with the fractional timestep distance $0 < X_i < 1$. In Fig. 4.36, t_z represents one of the possible events illustrated in Fig. 3.4.

4.4 Summary

This chapter explained how to discretize both the electrical network and the control networks. Discretization produces equations solvable at time grid intervals. At the electrical network level, discretization was shown using two methods. One method was tunable integration for stand-alone branches and control blocks, and the other was root-matching integration method for branch pairs. Following the discussion on discretization, the chapter introduced linearization for six switch types. It was shown that regardless of the switch type, switch branches also conform to the generic branch model introduced at the beginning of the chapter.

At the control network level, discretization was demonstrated for state-variable equations, transfer functions, moving RMS windows, moving average *windows*, power flows, PID controllers, and PWM generators. These seven control blocks, while not a comprehensive list, showed, nonetheless, the pattern of how control network blocks are discretized individually, and how they can be solved during power system simulations. All these control blocks make their outputs available to the electrical network, which, subsequently, uses the control network solution at the next timestep.

The exchange of results between the electrical and control network was first mentioned in section 3.1. As underlined throughout this chapter, the electrical network solution returns the values of all mesh currents (in mesh formulations) or node voltages (in nodal formulations) at each timestep. This network solution vector is available to the control network. The control network, on the other hand, is the solution to all post-electrical network arithmetic operations, and it is responsible for solving state-variable equations, transfer functions, calculating RMS measurements, power flows, PWM firing signal outputs, among other things. The electrical and control network solutions can be (stably) solved sequentially if Δt is sufficiently small [47]—but they can be solved simultaneously as well [59,60,100].

There is pragmatic motive to the electrical and control networks sequentially. The motive is a careful balance between simulation performance and software design. While it is preferable to solve these two networks simultaneously from the standpoint of numerical accuracy and stability, doing so can increase the size of the network matrices require iterative solutions to account for non-linearities, and can significantly reduce simulation performance (i.e., increase frame time). The ever-present trade-off between performance and accuracy, as seen herein, often dictates how programs are designed. A clear testimony of differences in software design can be seen by comparing results of different solvers: in many instances, the results are in reasonable agreement but do not agree entirely [88,89].

Chapter 5
Power apparatus models

> Remember that all models are wrong; the practical question is how wrong do they have to be to not be useful.
>
> —*George E. P. Box and Norman R. Draper*[50]

Discretization of the electrical and control networks was discussed in Chapter 4. This chapter presents basic models to represent the power apparatus listed in Table 2.2. The interconnection of these models comprises the notional power system presented in Fig. 2.1 (*System* 4). Each power apparatus model is presented in both its continuous and discrete forms. The continuous representations show the power apparatus models at the branch level; the discrete representations show the same models using the discrete branches derived in the previous chapter. Since this book covers both mesh and nodal formulations, the discrete representation of each power apparatus model shows its corresponding mesh and nodal equations.

Readers will notice that the power apparatus models are presented enclosed by gray boxes. This enclosure is a functional and didactic visual aid as gray boxes help isolate the electrical part of each power apparatus and facilitate their abstraction as identifiable modular building blocks. In the context of this book, software modularity means that the power apparatus models included in the notional system under study are replaceable. The models presented here (enclosed in gray boxes) are considered "place holders" in the sense that their (potential) substitution with other more elaborate component models will not affect the partitioning approach nor the parallelization methodology brought forward in later chapters. The meaning of modularity, borrowed from the field of software engineering, is an important concept to keep in mind throughout the book.

5.1 Cables

For the system under analysis, power is distributed from the generators to the loads via cables at two voltage levels: at 120 and 450 VAC. Since in the notional power

[50] Quote taken from *Empirical Model-Building and Response Surfaces*, 1st edn., New York; Wiley & Sons, 1987, page 74, textbook authored by George E. P. Box and Normal R. Draper. The quotation of these renowned statisticians and professors extends to power system modeling as well.

system model single-phase loads are lumped behind transformers as single three-phase loads, single-phase cables are not considered here. At 450 VAC, on the other hand, three-phase cables are modeled as nominal-π segments *without* shunt admittances to the hull. The choice of using π-sections stems from the cable lengths, which in shipboard applications are short. Treating cables as nominal-π segments they provides sufficient complexity for most simulation conditions.

Although shipboards include several types of cables on board, the cable types assumed for the notional shipboard power system model (*System* 1–4 in Chapter 2) are types *LSSSGU* and *LSTSGU* (Table 320-1-4 in Reference 30 and Table XIII in References 27, 28).

As a first approximation, conductor types, ampacities, voltage regulation, material characteristics, flexibility, geometric shape, economics, operating temperature, insulation [37], stranding temperature, and number shipboard cable conductors [27,28,30] required to supply necessary current levels are not considered as they do not impact the results considerably. However, conductor selection is contingent upon these considerations [101] and can be narrowed by referring to manufacturer catalogs or recommended practices [102] complying with military specifications. This level of detail, however, is not the scope of this book, but it should be taken into account as cable impact (i.e., size, weight, cost, dielectric requirement, bending radii [103], among other variables) may indeed become significant.

Consider the simple cable model in Fig. 5.1 modeled as a nominal-π section with no stray connections to the ship hull. Unintentional, implicit, leakage, or stray capacitance from cables to the hull exist in practice [104], but they are neglected here.[51] The continuous representation of the cable model shows all branches in continuous form and shows annotations for all terminal voltages and currents. The two discrete representations below the continuous representation show the same branches in discretized form and provide a numbering for its meshes and nodes. (The superscript $k+1$ is neglected *ex professo* from most discrete representations in this chapter, for clarity.) As noticed from the mesh and nodal cable models, each branch conforms to the generic branch model of Fig. 4.2.

In the discrete mesh model R_{La} represents the discrete resistance upon discretizing the series RL branch for phase a. Similarly, R_{Cab1} represents the discrete resistance upon discretizing the parallel RC branch connected between phases a and b on the left side. In the nodal equation set, the same resistance is expressed as conductance to avoid the use of fractions such as $G_{Cab1} = 1/R_{Cab1}$. The expressions for the discrete impedance and historical sources are found from the summary tables elaborated in Figs. 4.18 and 4.19 in Chapter 4.

The discrete mesh and discrete nodal equations for the three-phase cable model are given in (5.1) and (5.2) (*infra*). As seen, eight meshes are required to represent the ungrounded cable model; this is more than the quantity of equations required by the

[51] Although stray capacitance to the hull exists, the capacitance between unshielded conductors is considered dominant due to the proximity between electrical conductors.

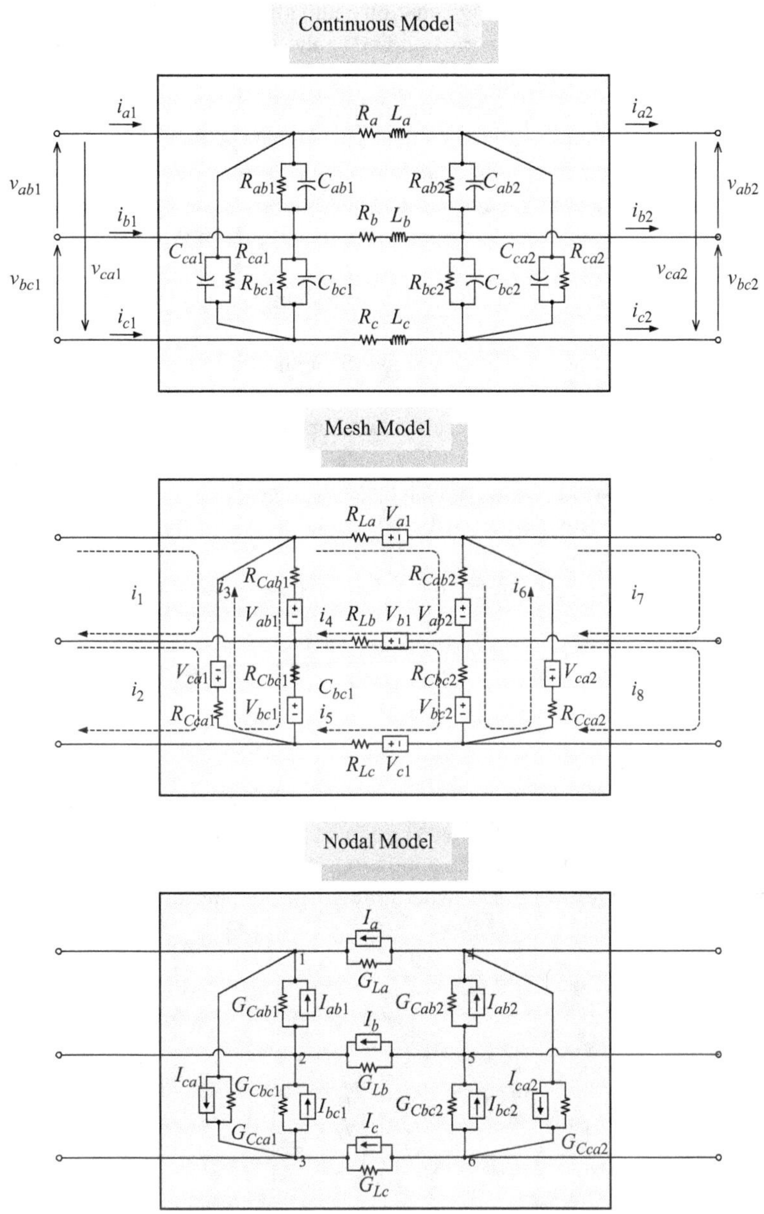

Fig. 5.1 Three-phase cable modeled as ungrounded nominal-π segment

nodal model (six). The difference in equation count affects the order of the network—especially because there are over one hundred cables in the notional shipboard power system model considered in Fig. 2.1.

In solvers that use state-variable formulations (such as *Simulink*), the model order is determined by the number of independent state-variables. Of the six capacitors and three inductances in the three-phase cable model, only four capacitors' voltages and two inductors' currents are independent. This independence reduces the equation count in programs based on state-variables. It may be tempting to conclude that state-variable formulations are more efficient than mesh or nodal ones. However, as shown in Fig. 2.8, discrete matrices based on state-variables formulations may be dense in comparison to either mesh or nodal formulations.

Additionally, it is noted that equations (5.1) and (5.2) assumed zero inputs at the cable terminals; that is, they assume 0 V impressions in the mesh case and 0 A injections in the nodal case. This zero-input approach of viewing power apparatus models promotes software modularity because it regards them as isolated, replaceable "miniature networks". The right-hand side input vectors in these equations show the historical terms that must be computed at each timestep of the simulation according to the tables in Figs. 4.18 and 4.19. It is noted that when events occur between time grid divisions, the values of each historical source undergoes interpolation.

5.2 Static loads

For the power system under contemplation, the loads considered here are modeled as three-phase, static, 450 VAC loads. Each of these loads encloses three series RL branches connected in delta. The continuous and discrete representations for the load model are shown in Fig. 5.2, where the corresponding discrete equations are given by (5.3) and (5.4).

Mesh

$$\begin{bmatrix} R_{Lab} & \cdot & R_{Lab} \\ \cdot & R_{Lbc} & R_{Lbc} \\ R_{Lab} & R_{Lbc} & (R_{Lab}+R_{Lbc}+R_{Lca}) \end{bmatrix} \begin{bmatrix} i_1 \\ i_2 \\ i_3 \end{bmatrix} = \begin{bmatrix} -V_{Lab} \\ -V_{Lbc} \\ -V_{Lab}-V_{Lbc}-V_{Lca} \end{bmatrix} \tag{5.3}$$

Nodal

$$\begin{bmatrix} G_{Lab}+G_{Lca} & -G_{Lab} & -G_{Lca} \\ -G_{Lab} & G_{Lab}+G_{Lbc} & -G_{Lbc} \\ -G_{Lca} & -G_{Lbc} & G_{Lbc}+G_{Lca} \end{bmatrix} \begin{bmatrix} v_1 \\ v_2 \\ v_3 \end{bmatrix} = \begin{bmatrix} I_{ab}-I_{ca} \\ I_{bc}-I_{ab} \\ I_{ca}-I_{bc} \end{bmatrix} \tag{5.4}$$

Mesh

$$\mathbf{R}_{CBL}\mathbf{i}_{CBL} = \mathbf{e}_{CBL} \tag{5.1}$$

$$\mathbf{R}_{CBL} = \begin{bmatrix} R_{Cab1} & \cdot & R_{Cab1} & -R_{Cab1} & \cdot & \cdot & \cdot & \cdot \\ \cdot & R_{Cbc1} & R_{Cbc1} & \cdot & -R_{Cbc1} & \cdot & \cdot & \cdot \\ R_{Cab1} & R_{Cbc1} & \begin{pmatrix} R_{Cab1} + R_{Cbc1} \\ + R_{Cca1} \end{pmatrix} & -R_{Cab1} & -R_{Cbc1} & \cdot & \cdot & \cdot \\ -R_{Cab1} & \cdot & -R_{Cab1} & \begin{pmatrix} R_{Cab1} + R_a + R_{La} \\ + R_{Cab2} + R_b + R_{Lb} \end{pmatrix} & -R_a - R_{La} & R_{Cab2} & -R_C & \cdot \\ \cdot & -R_{Cbc1} & -R_{Cbc1} & -R_a - R_{La} & \begin{pmatrix} R_{Cab2} + R_b + R_{Lb} \\ + R_{Cbc2} + R_c + R_{Lc} \end{pmatrix} & R_{Cbc2} & \cdot & -R_C \\ \cdot & \cdot & \cdot & R_{Cab2} & R_{Cbc2} & \begin{pmatrix} R_{Cab2} + R_{Cbc2} \\ + R_{Cca2} \end{pmatrix} & -R_{Cab2} & -R_{Cbc2} \\ \cdot & \cdot & \cdot & -R_C & \cdot & -R_{Cab2} & R_{Cab2} & \cdot \\ \cdot & \cdot & \cdot & \cdot & -R_C & -R_{Cbc2} & \cdot & R_{Cbc2} \end{bmatrix}$$

$$\mathbf{i}_{CBL} = \begin{bmatrix} i_1 \\ i_2 \\ i_3 \\ i_4 \\ i_5 \\ i_6 \\ i_7 \\ i_8 \end{bmatrix} \qquad \mathbf{e}_{CBL} = \begin{bmatrix} -V_{Cab1} \\ -V_{Cbc1} \\ -V_{Cab1} - V_{Cbc1} - V_{Cca1} \\ -V_{La} - V_{Cab2} + V_{Lb} + V_{Cab1} \\ -V_{Lb} - V_{Cbc2} + V_{Lc} + V_{Cbc1} \\ -V_{Cab2} - V_{Cbc2} - V_{Cca2} \\ V_{Cab2} \\ V_{Cbc2} \end{bmatrix}$$

Nodal

$$\mathbf{G}_{CBL}\mathbf{v}_{CBL} = \mathbf{j}_{CBL} \tag{5.2}$$

$$\mathbf{G}_{CBL} = \begin{bmatrix} \begin{pmatrix} G_{Cab1} + G_{Cca1} \\ +G_{La} \end{pmatrix} & -G_{Cab1} & -G_{Cca1} & -G_{La1} & \cdot & \cdot \\ \cdot & \begin{pmatrix} G_{Cab1} + G_{Cbc1} \\ +G_{Lb} \end{pmatrix} & -G_{Cbc1} & \cdot & -G_{Lb} & \cdot \\ \cdot & \cdot & \begin{pmatrix} G_{Cbc1} + G_{Cca1} \\ +G_{Lc} \end{pmatrix} & \cdot & \cdot & -G_{Lc} \\ -G_{La1} & \cdot & \cdot & \begin{pmatrix} G_{Cab2} + G_{Cca2} \\ +G_{La} \end{pmatrix} & -G_{Cab2} & -G_{Cca2} \\ \cdot & -G_{Lb} & \cdot & -G_{Cab2} & \begin{pmatrix} G_{Cab2} + G_{Cbc2} \\ +G_{Lb} \end{pmatrix} & -G_{Cbc2} \\ \cdot & \cdot & -G_{Lc} & -G_{Cca2} & -G_{Cbc2} & \begin{pmatrix} G_{Cbc2} + G_{Cca2} \\ +G_{Lc} \end{pmatrix} \end{bmatrix}$$

$$\mathbf{v}_{CBL} = \begin{bmatrix} v_1 \\ v_2 \\ v_3 \\ v_4 \\ v_5 \\ v_6 \end{bmatrix} \qquad \mathbf{j}_{CBL} = \begin{bmatrix} I_{ab1} - I_{ca1} + I_a \\ I_{bc1} - I_{ab1} + I_b \\ I_{ca1} - I_{bc1} + I_c \\ I_{ab2} - I_{ca1} - I_a \\ I_{bc2} - I_{ab1} - I_b \\ I_{ca2} - I_{bc1} - I_c \end{bmatrix}$$

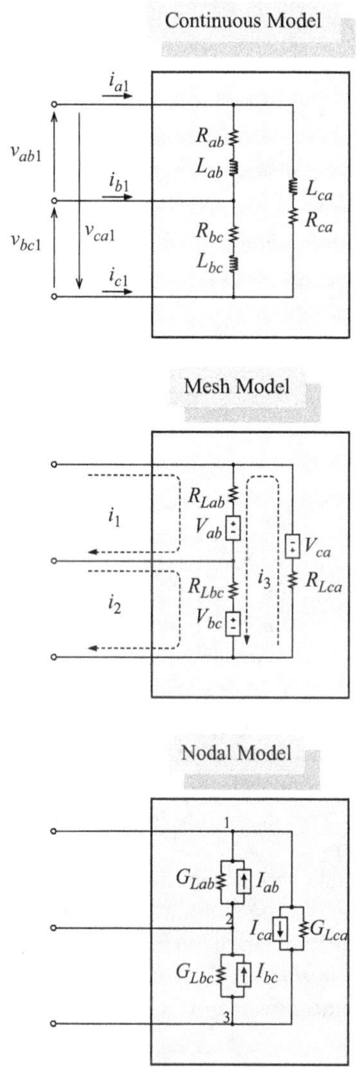

Fig. 5.2 Three-phase load model

Since the RL values of a load are (typically) unknown in practice, these RL values are computed from the load's nameplate data. For example, the R and L values (in each phase) of a load can be computed from the per-phase voltage and power using (5.5)

$$R_p = \frac{P_p V_p^2}{P_p^2 + Q_p^2}(\Omega) \qquad L_p = \frac{Q_p V_p^2}{2\pi f(P_p^2 + Q_p^2)}(\text{H}) \tag{5.5}$$

where:
R_p = branch resistance (ohms)
L_p = branch inductance (H)
f = system frequency (Hz)
P_p = per-phase *real* power consumption (W)
Q_p = per-phase *reactive* power consumption (Vars)
V_p = rated voltage across RL branch (RMS volts).

It is re-emphasized that the equations in (3.65) and (3.66) should be viewed as replaceable. Additionally, and similar to three-phase cables (and all power apparatus models), the right-hand side input vectors in (3.65) and (3.66) include historical terms that *must* be updated at each timestep of the simulation according to the tables in Figs. 4.18 and 4.19. As can be noticed, power apparatus model can be deemed as a "miniature network" described by equations in the form $\mathbf{A} \cdot \mathbf{x} = \mathbf{b}$.

5.3 Protective devices

The treatment of protective devices herein is purposefully brief. An in-depth description of protective device operation, control, and their coordination is a broad topic and outside the scope of this book. However, a brief description of protective devices is important because, when counted together,[52] they are the dominant power apparatus in the notional power system shown in Fig. 2.1.

Protective devices protect power systems from undesirable electrical conditions. This protection may be in the form of materials, devices, or measures that help prevent components (and personnel) from damage or harmful conditions. Therefore, the basic function of protective devices is to interrupt electric service. This interruption, in addition to being vital, should occur within milliseconds of detecting the presence of an undesirable condition. Two examples of an undesirable electrical conditions are short-circuits (faults) and low-voltage conditions. When these conditions emerge, protective devices isolate the problem and/or disconnect equipment to prevent damages. After corrective action is taken to remediate the undesirable electrical condition, protective devices may re-close to restore service to un-energized portions of the power system.

This section presents basic models for three different protective devices. These protective devices are circuit breakers (83), bus transfer switches (28), and low-voltage protective devices (19). Although the *control* logic of each protective devices is different, their electrical models are the same. The electrical model of all protective devices considered here is a three-phase breaker. Circuit breakers and low-voltage protective devices are modeled, each, as one three-phase breaker. Bus transfers, which tie three *paths* of electric service, are modeled as three three-phase breakers. The total three-phase breaker count is, then, $83 + (28 \times 3) + 19 = 186$

[52] Circuit breakers, bus transfers, and low-voltage protective devices are all protective devices.

three-phase breakers. Although each three-phase breaker consists of three switches each,[53] these switches do not toggle often and do not constitute an effective computational burden (as do power electronic valves).

Circuit breakers and low-voltage protective devices differ in the way they operate. Over-current circuit breakers, for example, operate when high currents are detected and sustained. Low-voltage protective devices, on the other hand, operate after low-voltage conditions are detected and sustained.

Bus transfers are switchgear that transfer loads (on their load path) from one bus (e.g., the normal path) to another bus (e.g., the alternate path). Bus transfers, however, cannot make this transfer when both supply buses undergo low voltage conditions. In this situation, bus transfers disconnect their load from the electric service. Bus transfers, similar to low-voltage protective devices, initiate action due to low-voltage conditions.

The models for each aforementioned protective device are presented next. Additional information on protective devices for AC-radial shipboard power systems can be found in References 105 and 106. Readers interested in more general information on power system protection are directed to References 107 and 108.

5.3.1 Circuit breakers

The basic function of a circuit breaker[54] is to isolate the power apparatus connected to the breaker. Circuit breakers automatically interrupt a circuit by separating its contacts during abnormal conditions. The circuit breaker position (i.e., open or closed) of an over-current circuit breaker is controlled by its relay logic, which produces a commanding action in response to sustained over currents and coordination with other relays protecting the same network area.

For the notional system at hand, the model used for circuit breakers is shown in Fig. 5.3. As circuit breakers do not have branches with differential equations, circuit breakers are modeled as purely resistive power apparatus. The values of these resistive branches, however, depend on the state of the circuit breaker and change according to (5.6).

$$R_i \rightarrow \begin{cases} = 1\ \text{m}\Omega, & \text{when phase } i \text{ is closed} \\ = 1\ \text{M}\Omega, & \text{when phase } i \text{ is open} \end{cases} \tag{5.6}$$

Similar to what has been discussed, many commercial power system simulators also model circuit breakers as time-varying resistances; however, there is a caveat

[53] The 186 three-phase breakers add $186 \times 3 = 558$ switches to the model. This number of switches, when added to the existing 342 power electronic switches in the system, results in a total of 900 switches included in the notional shipboard power system model displayed in Fig. 2.1.

[54] There are many sets of circuit breaker definitions and terms used in industry. For instance, there are definitions that revolve around voltages, currents, and still others that refer to various operating characteristics. Other definitions can be found in the standards, such as ANSI C37.03-1964, "American Standard Definitions for AC High Voltage Circuit Breakers," ANSI, New York, 1964.

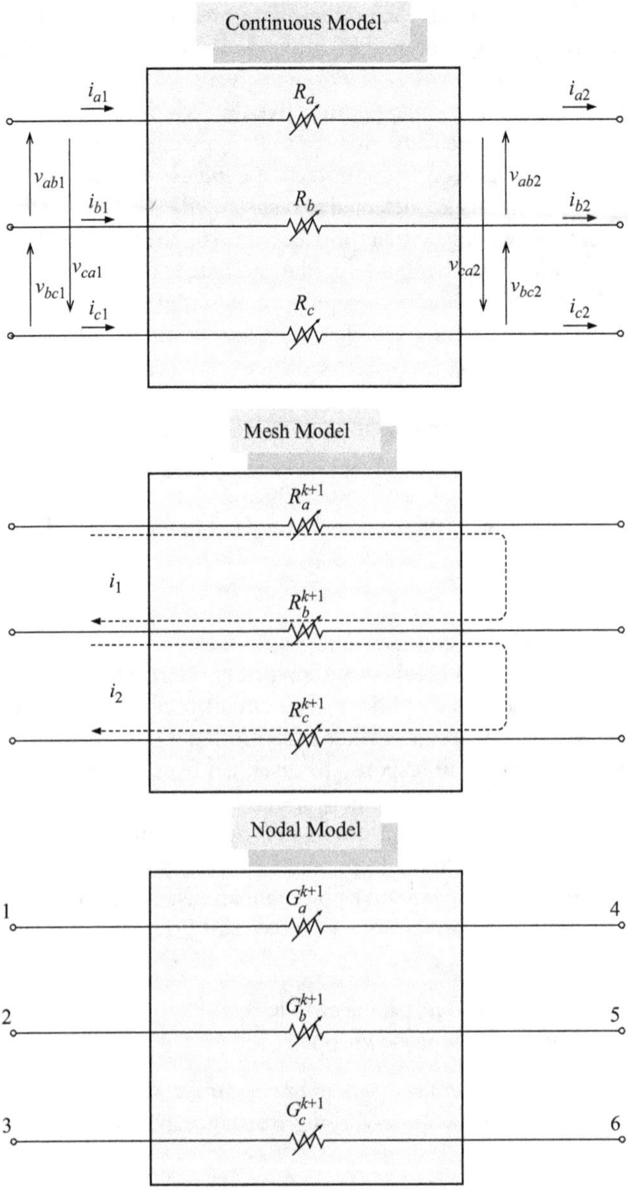

Fig. 5.3 Circuit breaker modeled as a generic three-phase breaker

in doing so: galvanic continuity exists in the open *and* closed positions. This galvanic continuity produces leakage currents in proportion to the system voltage level. To reinforce what was said in section 4.2.1.1, the over-use of large resistor branches can lead to unintended differences in simulation results. A better implementation of an interrupter (applicable in nodal and mesh formulations), is to re-structure the electrical network's immittance matrix.

The circuit breaker equations for the mesh and nodal models are given in (5.7) and (5.8), respectively. In both cases, the superscript $k+1$ is included to emphasize that these protective devices are treated as time-varying resistances. Furthermore, it is noticed that the right-hand side vectors do not have inputs. If arcing[55] is needed in a circuit breaker model, voltage sources may be inserted in series with the resistive elements to model the arcing stage [109–113].

Comparing the mesh and nodal equation sets, the mesh model requires two equations, whereas the nodal model requires six equations. While it appears advantageous to model circuit breakers using mesh equations, the disadvantage is that there is no (easy) way to make line-voltage measurements. In this sense, the nodal method is more flexible and versatile, and it also requires more equations. If voltage measurements are not required at circuit breakers, using a mesh formulation results in an electrical network of smaller order.

Mesh

$$\begin{bmatrix} R_a^{k+1} + R_b^{k+1} & -R_b^{k+1} \\ -R_b^{k+1} & R_b^{k+1} + R_c^{k+1} \end{bmatrix} \begin{bmatrix} i_1 \\ i_2 \end{bmatrix} = \begin{bmatrix} \cdot \\ \cdot \end{bmatrix} \tag{5.7}$$

Nodal

$$\begin{bmatrix} G_a^{k+1} & \cdot & \cdot & -G_a^{k+1} & \cdot & \cdot \\ \cdot & G_b^{k+1} & \cdot & \cdot & -G_b^{k+1} & \cdot \\ \cdot & \cdot & G_c^{k+1} & \cdot & \cdot & -G_c^{k+1} \\ -G_a^{k+1} & \cdot & \cdot & G_a^{k+1} & \cdot & \cdot \\ \cdot & -G_b^{k+1} & \cdot & \cdot & G_b^{k+1} & \cdot \\ \cdot & \cdot & -G_c^{k+1} & \cdot & \cdot & G_c^{k+1} \end{bmatrix} \begin{bmatrix} v_1 \\ v_2 \\ v_3 \\ v_4 \\ v_5 \\ v_6 \end{bmatrix} = \begin{bmatrix} \cdot \\ \cdot \\ \cdot \\ \cdot \\ \cdot \\ \cdot \end{bmatrix} \tag{5.8}$$

5.3.2 Low-voltage protection

Low-voltage protective devices safeguard loads against sustained low-voltage conditions. In terms of the notional power system model under explanation, low-voltage protective devices are placed upstream of the motor loads. These low-voltage protective devices are modeled as three-phase breakers, but they have different pickup value and delay than circuit breakers do. That is, the pickup value of low-voltage protective devices is 405 V (90% of nominal voltage) and the delay is a few cycles [114] (e.g., 50 ms).

The low-voltage protective device model is shown in Fig. 5.4, where its corresponding equations are given in (5.9) and (5.10). Comparing Figs. 5.3 and 5.4, both the circuit breaker and low-voltage protective device are modeled as a three-phase

[55] It is more common to consider circuit breaker arcing in medium-voltage (2.4–69 kV) and high-voltage ($\geq$115 kV) applications rather than in low-voltage ($\leq$600 V) applications.

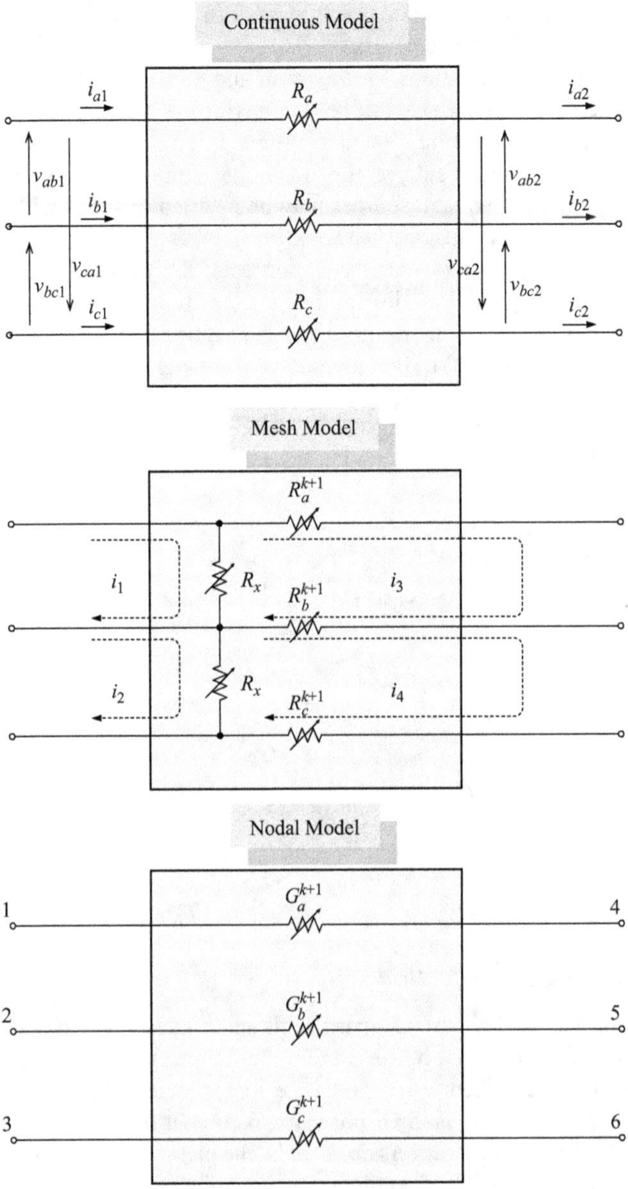

Fig. 5.4 Low-voltage protective device modeled as a generic three-phase breaker

breaker. Their differences, however, lie in the relay logic, count and placement, coordination, and in the topology of the mesh model.

Because voltage measurements are required for low-voltage protective devices, fictitious shunt resistances (noted R_x) are inserted in the mesh model. Adding these impedances increases the equation count of the mesh model from two to four—but this count is still less than the six equations required by the nodal model. It is also possible, however, to add a third resistance across phases c and a to compute v_{ca} and improve balancing, but doing so is not necessary since the expected unbalance is small and v_{ca} is readily available from $v_{ca} = -(v_{ab} + v_{bc})$.

Mesh

$$\begin{bmatrix} R_x & \cdot & -R_x & \cdot \\ \cdot & R_y & \cdot & -R_y \\ -R_x & \cdot & R_a^{k+1} + R_x + R_b^{k+1} & -R_b^{k+1} \\ \cdot & -R_y & -R_b^{k+1} & R_b^{k+1} + R_y + R_c^{k+1} \end{bmatrix} \begin{bmatrix} i_1 \\ i_2 \\ i_3 \\ i_4 \end{bmatrix} = \begin{bmatrix} \cdot \\ \cdot \\ \cdot \\ \cdot \end{bmatrix} \tag{5.9}$$

Nodal

$$\begin{bmatrix} G_a^{k+1} & \cdot & \cdot & -G_a^{k+1} & \cdot & \cdot \\ \cdot & G_b^{k+1} & \cdot & \cdot & -G_b^{k+1} & \cdot \\ \cdot & \cdot & G_c^{k+1} & \cdot & \cdot & -G_c^{k+1} \\ -G_a^{k+1} & \cdot & \cdot & G_a^{k+1} & \cdot & \cdot \\ \cdot & -G_b^{k+1} & \cdot & \cdot & G_b^{k+1} & \cdot \\ \cdot & \cdot & -G_c^{k+1} & \cdot & \cdot & G_c^{k+1} \end{bmatrix} \begin{bmatrix} v_1 \\ v_2 \\ v_3 \\ v_4 \\ v_5 \\ v_6 \end{bmatrix} = \begin{bmatrix} \cdot \\ \cdot \\ \cdot \\ \cdot \\ \cdot \\ \cdot \end{bmatrix} \tag{5.10}$$

5.3.3 Bus transfers

Bus transfers are switchgear assemblies that route power from either of two supply paths: from the normal path or from the alternate path. The choice of the supply path depends on the voltage condition of the normal path. If the normal path voltage is too low, bus transfer switches automatically change the supply path to the alternate supply side. When the normal path voltage is restored, bus transfers switch back to the normal path. If the voltage is low on both supply paths (or sides), bus transfers switches can disconnect their load downstream. The loads commonly protected by bus transfers are vital loads; therefore, there are not as many bus transfers as there are circuit breakers.

Bus transfers are modeled as three interconnected three-phase breakers as shown in Fig. 5.5. When there is a low-voltage condition detected on the normal path, the normal path breaker opens (after the pickup delay) and the alternate path closes (after the transfer delay).

As in the case of the low-voltage protective device, fictitious 1 MΩ shunt resistances (R_{xab} and R_{xbc}) make available the line voltages at all terminals: side 1, side 2, and side 3. In a 450-VAC system, the power losses in each fictitious resistance is $450^2/10^6 = 200$ mW. In a megawatt-level system, however, milliwatt-level power losses produced by these fictitious resistances are tolerable.

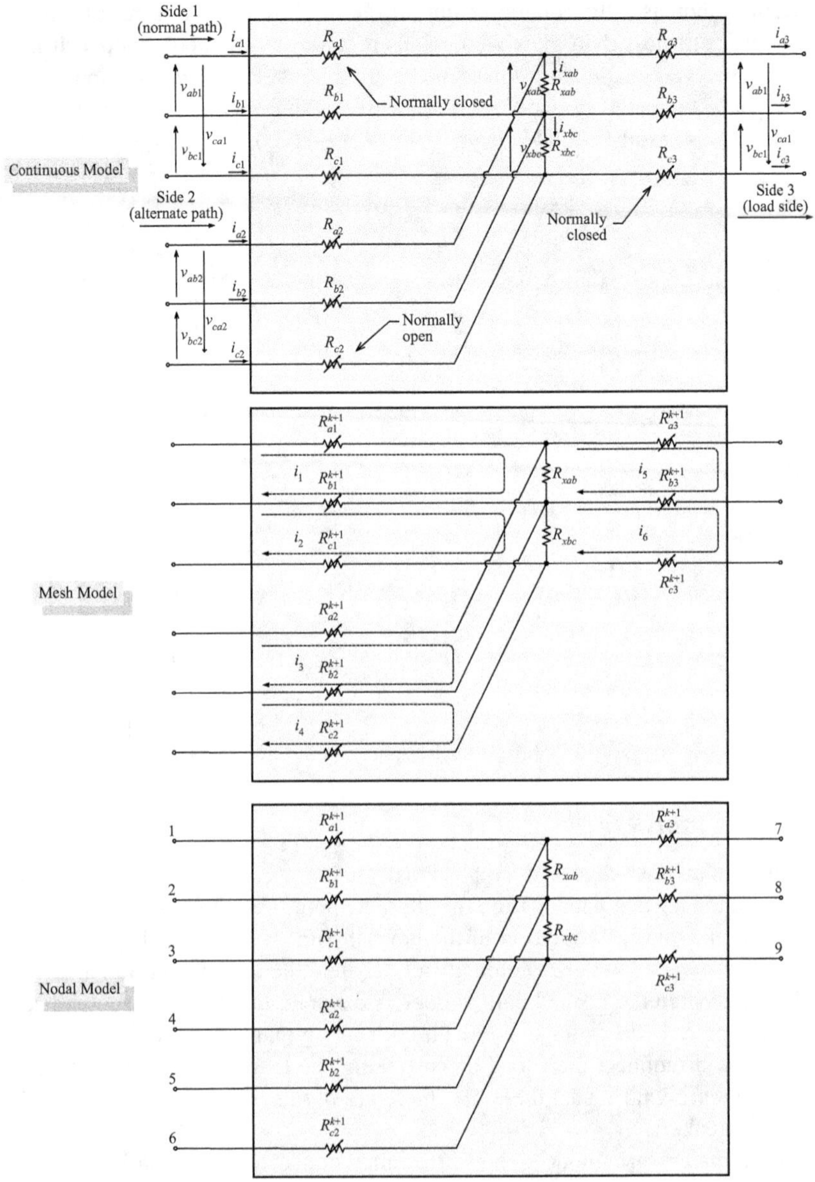

Fig. 5.5 Bus transfer model

The equations for the mesh and nodal models are written out in (5.11) and (5.12), respectively. Like in the case of the circuit breaker and low-voltage protective device above, the mesh model requires fewer equations. When there are numerous protective devices in a power system, mesh formulations keep the electrical network's equation count low and, therefore, reduces the computational burden of the solver.

5.4 Motor drive

Modeling motor drives is computationally intense due to the number of switches its converters have, and the number of times a solver must interpolate to address natural and forced-commutating switching actions. The computational burden exacerbates when many motor drives are included in a power system model.

An illustration of an induction motor drive included in the power system model under analysis is depicted in Fig. 5.6. Such an induction motor drive comprises several subcomponents, namely:

- front-end three-phase rectifier (6 diodes)
- DC-link filter
- voltage-source inverter driven from a PWM controller (6 IGBTs, 6 feedback diodes)
- motor stator and rotor windings
- rotor shaft including a mechanical load.

As was seen from Table 2.2, 19 induction motor drives, as the one depicted in Fig. 5.6, are included in the notional power system diagram displayed in Fig. 2.1. This inclusion makes the simulation of such power systems a rather burdensome task to carry out. The models of each aforementioned motor drive subcomponent are presented next.

5.4.1 Rectifier

The front-end rectifiers of all motor loads are modeled in the notional power system as line-commutated, six pulse rectifiers. The front-end rectifier model is shown in Fig. 5.7 and its discrete equations are given in (5.13) and (5.14). Referring to Fig. 4.20, the rectifier diodes are switches of type 4. Although snubber circuits are not required to model rectifiers, they are included in the motor drive rectifiers to preserve correspondence with how switches are treated in *SimPowerSystems*. In passing, it should be highlighted that switch modeling [90] is outside the scope of this book, but average-model techniques also exist to reduce computational burden [48,115].

A comparison of simulating a stand-alone diode in *Simulink* with the multicore solver developed for this book was shown in Fig. 4.24. Now, consider the case of the three-phase, six-diode rectifier shown in Fig. 5.8. The simulation results are shown in Fig. 5.9, where the left-hand side shows the overlay of the DC load's voltage and current. The right-hand side shows the differences of each overlay in logarithmic scale. While the overall results are in reasonable agreement, a closer examination and contrast of the two solvers reveals that a large absolute difference appears at the beginning of the simulation as shown in Fig. 5.10.

Mesh

$$\left[\begin{array}{cc|cc|cc}
\begin{pmatrix} R_{a1}^{k+1} + R_{xab} \\ + R_{b1}^{k+1} \end{pmatrix} & -R_{b1}^{k+1} & R_{xab} & \cdot & -R_{xab} & \cdot \\
-R_{b1}^{k+1} & \begin{pmatrix} R_{b1}^{k+1} + R_{xbc} \\ + R_{c1}^{k+1} \end{pmatrix} & \cdot & R_{xbc} & \cdot & -R_{xbc} \\
\hline
R_{xab} & \cdot & \begin{pmatrix} R_{a2}^{k+1} + R_{xab} \\ + R_{b2}^{k+1} \end{pmatrix} & -R_{b2}^{k+1} & -R_{xab} & \cdot \\
\cdot & R_{xbc} & -R_{b2}^{k+1} & \begin{pmatrix} R_{b2}^{k+1} + R_{xbc} \\ + R_{c2}^{k+1} \end{pmatrix} & \cdot & -R_{xbc} \\
\hline
-R_{xab} & \cdot & -R_{xab} & \cdot & \begin{pmatrix} R_{a3}^{k+1} + R_{xab} \\ + R_{b3}^{k+1} \end{pmatrix} & -R_{b3}^{k+1} \\
\cdot & -R_{xbc} & \cdot & -R_{xbc} & -R_{b3}^{k+1} & \begin{pmatrix} R_{b3}^{k+1} + R_{xbc} \\ + R_{c3}^{k+1} \end{pmatrix}
\end{array}\right]
\begin{bmatrix} i_1 \\ i_2 \\ i_3 \\ i_4 \\ i_5 \\ i_6 \end{bmatrix}
=
\begin{bmatrix} \cdot \\ \cdot \\ \cdot \\ \cdot \\ \cdot \\ \cdot \end{bmatrix}
\tag{5.11}$$

Nodal

$$\left[\begin{array}{ccc:ccc:ccc:ccc}
G_{a1}^{k+1} & & & & & & & & & -G_{a1}^{k+1} & & \\
& G_{b1}^{k+1} & & & & & & & & & -G_{b1}^{k+1} & \\
& & G_{c1}^{k+1} & & & & & & & & & -G_{c1}^{k+1} \\
\hdashline
& & & G_{a2}^{k+1} & & & & & & -G_{a2}^{k+1} & & \\
& & & & G_{b2}^{k+1} & & & & & & -G_{b2}^{k+1} & \\
& & & & & G_{c2}^{k+1} & & & & & & -G_{c2}^{k+1} \\
\hdashline
& & & & & & G_{a3}^{k+1} & & & -G_{a3}^{k+1} & & \\
& & & & & & & G_{b3}^{k+1} & & & -G_{b3}^{k+1} & \\
& & & & & & & & G_{c3}^{k+1} & & & -G_{c3}^{k+1} \\
\hdashline
-G_{a1}^{k+1} & & & -G_{a2}^{k+1} & & & -G_{a3}^{k+1} & & & \begin{pmatrix} G_{a1}^{k+1} + G_{a2}^{k+1} \\ + G_{a3}^{k+1} + G_{xab} \end{pmatrix} & -G_{xab} & \\
& -G_{b1}^{k+1} & & & -G_{b2}^{k+1} & & & -G_{b3}^{k+1} & & -G_{xab} & \begin{pmatrix} G_{b1}^{k+1} + G_{b2}^{k+1} \\ + G_{b3}^{k+1} + G_{xab} + G_{xbc} \end{pmatrix} & -G_{xbc} \\
& & -G_{c1}^{k+1} & & & -G_{c2}^{k+1} & & & -G_{c3}^{k+1} & & -G_{xbc} & \begin{pmatrix} G_{c1}^{k+1} + G_{c2}^{k+1} \\ + G_{c3}^{k+1} + G_{xbc} \end{pmatrix}
\end{array}\right]
\begin{bmatrix} v_1 \\ v_2 \\ v_3 \\ v_4 \\ v_5 \\ v_6 \\ v_7 \\ v_8 \\ v_9 \\ v_{10} \\ v_{11} \\ v_{12} \end{bmatrix} = \begin{bmatrix} \cdot \\ \cdot \\ \cdot \\ \cdot \\ \cdot \\ \cdot \\ \cdot \\ \cdot \\ \cdot \\ \cdot \\ \cdot \\ \cdot \end{bmatrix} \tag{5.12}$$

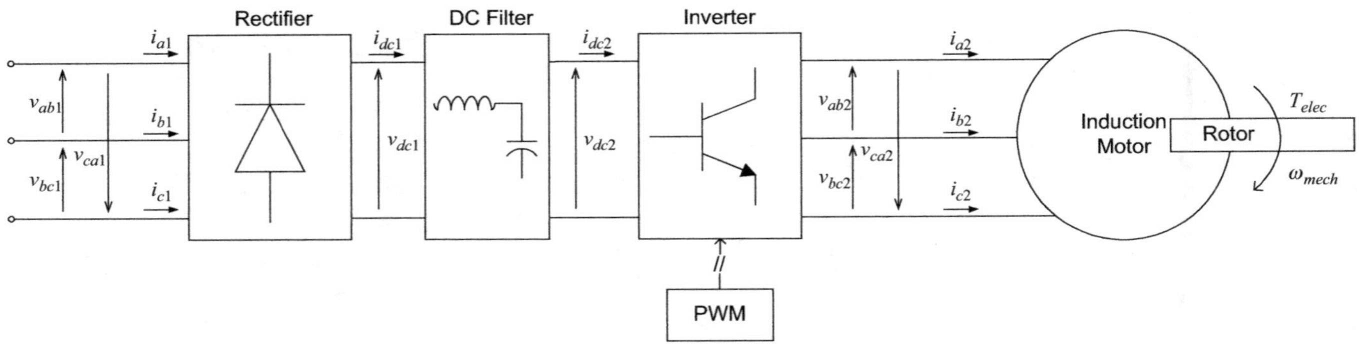

Fig. 5.6 Induction motor drive model

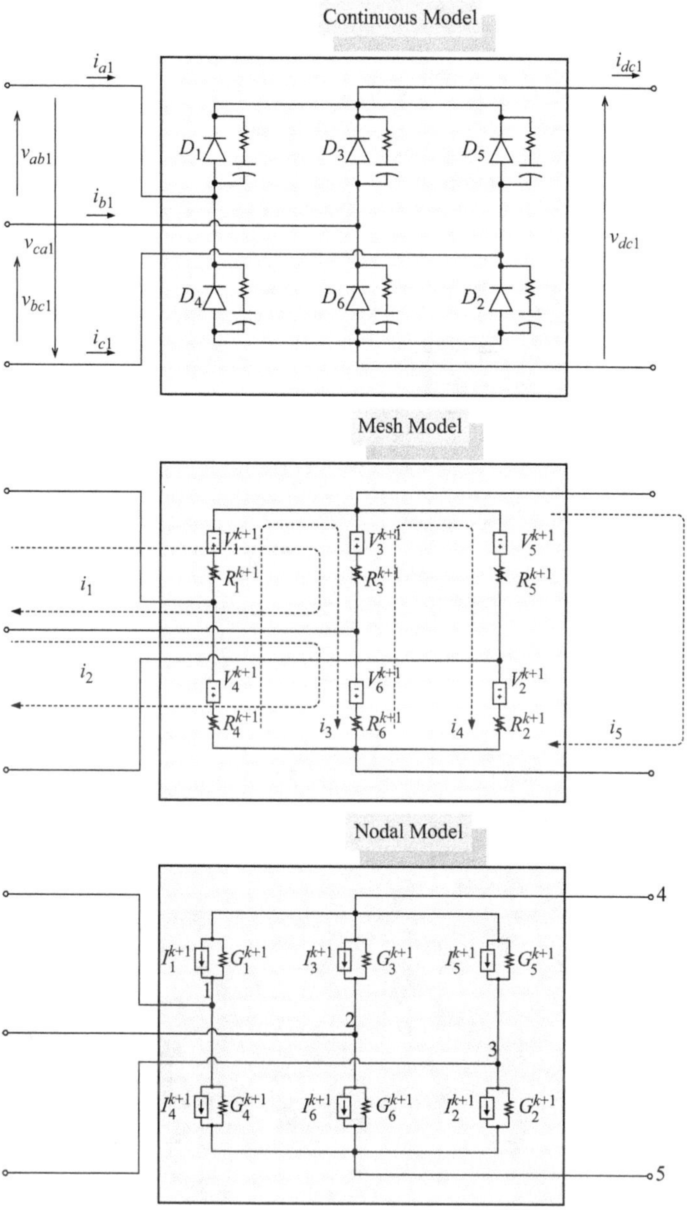

Fig. 5.7 Rectifier model

In this close-up, the difference occurs at $t = 0$, where *Simulink* produced a voltage and current output at $t = 0$ and the multicore solver did *not*. The reason for *not* producing an output at $t = 0$ is explained by the particular choice (decision-making) in designing the solver, which doubtless to say, can be reconsidered at any time.

Mesh

$$\begin{bmatrix} R_1^{k+1}+R_3^{k+1} & -R_3^{k+1} & R_1^{k+1}+R_3^{k+1} & -R_3^{k+1} & \cdot \\ -R_3^{k+1} & R_3^{k+1}+R_5^{k+1} & -R_3^{k+1} & R_3^{k+1}+R_5^{k+1} & -R_5^{k+1} \\ R_1^{k+1}+R_3^{k+1} & -R_3^{k+1} & \begin{pmatrix} R_1^{k+1}+R_3^{k+1} \\ +R_4^{k+1}+R_6 \end{pmatrix} & -R_3^{k+1}-R_6^{k+1} & \cdot \\ -R_3^{k+1} & R_3^{k+1}+R_5^{k+1} & -R_3^{k+1}-R_6^{k+1} & \begin{pmatrix} R_2^{k+1}+R_3^{k+1} \\ +R_5^{k+1}+R_6^{k+1} \end{pmatrix} & -R_2^{k+1}-R_5^{k+1} \\ \cdot & -R_5^{k+1} & \cdot & -R_2^{k+1}-R_5^{k+1} & R_2^{k+1}+R_5^{k+1} \end{bmatrix} \begin{bmatrix} i_1 \\ i_2 \\ i_3 \\ i_4 \\ i_5 \end{bmatrix}$$

$$= \begin{bmatrix} -V_1^{k+1}+V_3^{k+1} \\ -V_3^{k+1}+V_5^{k+1} \\ -V_1^{k+1}+V_3^{k+1}+V_6^{k+1}-V_4^{k+1} \\ -V_3^{k+1}+V_5^{k+1}+V_2^{k+1}-V_6^{k+1} \\ -V_2^{k+1}-V_5^{k+1} \end{bmatrix} \tag{5.13}$$

Nodal

$$\begin{bmatrix} G_1^{k+1} + G_4^{k+1} & \cdot & \cdot & -G_1^{k+1} & -G_4^{k+1} \\ \cdot & G_3^{k+1} + G_6^{k+1} & \cdot & -G_3^{k+1} & -G_6^{k+1} \\ \cdot & \cdot & G_2^{k+1} + G_5^{k+1} & -G_5^{k+1} & -G_2^{k+1} \\ -G_1^{k+1} & -G_3^{k+1} & -G_5^{k+1} & \begin{pmatrix} G_1^{k+1} + G_3^{k+1} \\ + G_5^{k+1} \end{pmatrix} & \cdot \\ -G_4^{k+1} & \cdot - G_6^{k+1} & -G_2^{k+1} & \cdot & \begin{pmatrix} G_2^{k+1} + G_4^{k+1} \\ + G_6^{k+1} \end{pmatrix} \end{bmatrix} \begin{bmatrix} v_1 \\ v_2 \\ v_3 \\ v_4 \\ v_5 \end{bmatrix} = \begin{bmatrix} I_1^{k+1} - I_4^{k+1} \\ I_3^{k+1} - I_6^{k+1} \\ I_5^{k+1} - I_2^{k+1} \\ -I_1^{k+1} - I_3^{k+1} - I_5^{k+1} \\ I_2^{k+1} + I_4^{k+1} + I_6^{k+1} \end{bmatrix} \quad (5.14)$$

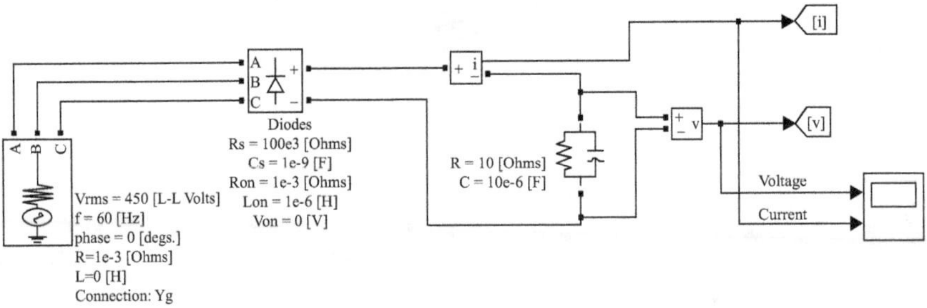

Fig. 5.8 Three-phase six-diode rectifier circuit to show differences in solvers

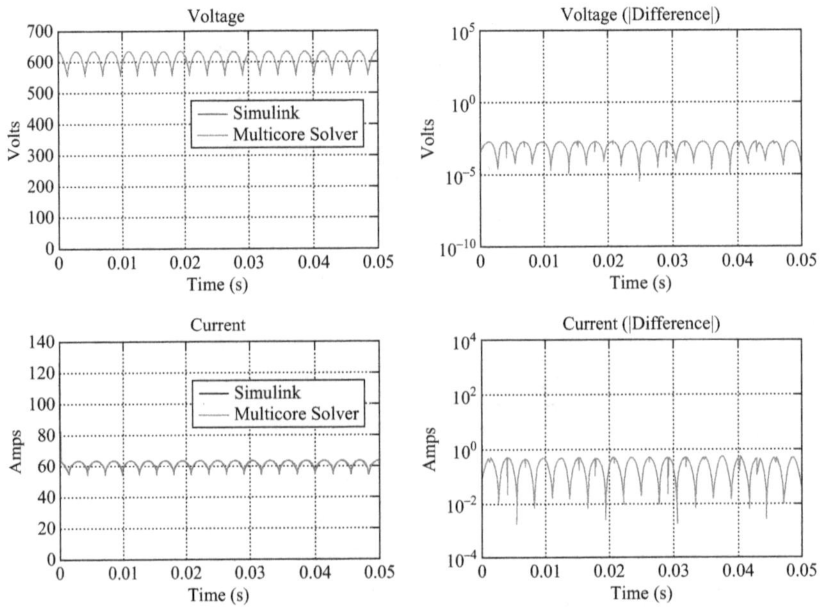

Fig. 5.9 Overlay of rectifier's load voltage and current. Difference shown in logarithmic scale on the right

This example clearly corroborates the fact that whenever there are differences in the design of two solvers, the expected results from testing them, like in this situation, will manifest such differences [116–118].

5.4.2 DC filter

Filters placed at DC links improve power quality by reducing voltage ripple and mitigating harmonic content [94]. In this regard, the DC filter model used for the motor loads included in the notional power system under analysis is shown Fig. 5.11. This is a simple second order filter with equations given in (5.15) and (5.16).

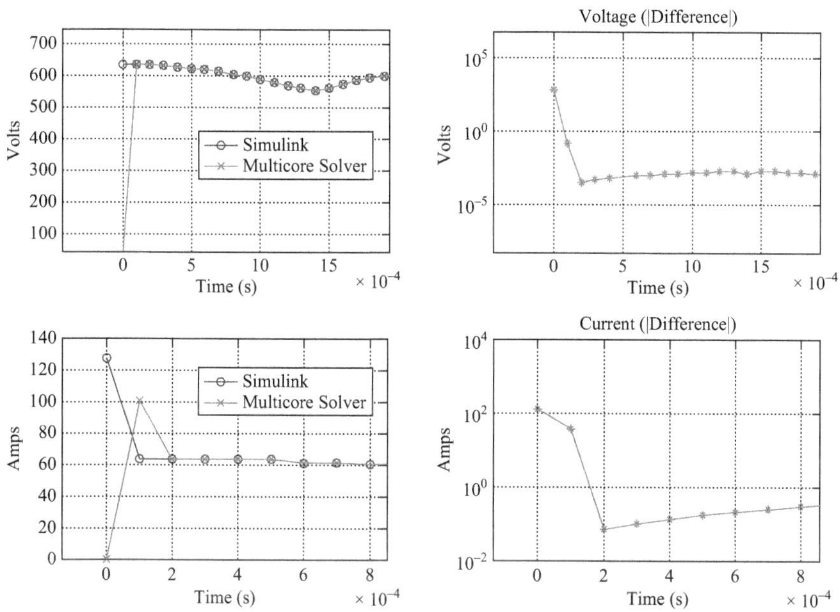

Fig. 5.10 Close-up of first simulation point of two solvers for a three-phase rectifier circuit

Mesh

$$\begin{bmatrix} R_{Ldc} + R_{Cdc} & -R_{Cdc} \\ -R_{Cdc} & R_{Cdc} \end{bmatrix} \begin{bmatrix} i_1 \\ i_2 \end{bmatrix} = \begin{bmatrix} -V_{Ldc} - V_{Cdc} \\ -V_{Ldc} \end{bmatrix} \tag{5.15}$$

Nodal

$$\begin{bmatrix} G_{Ldc} & -G_{Ldc} & \cdot \\ -G_{Ldc} & (G_{Ldc} + G_{Cdc}) & -G_{Cdc} \\ \cdot & -G_{Cdc} & G_{Cdc} \end{bmatrix} \begin{bmatrix} v_1 \\ v_2 \\ v_3 \end{bmatrix} = \begin{bmatrix} I_{Ldc} \\ -I_{Ldc} + I_{Ldc} \\ -I_{Cdc} \end{bmatrix} \tag{5.16}$$

Adding this DC filter to the output of the three-phase rectifier smoothens the DC bus voltage and capacitor charging current. To illustrate how this DC filter works, consider the three-phase rectifier with added DC filter and AC-side line inductance as shown in Fig. 5.12. The DC-side and AC-side simulation results are shown in Figs. 5.13 and 5.14, respectively, when simulated with a timestep of $\Delta t = 10\ \mu\text{s}$ using the nodal method. (To note, the mesh method gives the same outcome.) The left-hand side of these two latter figures shows the results produced by the *Simulink* solver, and the right-hand side shows the results produced using the multicore solver developed for this book. In contrast to Fig. 4.12, the charging current now reduces in peak value and increases in duration.

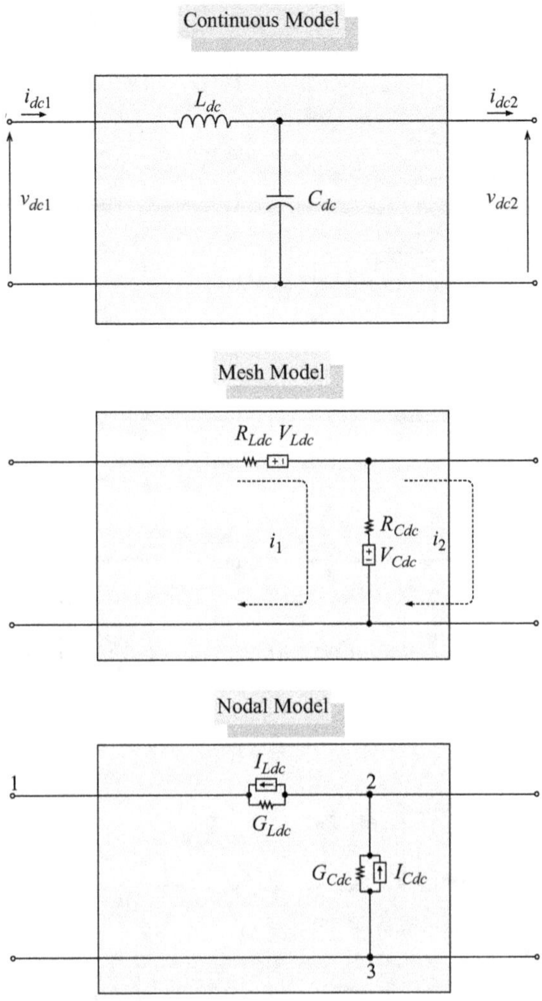

Fig. 5.11 DC filter model

5.4.3 Inverter

For the system under consideration, it suffices to model inverters (in reality, a converter) as PWM-controlled IGBTs[56] including feedback diodes and snubbers. Fortunately, six-pulse inverters and six-pulse rectifiers share the *same* model differing only in their switch types, switch-toggle control, and power flow direction. This convenience in modeling allows re-using the rectifier model as an inverter in software code.

[56] To recall, these acronyms stand for pulse-width modulation (PWM) and insulated-gate bipolar transistor (IGBT)

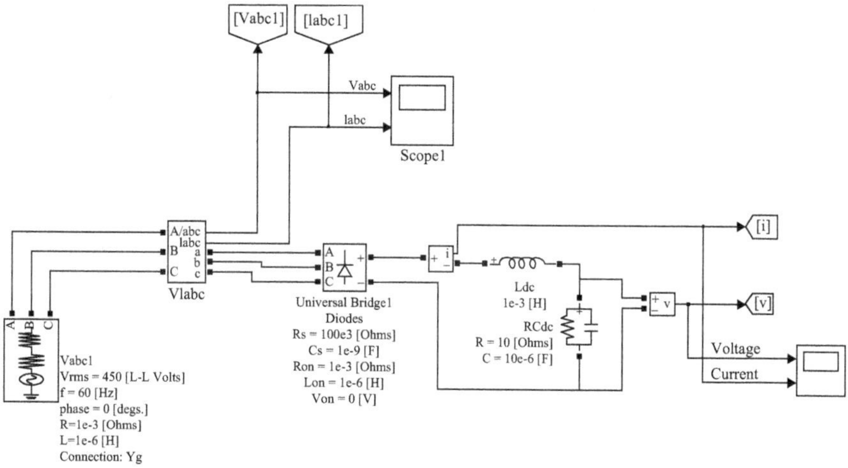

Fig. 5.12 Three-phase six-diode rectifier circuit with added DC filter and AC-side line inductance

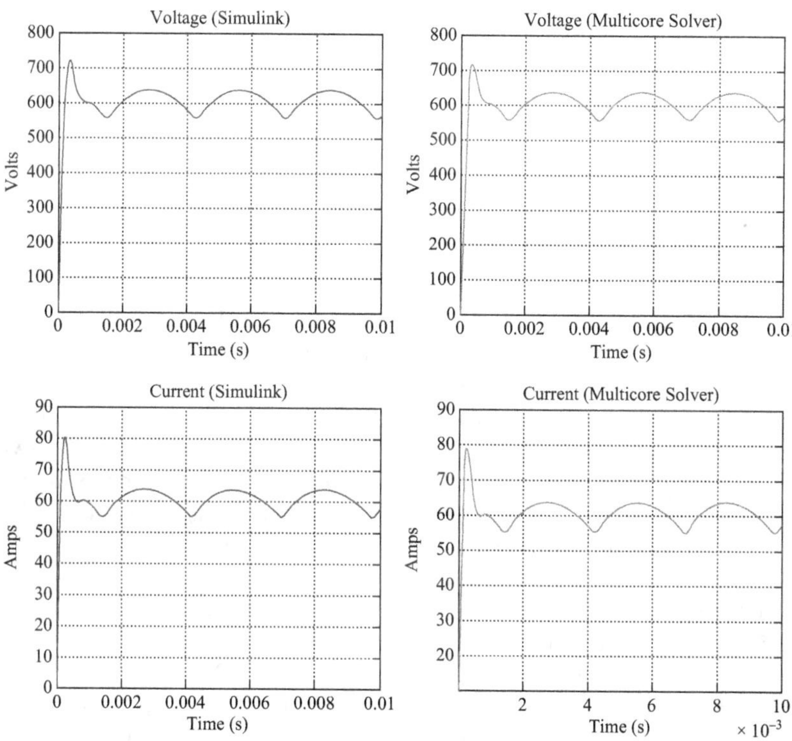

Fig. 5.13 DC-side startup characteristics of a rectifier with external DC filter. Left: Simulink results. Right: multicore results

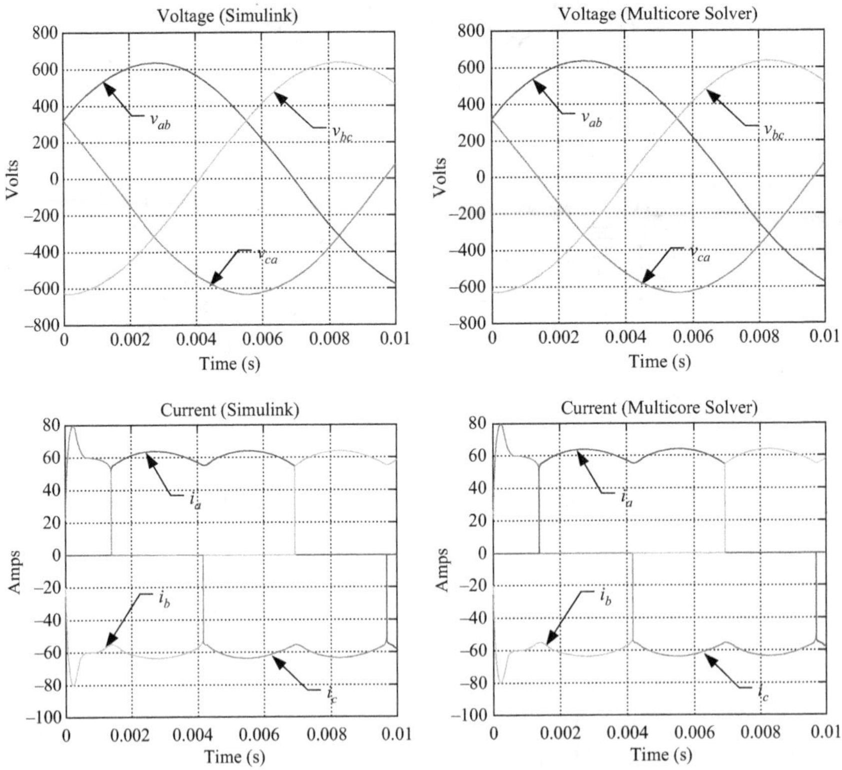

Fig. 5.14 AC-side startup characteristics of a rectifiers with external DC filter. Left: Simulink results. Right: multicore results

The inverter model for the motor drives of the notional power system is shown in Fig. 5.15. Comparing the inverter with the rectifier model in Fig. 5.7, the terminals of the inverter are oriented in the direction of the power flow: from the DC to the AC side. Although the discrete switches in Fig. 5.15 appear unchanged, inverters use switch type 6 in Fig. 4.20 (rectifiers use type 4). Additionally, IGBTs switch toggle according to an external control signal based on the reference and carrier comparisons shown in Fig. 4.35 in the previous chapter. Since the rectifier mesh and nodal equations ((5.13) and (5.14)) are similar to the equations for the inverter, they are intentionally omitted.

Simulating inverter circuits are a good way to test switch models and interpolation routines. Consider the inverter circuit shown in Fig. 5.16, and the results are shown in Fig. 5.17 for a resistive load ($P = 10$ kW) and in Fig. 5.18 for an RL load ($P = 10$ kW, $Q_L = 10$ kVar). For the resistive load case, a timestep of $\Delta t = 100$ μs was used with the nodal method. The left side in Fig. 5.17 shows the overlay of the inverter output voltage (*ab*) and line current (phase *a*). The right side shows the difference in each overlay. As can be gathered, the results are in reasonable agreement, but finite differences exist.

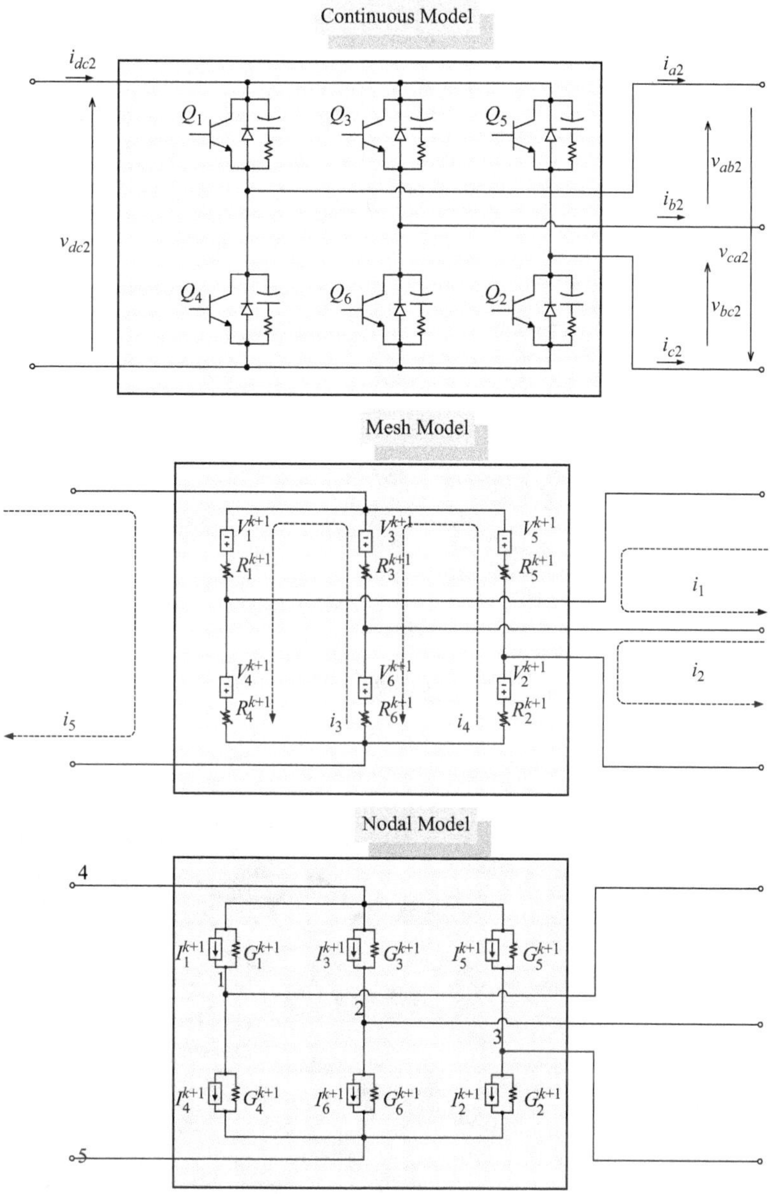

Fig. 5.15 Inverter model

The simulation results in Fig. 5.18 realized in terms of the above discussion are produced at much higher fidelity by using a timestep of $\Delta t = 1$ μs. Reducing the timestep allows comparing simulation results at different timestep sizes and check for abnormalities that could otherwise be over-looked. For example, small timesteps

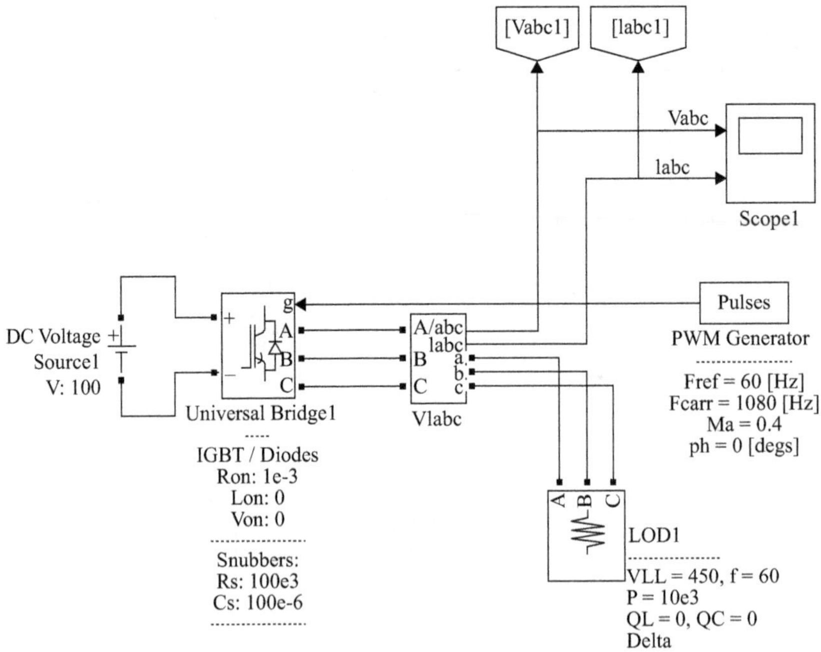

Fig. 5.16 Three-phase inverter circuit supplying a delta-connected resistive load

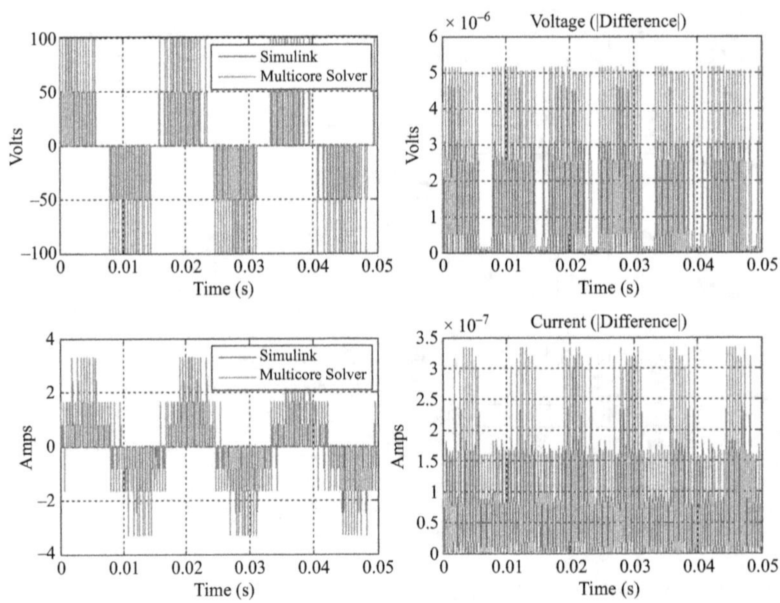

Fig. 5.17 Overlay of inverter voltage ab and line current a (resistive load, $\Delta t = 100\ \mu s$, nodal)

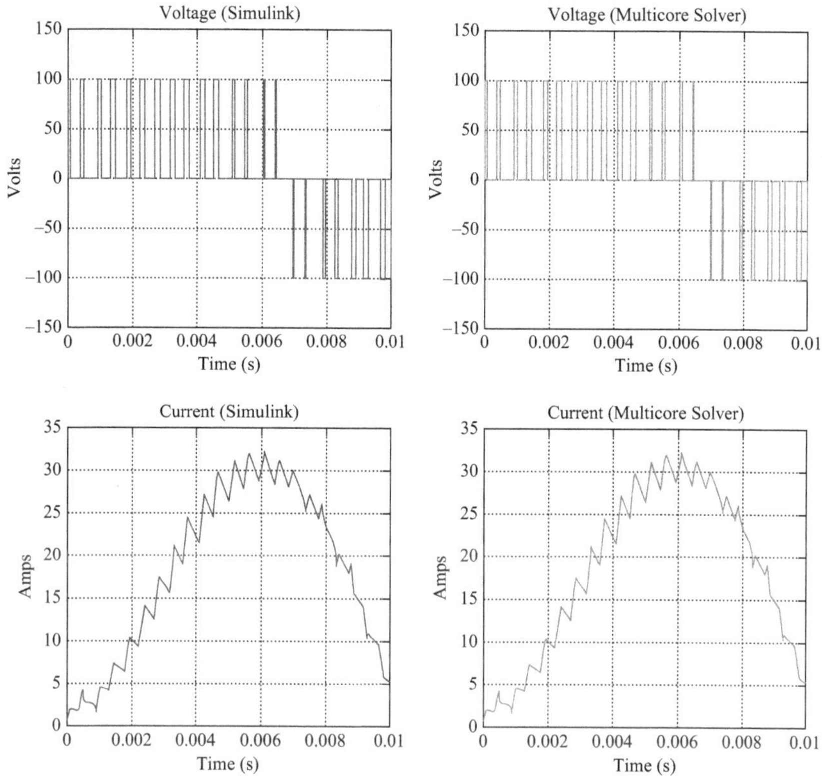

Fig. 5.18 Comparison of inverter voltage ab and line current a (RL load, $\Delta t = 1$ μs, mesh)

such as $\Delta t = 1$ μs requires more memory to store simulation data. In all fairness, this realization can be challenging for efficient computer memory management, as well as it can bring about computer memory leaks and increase simulation runtime. Small timesteps sacrifices model performance.

Larger timesteps, on the other hand, can be useful not only to accelerate simulations, but also to test interpolation routines as they allow multiple events to occur between timesteps as demonstrated earlier by Fig. 3.4. At the end of each timestep in a simulation task, the interpolation routine should be able to detect *all* intermediate events occurring between any two time grid divisions.

Lastly, modeling inverters is more challenging than modeling rectifiers as inverters have both naturally commutated and forced-commutated switches. Naturally commutated switches are controlled by each switch's voltage and current; forced-commutated switches are controlled by the results of the control network. The presence of both types of commutations increases the time spent in the interpolation procedure.

5.4.4 Motor

Modeling induction motors is a broad subject and well beyond the scope of this book. For avid readers wishing to learn more on this subject, it should be mentioned that there are several ways to model these machines, including the $0dq$ model [48], the phase-domain model [119], and the steady-state model [120]. Nonetheless, as the simulation focus of this book is at the *system* level, as opposed to at the motor-winding level, it suffices to model motors as approximate per-phase equivalents [57,94]. Treatment of winding-level motor analyses can be found in References 48, 121 and 122.

The continuous representation of the delta-connected motor windings is shown in Fig. 5.19. Although delta-connected stators are rare, they exist in shipboard applications. Consider the discretized induction motor model in Fig. 5.20. As noticed from the rotor side, the rotor slip s makes the motor model time-varying. Time-varying components increase the computational burden of the solver. Thus, to reduce this burden, motors can be modeled as having a constant speed using a fixed slip value $0 < s < 1$. This technique converts the induction motor model into a three-phase static load. The motor's discrete mesh and nodal equations are given in (5.17) and (5.18), respectively.

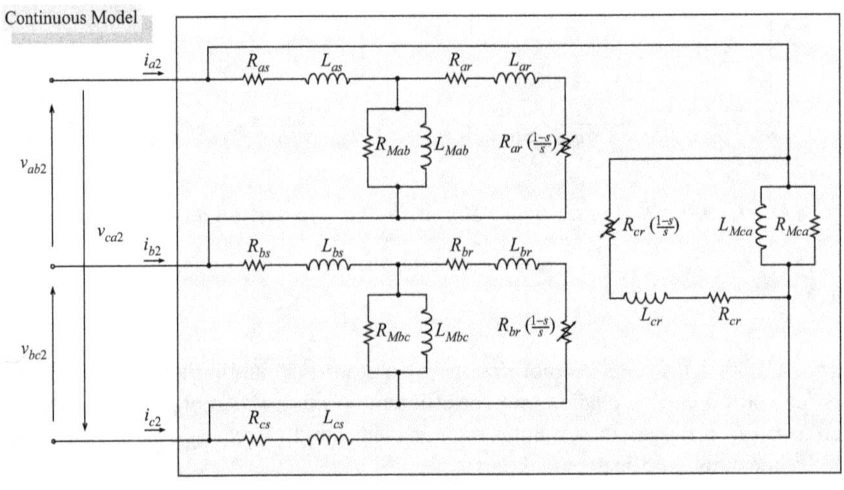

Fig. 5.19 Induction motor model (continuous representation)

Simulating a motor drive is also a good way to test switch models and interpolation routines. For example, consider the motor drive shown in Fig. 5.21, where the outcomes are shown in Fig. 5.22 for an intentionally large timestep of $\Delta t = 100\ \mu s$. The left-hand side in Fig. 5.22 shows an overlay of the three-phase source line voltage (ab) and its line current (phase a). The right-hand side shows the difference in each overlay. As can be seen, the results appear to be within reasonable agreement to one another.

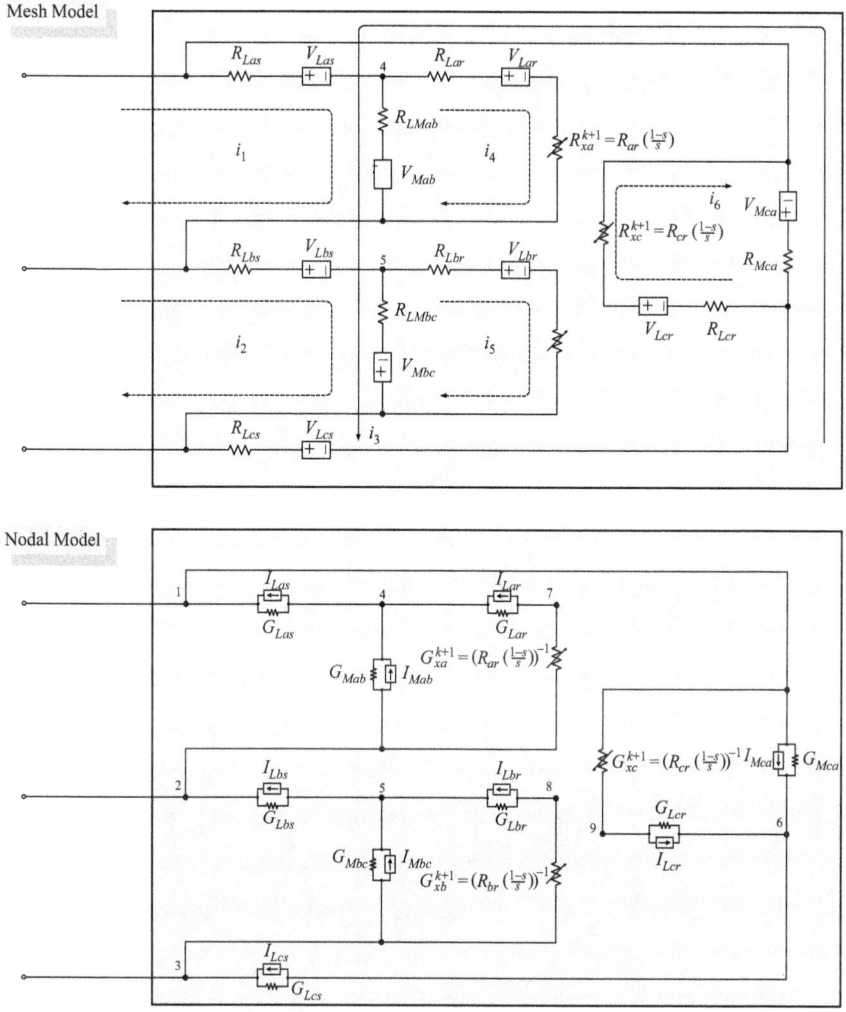

Fig. 5.20 Induction motor model (discrete representation in mesh and nodal variables)

5.4.5 Rotor

The shafts driven by induction motors can be modeled using Newton's law of rotational motion. Consider the rotor shaft of an induction motor depicted in Fig. 5.23, and its corresponding equation in (5.19).

$$T_{elec} - T_{mech} = J\frac{d\omega_{mech}}{dt} + \omega_{mech}D \tag{5.19}$$

Mesh

$$\left[\begin{array}{ccc|ccc} R_{Las}+R_{LMab} & \cdot & R_{Las}+R_{LMab} & -R_{LMab} & \cdot & \cdot \\ \cdot & R_{Lbs}+R_{LMbc} & R_{Lbs}+R_{LMbc} & \cdot & -R_{LMbc} & \cdot \\ R_{Las}+R_{LMab} & R_{Lbs}+R_{LMbc} & \begin{pmatrix} R_{Las}+R_{Lbs}+R_{Lcs} \\ +R_{LMab}+R_{LMbc}+R_{LMca} \end{pmatrix} & -R_{LMab} & -R_{LMbc} & -R_{LMca} \\ \hline -R_{LMab} & \cdot & -R_{LMab} & R_{LMab}+R_{Lar}+R_{xa}^{k+1} & \cdot & \cdot \\ \cdot & -R_{LMbc} & -R_{LMbc} & \cdot & R_{LMbc}+R_{Lbr}+R_{xb}^{k+1} & \cdot \\ \cdot & \cdot & -R_{LMca} & \cdot & \cdot & R_{LMca}+R_{Lcr}+R_{xc}^{k+1} \end{array}\right] \begin{bmatrix} i_1 \\ i_2 \\ i_3 \\ i_4 \\ i_5 \\ i_6 \end{bmatrix}$$

$$= \left[\begin{array}{c} -V_{Las}-V_{LMab} \\ -V_{Lbs}-V_{LMbc} \\ \begin{pmatrix} -V_{Las}-V_{Lbs}-V_{Lcs} \\ -V_{LMab}-V_{LMbc}-V_{LMca} \end{pmatrix} \\ \hline V_{LMab}-V_{Lsrab} \\ V_{LMbc}-V_{Lsrbc} \\ V_{LMca}-V_{Lsrca} \end{array}\right] \tag{5.17}$$

Nodal

$$\left[\begin{array}{ccc|ccc|ccc}
\begin{pmatrix}G_{Las}+G_{Mca}\\+G_{xc}^{k+1}\end{pmatrix} & \cdot & \cdot & -G_{Las} & \cdot & -G_{Mca} & \cdot & \cdot & -G_{xc}^{k+1}\\
\cdot & \begin{pmatrix}G_{Lbs}+G_{Mab}\\+G_{ax}^{k+1}\end{pmatrix} & \cdot & -G_{Mab} & -G_{Lbs} & \cdot & -G_{xa}^{k+1} & \cdot & \cdot\\
\cdot & \cdot & \begin{pmatrix}G_{Lcs}+G_{Mbc}\\+G_{bx}^{k+1}\end{pmatrix} & \cdot & -G_{Mbc} & -G_{Lcs} & \cdot & -G_{xb}^{k+1} & \cdot\\
\hline
-G_{Las} & -G_{Mab} & \cdot & \begin{pmatrix}G_{Mab}+G_{Las}\\+G_{Lar}\end{pmatrix} & \cdot & \cdot & \cdot & \cdot & \cdot\\
\cdot & -G_{Lbs} & -G_{Mbc} & \cdot & \begin{pmatrix}G_{Mbc}+G_{Lbs}\\+G_{Lbr}\end{pmatrix} & \cdot & \cdot & \cdot & \cdot\\
-G_{Mca} & \cdot & -G_{Lcs} & \cdot & \cdot & \begin{pmatrix}G_{Mca}+G_{Lcs}\\+G_{Lcr}\end{pmatrix} & \cdot & \cdot & \cdot\\
\hline
\cdot & -G_{xa}^{k+1} & \cdot & \cdot & \cdot & \cdot & G_{Lar}+G_{xa}^{k+1} & \cdot & \cdot\\
\cdot & \cdot & -G_{xb}^{k+1} & \cdot & \cdot & \cdot & \cdot & G_{Lr}+G_{xb}^{k+1} & \cdot\\
-G_{xc}^{k+1} & \cdot & \cdot & \cdot & \cdot & \cdot & \cdot & \cdot & G_{Lcr}+G_{xc}^{k+1}
\end{array}\right]
\begin{bmatrix}v_1\\v_2\\v_3\\v_4\\v_5\\v_6\\v_7\\v_8\\v_9\end{bmatrix}$$

$$= \left[\begin{array}{c}
I_{Las}-I_{Mca}\\
I_{Lbs}-I_{Mab}\\
I_{Lcs}-I_{Mbc}\\
\hline
-I_{Lba}+I_{Mab}+I_{Lar}\\
-I_{Lbc}+I_{Mbc}+I_{Lbr}\\
-I_{Lca}+I_{Mca}+I_{Lcr}\\
\hline
-I_{Lar}\\
-I_{Lbr}\\
-I_{Lcr}
\end{array}\right] \tag{5.18}$$

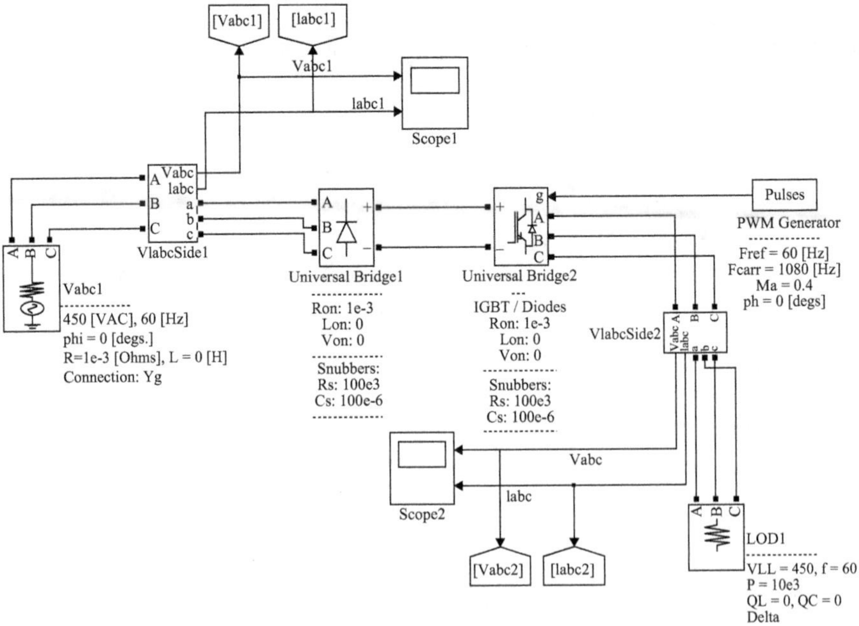

Fig. 5.21 Three-phase induction motor drive circuit (resistive load, $\Delta t = 100\ \mu s$)

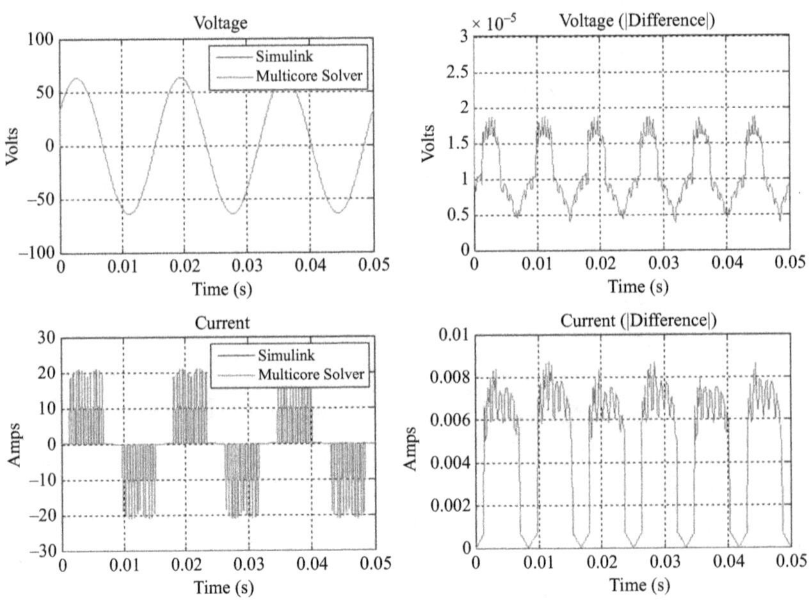

Fig. 5.22 Overlay of motor drive voltage ab and its line current on phase a

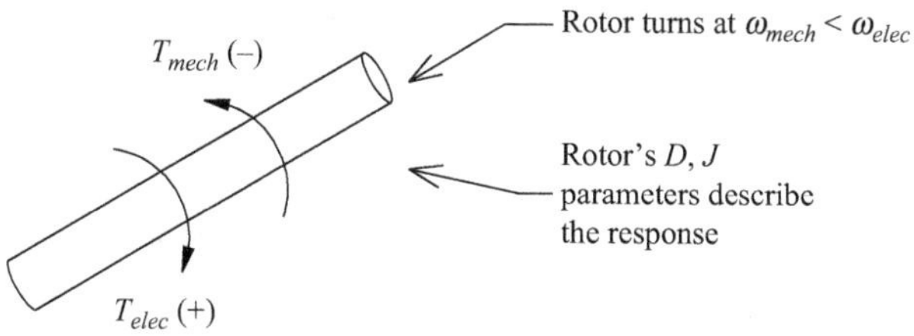

Fig. 5.23 Induction motor rotor shaft

where:

D = rotor damping coefficient (N-m-s)
J = rotor moment of inertia (kg-m^2)
T_{elec} = total (three-phase) electromagnetic torque (N-m)
T_{mech} = mechanical load torque (N-m)
ω_{elec} = electrical frequency (rad/s)
ω_{mech} = mechanical speed (rad/s).

Equation (5.19) is a first-order equation, which can be solved either via tunable integration or by root matching. Because the rotor angle θ_{mech} is conveniently a quantity of interest, (5.19) can be augmented with the rotor's angular position equation and solved via second-order state-variable formulation [51], as given by (5.20). (The solution of expression (5.20) is found from (4.39).)

$$\frac{d}{dt}\begin{bmatrix}\theta_{mech}\\ \omega_{mech}\end{bmatrix} = \begin{bmatrix}\cdot & 1\\ \cdot & \dfrac{-D}{J}\end{bmatrix}\begin{bmatrix}\theta_{mech}\\ \omega_{mech}\end{bmatrix} + \begin{bmatrix}\cdot\\ \dfrac{1}{J}\end{bmatrix}(T_{elec} - T_{mech}) \tag{5.20}$$

The per-phase electromagnetic torque produced by the induction motor is found from (5.21), where the time-varying variables are annotated with $k+1$ for clarity.

$$T_{elec/phase}^{k+1} = \frac{R_r (V_s^{k+1})^2}{s^{k+1}\omega_{elec}^{k+1}(R_{sr}^2 + (\omega_{elec}^{k+1})^2 L_{sr}^2)} \tag{5.21}$$

where:

L_{sr} = per-phase sum of stator and rotor inductance (H)
R_r = rotor resistance (Ω) (e.g., R_{ar} in Fig. 5.19 is the rotor resistance for phase *ab*)
R_{sr} = per-phase sum of stator and rotor resistance (Ω) (e.g., $R_{srab} = R_{stator} + R_{rotor}/s$ for phase ab)
s^{k+1} = rotor slip at timestep $k+1$ $\left(s^{k+1} = \left(\omega_{elec}^{k+1} - \frac{p}{2}\omega_{mech}^{k+1}\right)/\omega_{elec}^{k+1}\right)$, where p is the total number of motor poles

$T^{k+1}_{elec/phase}$ = torque developed in each phase (N-m)
V^{k+1}_{s} = per-phase stator voltage (RMS volts)
ω^{k+1}_{elec} = synchronous frequency at timestep $k+1$ (rad/s).

Summing the torque developed in each phase gives the three-phase torque T^{k+1}_{elec}, which is one of the two inputs of expression (5.20). (T_{mech} is assumed constant for simplicity.) Once ω^{k+1}_{mech} and s^{k+1} are known, the rotor resistance in the motor's discrete equations must be updated unless the motor is treated as a constant speed model.

Two aspects on the calculation of the slip s are observed. First, the rotor slip must be calculated at *each* timestep, which makes motors time-varying components—even if using reduced-order per-phase approximate models. Second, the electrical (synchronous) frequency ω^{k+1}_{elec} (typically 120π rad/s) in finite-inertia power systems may vary during disturbances. Therefore, slip calculations should include frequency measurements as well.

5.5 Transformers

The primary power-distribution voltage level in US Navy ships is 450 VAC at 60 Hz. Lighting distribution circuits, however, receive power at 120 VAC, 60 Hz [30]. These lighting transformers are modeled as three single-phase transformers connected in closed delta.

Fig. 5.24 shows the transformer model in its continuous representation. Three single-phase transformer units are connected in closed delta on both the primary and secondary sides. As noted from the voltage and current of the magnetizing branches, the secondary-side voltage is expressed in terms of the primary-side voltage, and the primary-side current is expressed in terms of the secondary-side current. These dependencies can be modeled with dependent voltage and current sources or by eliminating the dependencies. The use of dependent sources, however, removes symmetric positive-definiteness from **A** and is not covered here.

The other approach to account for these dependencies is to, first, define and then eliminate them from the transformer's equations. More specifically, three mesh currents can be eliminated from the mesh model and three voltages can be removed from the nodal model. This approach retains symmetric positive-definiteness in **A** and reduces the model's order.

Mesh model

Consider the transformer's mesh model in Fig. 5.25. It's mesh equations are given in (5.22), where $\mathbf{R}_{XFM}$ represents the mesh-coefficient matrix and $\mathbf{e}_{XMF}$ is the right-hand side vector. The currents are enumerated as follows:

- $\{i_1, i_2, i_3\}$ primary side mesh currents (independent)
- $\{i_4, i_5, i_6\}$ secondary side mesh currents (independent)
- $\{i_7, i_8, i_9\}$ magnetizing mesh currents (dependent)

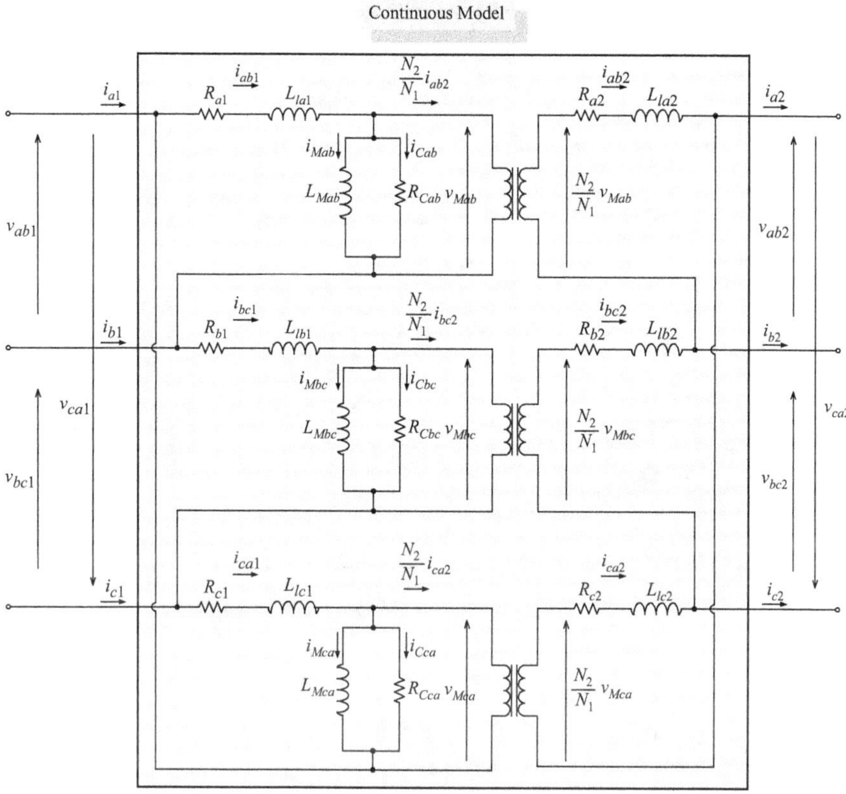

Fig. 5.24 Three-phase transformer model (continuous)

Mesh currents $\{i_1, i_2, i_3\}$ are enumerated first to maintain consistency with all the other power apparatus models. Mesh currents $\{i_4, i_5, i_6\}$ are enumerated next and marked as independent currents, so that their values are returned as part of the electrical network solution. Mesh currents $\{i_7, i_8, i_9\}$ were enumerated last both because they are dependent and because their removal at the outset does not require renumbering the independent meshes i_1 through i_6.

The relationship between independent and dependent mesh currents is given in (5.23). Transforming (5.22) with (5.23) results in the new reduced set (i.e., dimensions of 6×6 instead of 9×9) of independent mesh equations in (5.24), where $\mathbf{i}_{new}$ is the right-hand side vector consisting of independent mesh currents i_1 through i_6, and $\mathbf{C}$ is a transformation tensor. The new matrix $\mathbf{R}_{XFMnew}$ in (5.24) is the actual mesh coefficient matrix corresponding to the transformer mesh model (not the one given by (5.22)). (This new matrix, similar to the coefficient matrices of all power apparatus will be used in the next chapter to form the mesh resistance matrix of each electrical subsystem.) During simulation, the dependent (eliminated) mesh currents $\{i_7, i_8, i_9\}$ are computed using the last three rows of (5.23). These dependent mesh

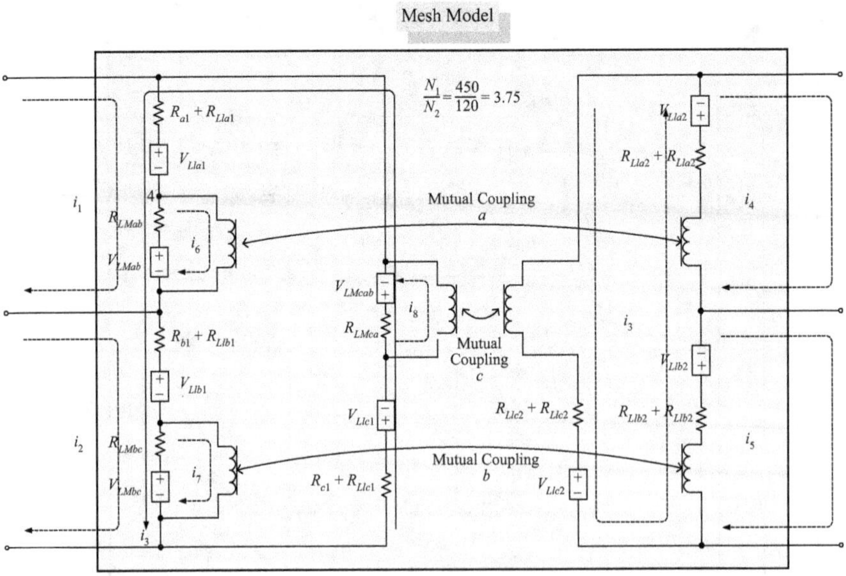

Fig. 5.25 Three-phase transformer model (mesh)

currents are required to compute the historical terms of the magnetizing parallel RL branches.

$$\begin{bmatrix} i_1 \\ i_2 \\ i_3 \\ \hline i_4 \\ i_5 \\ i_6 \\ \hline i_7 \\ i_8 \\ i_9 \end{bmatrix} = \underbrace{\left[\begin{array}{ccc|cccc} 1 & \cdot & \cdot & \cdot & \cdot & \cdot & \cdot \\ \cdot & 1 & \cdot & \cdot & \cdot & \cdot & \cdot \\ \cdot & \cdot & 1 & \cdot & \cdot & \cdot & \cdot \\ \hline \cdot & \cdot & \cdot & 1 & \cdot & \cdot & \cdot \\ \cdot & \cdot & \cdot & \cdot & 1 & \cdot & \cdot \\ \cdot & \cdot & \cdot & \cdot & \cdot & 1 & \cdot \\ \cdot & \cdot & \cdot & \cdot & \cdot & \cdot & 1 \\ \hline \cdot & \cdot & \cdot & \frac{N_2}{N_1} & \frac{N_2}{N_1} & \cdot & \cdot \\ \cdot & \cdot & \cdot & \frac{N_2}{N_1} & \cdot & \frac{N_2}{N_1} & \cdot \\ \cdot & \cdot & \cdot & \frac{N_2}{N_1} & \cdot & \cdot & \frac{N_2}{N_1} \end{array}\right]}_{\mathrm{C}} \underbrace{\begin{bmatrix} i_1 \\ i_2 \\ i_3 \\ \hline i_4 \\ i_5 \\ i_6 \end{bmatrix}}_{\mathbf{i}_{new}} \tag{5.23}$$

$$\underbrace{(\mathbf{C}^{\mathrm{T}}\mathbf{R}_{XFM}\mathbf{C})}_{\mathbf{R}_{XFM\mathrm{new}}}\mathbf{i}_{new} = \underbrace{\mathbf{C}^{\mathrm{T}}\mathbf{e}_{XFM}}_{\mathbf{e}_{XFM\mathrm{new}}} \tag{5.24}$$

Mesh

$$
\underbrace{\left[\begin{array}{ccc:ccc:ccc}
R_{La1}+R_{MCab} & \cdot & R_{La1}+R_{MCab} & \cdot & \cdot & \cdot & -R_{MCab} & & \\
\cdot & R_{Lb1}+R_{MCbc} & R_{Lb1}+R_{MCbc} & \cdot & \cdot & \cdot & & -R_{MCbc} & \\
R_{La1}+R_{MCab} & R_{Lb1}+R_{MCbc} & \begin{pmatrix} R_{La1}+R_{Cab} \\ R_{Lb1}+R_{Cbc} \\ R_{Lc1}+R_{Cca} \end{pmatrix} & \cdot & \cdot & \cdot & -R_{MCab} & -R_{MCbc} & -R_{MCca} \\
\hdashline
\cdot & \cdot & \cdot & \begin{pmatrix} R_{Lc2}+R_{Lb2} \\ +R_{La2} \end{pmatrix} & R_{La2} & R_{Lb2} & \cdot & \cdot & \cdot \\
\cdot & \cdot & \cdot & R_{La2} & R_{La2} & \cdot & \cdot & \cdot & \cdot \\
\cdot & \cdot & \cdot & R_{Lb2} & \cdot & R_{Lb2} & \cdot & \cdot & \cdot \\
\hdashline
-R_{MCab} & \cdot & -R_{MCab} & \cdot & \cdot & \cdot & R_{MCab} & & \cdot \\
\cdot & -R_{MCbc} & -R_{MCbc} & \cdot & \cdot & \cdot & \cdot & R_{MCbc} & \cdot \\
\cdot & \cdot & -R_{MCca} & \cdot & \cdot & \cdot & \cdot & \cdot & R_{MCca}
\end{array}\right]}_{\mathbf{R}_{XFM}}
\begin{bmatrix} i_1 \\ i_2 \\ i_3 \\ \hline i_4 \\ i_5 \\ i_6 \\ \hline i_7 \\ i_8 \\ i_9 \end{bmatrix}
=
\underbrace{\left[\begin{array}{c}
-V_{La1}-V_{MCab} \\
-V_{Lb1}-V_{MCbc} \\
\begin{pmatrix} -V_{La1}-V_{Lb1}-V_{Lc1}\ldots \\ \ldots-V_{MCab}-V_{MCbc}-V_{MCca} \end{pmatrix} \\
\hdashline
-V_{La2}-V_{Lb2}-V_{Lc2} \\
-V_{La2} \\
-V_{Lb2} \\
\hdashline
V_{MCab} \\
V_{MCbc} \\
V_{MCca}
\end{array}\right]}_{\mathbf{e}_{XFM}}
\tag{5.22}
$$

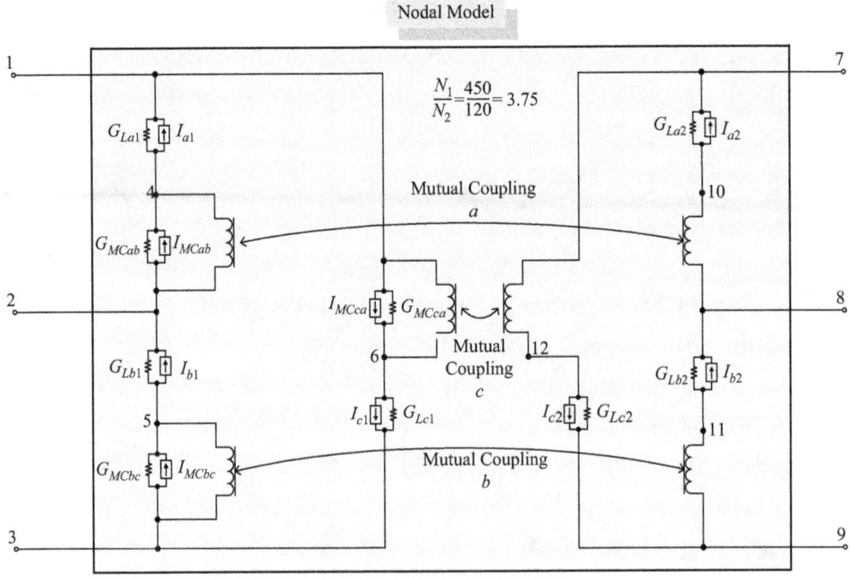

Fig. 5.26 Three-phase transformer model (nodal)

Nodal model

In a similar fashion, consider the transformer's nodal model shown in Fig. 5.26, where its corresponding nodal equations are given in (5.25). Similar to the mesh model case, nodes 10, 11, and 12 are enumerated last because they are dependent. The coefficient matrix in the nodal case is denoted as $\mathbf{G}_{XFM}$, and the right-hand side vector as $\mathbf{j}_{XFM}$. The nodes are enumerated as follows:

- $\{v_1, v_2, v_3, v_4, v_5, v_6\}$ primary side voltages (independent)
- $\{v_7, v_8, v_9\}$ secondary side voltages (independent)
- $\{v_{10}, v_{11}, v_{12}\}$ magnetizing voltages (dependent)

The relationship between independent and dependent node voltages is given by (5.26). Using (5.26) to reduce (5.25) (i.e., from 12×12 to 9×9) results in the new node voltage equations in (5.27), where $\mathbf{v}_{new}$ is a vector of independent voltages and $\mathbf{C}$ is a transformation tensor [123–125] that uses the winding ratio N_2/N_1 to express the relationship between dependencies. The matrix $\mathbf{G}_{XFMnew}$ is the nodal coefficient matrix corresponding to the transformer's nodal model. It should be kept in mind that, during simulation, the dependent node voltages are computed using the last three rows in (5.26), which are required to update the historical current source in the magnetizing parallel RL branches.

Nodal

$$
\underbrace{\left[\begin{array}{cccccc|ccc|ccc}
G_{La1}+G_{MCca} & \cdot & \cdot & -G_{La1} & \cdot & -G_{MCa} & \cdot & \cdot & \cdot & \cdot & \cdot & \cdot \\
\cdot & G_{Lb1}+G_{MCcb} & \cdot & -G_{MCab} & -G_{Lb1} & \cdot & \cdot & \cdot & \cdot & \cdot & \cdot & \cdot \\
\cdot & \cdot & G_{Lc1}+G_{MCbc} & \cdot & -G_{MCbc} & -G_{Lc1} & \cdot & \cdot & \cdot & \cdot & \cdot & \cdot \\
-G_{La1} & -G_{MCab} & \cdot & G_{La1}+G_{MCab} & \cdot & \cdot & \cdot & \cdot & \cdot & \cdot & \cdot & \cdot \\
\cdot & -G_{Lb1} & -G_{MCbc} & \cdot & G_{Lb1}+G_{MCbc} & \cdot & \cdot & \cdot & \cdot & \cdot & \cdot & \cdot \\
-G_{MCa} & \cdot & -G_{Lc1} & \cdot & \cdot & G_{Lc1}+G_{MCca} & \cdot & \cdot & \cdot & \cdot & \cdot & \cdot \\
\hline
\cdot & \cdot & \cdot & \cdot & \cdot & \cdot & G_{La2} & \cdot & \cdot & -G_{La2} & \cdot & \cdot \\
\cdot & \cdot & \cdot & \cdot & \cdot & \cdot & \cdot & G_{Lb2} & \cdot & \cdot & -G_{Lb2} & \cdot \\
\cdot & \cdot & \cdot & \cdot & \cdot & \cdot & \cdot & \cdot & G_{Lc2} & \cdot & \cdot & -G_{Lc2} \\
\hline
\cdot & \cdot & \cdot & \cdot & \cdot & \cdot & -G_{La2} & \cdot & \cdot & G_{La2} & \cdot & \cdot \\
\cdot & \cdot & \cdot & \cdot & \cdot & \cdot & \cdot & -G_{Lb2} & \cdot & \cdot & G_{Lb2} & \cdot \\
\cdot & \cdot & \cdot & \cdot & \cdot & \cdot & \cdot & \cdot & -G_{Lc2} & \cdot & \cdot & G_{Lc2}
\end{array}\right]}_{\mathbf{G}_{XFM}}
$$

$$
\left[\begin{array}{c} v_1 \\ v_2 \\ v_3 \\ v_4 \\ v_5 \\ v_6 \\ \hline v_7 \\ v_8 \\ v_9 \end{array}\right]
= \underbrace{\left[\begin{array}{c} I_{a1}-I_{MCa} \\ I_{b1}-I_{MCab} \\ I_{c1}-I_{MCbc} \\ I_{MCab}-I_{a1} \\ I_{MCbc}-I_{b1} \\ I_{MCca}-I_{c1} \\ \hline -I_{a2} \\ -I_{b2} \\ -I_{c2} \\ \hline I_{a2} \\ I_{b2} \\ I_{c2} \end{array}\right]}_{\mathbf{j}_{XFM}} \qquad (5.25)
$$

$$\begin{bmatrix} v_1 \\ v_2 \\ v_3 \\ v_4 \\ v_5 \\ v_6 \\ v_7 \\ v_8 \\ v_9 \\ \hdashline v_{10} \\ v_{11} \\ v_{12} \end{bmatrix} = \underbrace{\begin{bmatrix} 1 & \cdot & \cdot & \cdot & \cdot & \cdot & \cdot & \cdot & \cdot \\ \cdot & 1 & \cdot & \cdot & \cdot & \cdot & \cdot & \cdot & \cdot \\ \cdot & \cdot & 1 & \cdot & \cdot & \cdot & \cdot & \cdot & \cdot \\ \cdot & \cdot & \cdot & 1 & \cdot & \cdot & \cdot & \cdot & \cdot \\ \cdot & \cdot & \cdot & \cdot & 1 & \cdot & \cdot & \cdot & \cdot \\ \cdot & \cdot & \cdot & \cdot & \cdot & 1 & \cdot & \cdot & \cdot \\ \cdot & \cdot & \cdot & \cdot & \cdot & \cdot & 1 & \cdot & \cdot \\ \cdot & \cdot & \cdot & \cdot & \cdot & \cdot & \cdot & 1 & \cdot \\ \cdot & \cdot & \cdot & \cdot & \cdot & \cdot & \cdot & \cdot & 1 \\ \hdashline \cdot & -\frac{N_2}{N_1} & \cdot & \frac{N_2}{N_1} & \cdot & \cdot & \cdot & 1 & \cdot \\ \cdot & \cdot & -\frac{N_2}{N_1} & \cdot & \frac{N_2}{N_1} & \cdot & \cdot & \cdot & 1 \\ -\frac{N_2}{N_1} & \cdot & \cdot & \cdot & \cdot & \frac{N_2}{N_1} & 1 & \cdot & \cdot \end{bmatrix}}_{\mathbf{C}} \underbrace{\begin{bmatrix} v_1 \\ v_2 \\ v_3 \\ v_4 \\ v_5 \\ v_6 \\ v_7 \\ v_8 \\ v_9 \end{bmatrix}}_{\mathbf{v}_{new}} \tag{5.26}$$

$$\underbrace{(\mathbf{C}^{\mathrm{T}}\mathbf{G}_{XFM}\mathbf{C})}_{\mathbf{G}_{XFM\text{new}}}\mathbf{v}_{new} = \underbrace{\mathbf{C}^{\mathrm{T}}\mathbf{j}_{XFM}}_{\mathbf{j}_{XFM\text{new}}} \tag{5.27}$$

5.6 Generation

AC-synchronous generators convert mechanical energy provided by a prime mover into electrical energy delivered through the power distribution network. The intended recipients of electrical energy are the system loads; however, electrical energy is also lost during its distribution. The rotational energy conversion occurs at a fixed speed called the synchronous frequency, and is normally 60 Hz.

As it appears, there is more literature on synchronous generators than on any other power apparatus type [126]. Therefore, this section is not intended to serve as a guide on synchronous generators.[57] Such topics are outside the scope of this book—albeit, an intricate and detailed subject that requires a chapter of its own. As the reader will reminisce from earlier chapters, this book is not about power system *modeling*; it is about power system *partitioning* and parallelization to reduce the simulation runtime of power system transient simulations running on *Windows*-based desktop computers. This section only scratches the surface of a very deep topic on synchronous generators. A few textbooks and research articles about AC synchronous generator theory, design, and modeling that may be of interest to readers are included in References 48, 52, 122, and 127–145.

For the notional shipboard power system model illustrated in Fig. 2.1, both generators are gas-turbine-driven synchronous machines with delta-connected stator

[57] Readers interested in AC generator modeling should consult the industry standard IEEE 1110, *Guide for Synchronous Generator Modeling Practices and Applications in Power System Stability Analyses*.

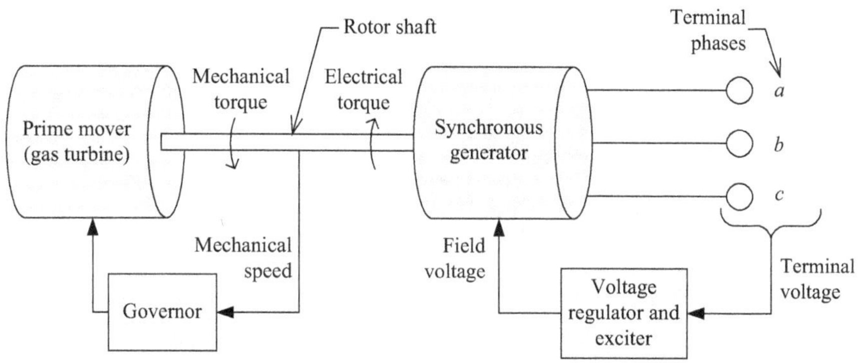

Fig. 5.27 Depiction of a prime-mover, synchronous generator, and their controllers

windings, rated at 450 VAC, 2.5 MW, 3.125 MVA, 900 RPM (60 Hz), 0.8 PF; they are also eight-pole machines and use the parameters presented in Reference 137. In a similar way to induction motor drives examined earlier, each generator in the notional power system consists of subcomponents as shown in Fig. 5.27: a prime-mover and governor, a rotor shaft, a voltage regulator and exciter, machine windings, and those mechanical parts required to hold the machine together and ensure proper alignment, cooling, and other aspects of operation.

When modeling the dynamics of a synchronous generator, each of these sub-components must be modeled and parameterized adequately. The generator rotor dynamics are normally based on the swing equation ((7.82) in Reference 51). The prime mover and governor models are based on Reference 137, and the voltage regulator and exciter are based on the IEEE Type II excitation system [99,146]. The stator and rotor winding models are based on the six-winding time-varying inductances model in References 122, 132, and 135, which stator is connected in delta (rather unusual). The rotor windings include a field, d-axis damper, and q-axis damper windings. The six windings in the stator and rotor windings are magnetically coupled through time-varying inductances. Each of the aforementioned subsystems requires not only careful modeling, but also careful *implementation* in a solver. Even when models are implemented properly, numerical instabilities may arise as discussed in References 38 and 147.

Consistent with the scope of this book, synchronous generators are not treated here. Instead, for the system under analysis, these power apparatus are modeled as delta-connected three-phase voltage sources behind RL impedances. It should be pointed out that replacement in the future of these three-phase voltage sources will not affect the partitioning method presented in this book. AC synchronous generators, as with induction machines, can be modeled in the *0dq* or *abc* domains. The *abc*-frame model presented in References 122, 132, and 135 works well with the partitioned mesh and nodal methods formulations illustrated in this book.

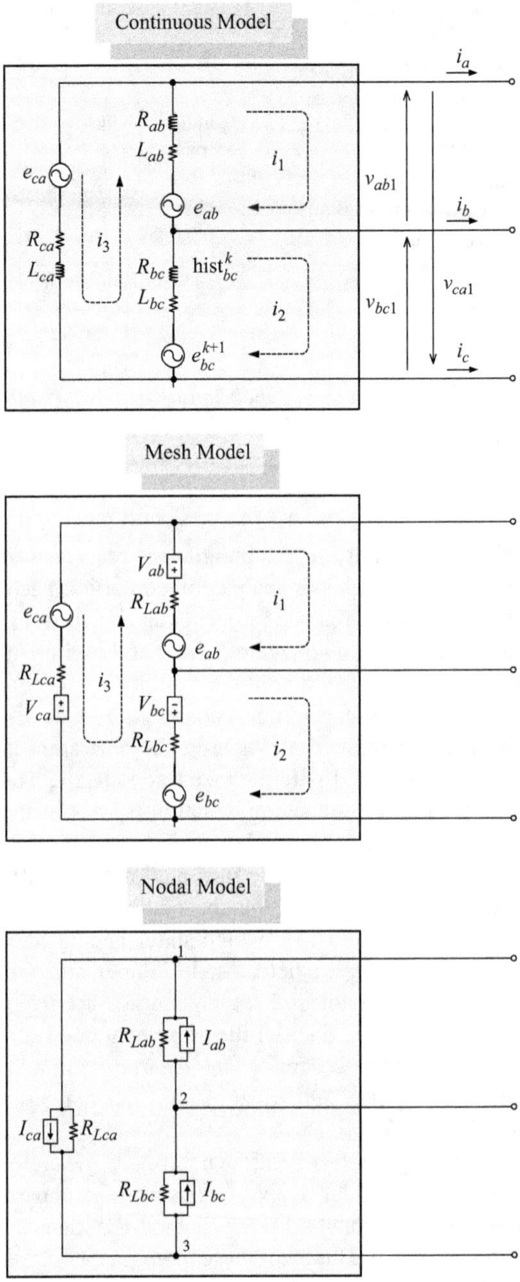

Fig. 5.28 Three-phase voltage source model

Consider the three-phase voltage source shown in Fig. 5.28. Its corresponding mesh and nodal equations are shown in (5.28) and (5.29), respectively. In the mesh model, the EMF values in each phase are superimposed to the historical sources of the series RL branches. In the nodal model, on the other hand, the voltage sources are first converted to Norton equivalents and then combined with the series RL branch in the current injection term.

Mesh

$$\begin{bmatrix} R_{Lab} & \cdot & R_{Lab} \\ \cdot & R_{Lbc} & R_{Lbc} \\ R_{Lab} & R_{Lbc} & (R_{Lab} + R_{Lbc} + R_{Lca}) \end{bmatrix} \begin{bmatrix} i_1 \\ i_2 \\ i_3 \end{bmatrix} = \begin{bmatrix} -V_{Lab} \\ -V_{Lbc} \\ -V_{Lab} - V_{Lbc} - V_{Lca} \end{bmatrix} \tag{5.28}$$

Nodal

$$\begin{bmatrix} G_{Lab} + G_{Lca} & -G_{Lab} & -G_{Lca} \\ -G_{Lab} & G_{Lab} + G_{Lbc} & -G_{Lbc} \\ -G_{Lca} & -G_{Lbc} & G_{Lbc} + G_{Lca} \end{bmatrix} \begin{bmatrix} v_1 \\ v_2 \\ v_3 \end{bmatrix} = \begin{bmatrix} I_{ab} - I_{ca} \\ I_{bc} - I_{ab} \\ I_{ca} - I_{bc} \end{bmatrix} \tag{5.29}$$

5.7 Summary

Each power apparatus model presented in the chapter was discussed in its continuous and discrete forms. For the discrete form, the equations were given as both mesh and nodal sets. It was also stated that representing discrete power apparatus models as isolated equations with zero inputs stems from multi-terminal component theory [148]. This theory is a convenient approach to design solvers as such representation follows the principles of software modularity and object-oriented design.

This chapter presented basic "place holder" models for the power apparatus types listed in Chapter 2. Among them, the most compute-intensive power apparatus model presented was the induction motor drive discussed in section 5.4. The induction motor included two converters: a front-end rectifier and an inverter. The rectifier was modeled with six switches and the inverter with twice as many. It was explained that 18 switches per motor drive significantly increased the computational burden of the simulation run.

The workings and especial features of the family of power apparatus models presented in this chapter should not be difficult to grasp. Throughout the discussions, it was presumed that readers have working knowledge of power system components. Moreover, as may be implied by the quotation of Box and Draper atop, one may say that the gains accrued from using more-complex power apparatus models must be carefully weighed. More often than not, employing more-complex power apparatus models results in an increase in power system simulation runtime with little (or null) gains in detail. Should this be the case, their implementation may not be warranted. Whether users elect simple or such complex models, this choice does not change the partitioning methodology presented in this book.

To end, it should be remembered that an improper implementation of (simple or complex) power apparatus models can lead to numerical instabilities. Unsuitable implementations highlights why the design of modular software is of great importance to power system solvers. Further, but not less important, is the perception and usage of the gray-enclosed representations of power apparatus models. This perception reduces the analysis and implementation time of models in a solver required by using "miniature networks" that are easy to understand and troubleshoot. A well-designed power system solver should provide its users (and developers) the ability to swap-out these power apparatus models effortlessly (e.g., replace a three-phase voltage source with a time-varying synchronous generator).

Chapter 6
Network formulation

Chapter 5 introduced the power apparatus models included in the notional shipboard power system model accompanied by their respective illustrations and equations. This chapter explains how to form the electrical network immittance[58] matrix **A** for mesh and nodal formulations starting from said power apparatus equations. The approach presented here first block-diagonalizes the power apparatus equations and then interconnects them using a tensor transformation. While this approach is uncommon, it promotes code re-use as the same method to form the immittance matrix applies to both nodal and mesh formulation types. Moreover, using a tensor to form immittance matrices does not require graph theory, allows working with small equation sets in code (at the power apparatus level), and promotes software modularity as power apparatus are treated as interchangeable blocks.

The chapter starts by introducing multi-terminal component theory to illustrate how isolated power apparatus blocks interconnect to form a connected network. Then, it is shown how to block-diagonalize said power apparatus equations, how to form the tensor, and, finally, how to obtain the immittance matrices of the electrical network.

6.1 Multi-terminal components

Power systems may be viewed as an overall interconnection of power apparatus at various bus nodes. Power *apparatus*, on the other hand, refers to machines, cables, converters, transmission lines, loads, transformers, and so forth—that is, everything shown as enclosed by "gray boxes" in the previous chapter. *Bus* nodes, or buses, refer to a group of electrical nodes where two[59] or more power apparatus interconnect.

[58] Immittance matrix is a term that refers to either of several possible electrical network coefficient matrices. Depending on the context, it may refer to the nodal matrix or the mesh matrix. In general, immittance matrices are denoted as **A**; in a discrete nodal context, they are denoted as $\mathbf{G}_{nodal}$ and in a discrete mesh context, they are denoted as $\mathbf{R}_{mesh}$.

[59] Typically, three or more power apparatus interconnect at a bus. To avoid using the term *three-phase node* to refer to the connection of two power apparatus, and the term *bus* to refer to the connection of three or more power apparatus, *bus* is used for convenience in both situations.

More specifically, in power engineering buses refer to some geographical location where terminal phases of various power apparatus meet at a junction. For example, a three-phase bus is where the electrical phases *a*, *b*, and *c* of two or more power apparatus interconnect.

Treatment of power apparatus as isolated networks can be traced back to Kron's tensor analysis of networks [123–125] and to multi-terminal component (MTC) theory [148], where power apparatus are conceptually enclosed in "black boxes" called MTCs. Depending on the MTC type (single-phase or three-phase), there can be one or more terminal leads (connectors) on the input and/or output sides of an MTC. The concept of multi-terminal component obfuscates the internal power apparatus type and, instead, provides a view of some generic multi-terminal box.

An MTC with three-phase terminals on both the input and output sides is illustrated in Fig. 6.1. The top view represents the case for mesh formulations, while the bottom view represents the case for nodal formulations. The MTC (and its internal power apparatus) is the same for both cases; what is different is the way their terminals are shown. To illustrate, in the mesh case the terminals are short-circuited to provide internal meshes with closed circular paths. This is so because shorted terminals enable isolation between neighboring components and explicitly show that the

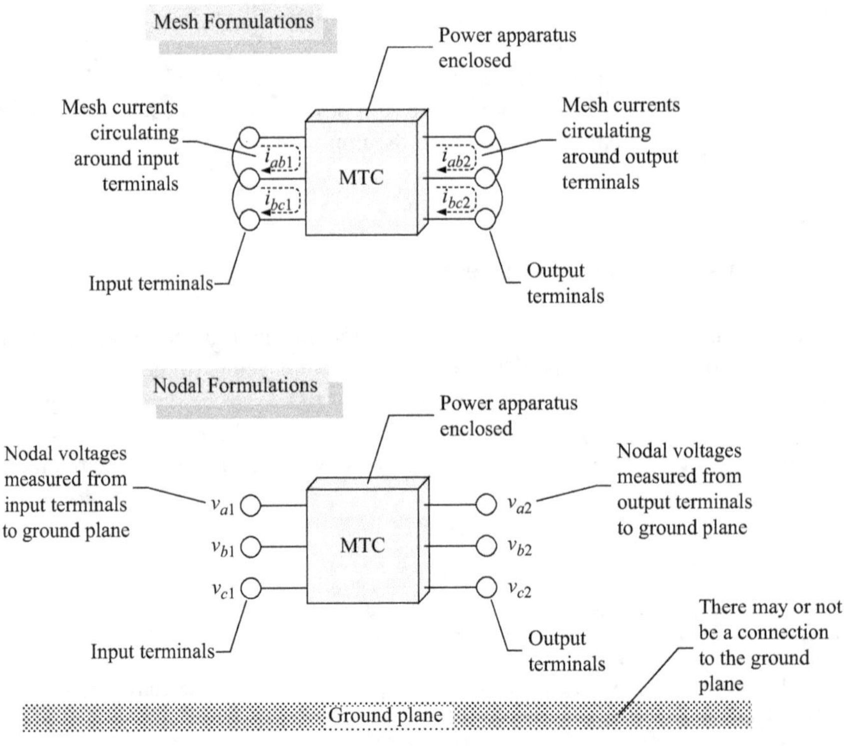

Fig. 6.1 A multi-terminal component illustrated for mesh and nodal formulations

MTCs have inputs (i.e., the terminal EMF or voltage impressions are zero). In the nodal case, the same MTC is shown with its terminals *open*. This is consistent with the concept of zero inputs as well, where zero-current injections at the terminals are equivalent to having open terminals.

Notice that Fig. 6.1 includes a ground plane at the bottom. Some power systems have explicit connections to this ground plane, while others do not. As seen from the power apparatus models in the previous chapter, no power apparatus was illustrated as having a connection to the ground plane—yet a ground plane (*datum* node) is required in nodal formulations.

6.2 Buses

Buses are electrical junctions where MTCs interconnect. Consider the two MTCs on the left side of Fig. 6.2. These MTCs enclose power apparatus of different types and are ready for an interconnection at the bus (shown as a black bar). Referring to the mesh representation, the mesh directionality is defined by bus arrows. Mesh currents *entering* the bus are shown by arrows toward the bus. Mesh currents *leaving* the bus are shown by arrows leaving the bus. With respect to the nodal representation, the bus nodes do not display directional arrows because node voltages are measured with respect to a common reference.

The right side of Fig. 6.2 shows the same MTCs *after* the interconnection takes place. After their interconnection, the terminal mesh currents flow *through* the bus (in the mesh case) and terminal voltages become bus voltages (in the nodal case). Essentially, this is how the interconnection tensor will be formed later in this chapter: at each bus, a set of equations will describe how the meshes or nodes of one MTC relate to the meshes or nodes of its neighbor MTCs.

Readers should know that power systems modeled by power utility companies can span 1,000 to 10,000 buses—and even to 50,000 buses in larger cases. The solution of these models is returned as phasors instead of by instantaneous voltages and currents. The inability to produce timely transient simulations for such large systems is an ongoing (and challenging) research topic. This book explains how to partition, parallelize, and solve large power systems so that their solution is produced in the *time domain* without having to use phasors.

The following sections rely heavily on the aforementioned concepts of MTCs and buses in order to give rise to the interconnection tensor. Once the power apparatus equations (presented in Chapter 4) are block-diagonalized, the purpose of the tensor is to transform the block-diagonalized power apparatus equations into the desired electrical network immittance matrix **A**.

6.3 Forming the mesh matrix

The mesh resistance matrix is the **A** (immittance) matrix of the electrical network and is a special case of the mesh impedance matrix. It is termed *resistance* matrix

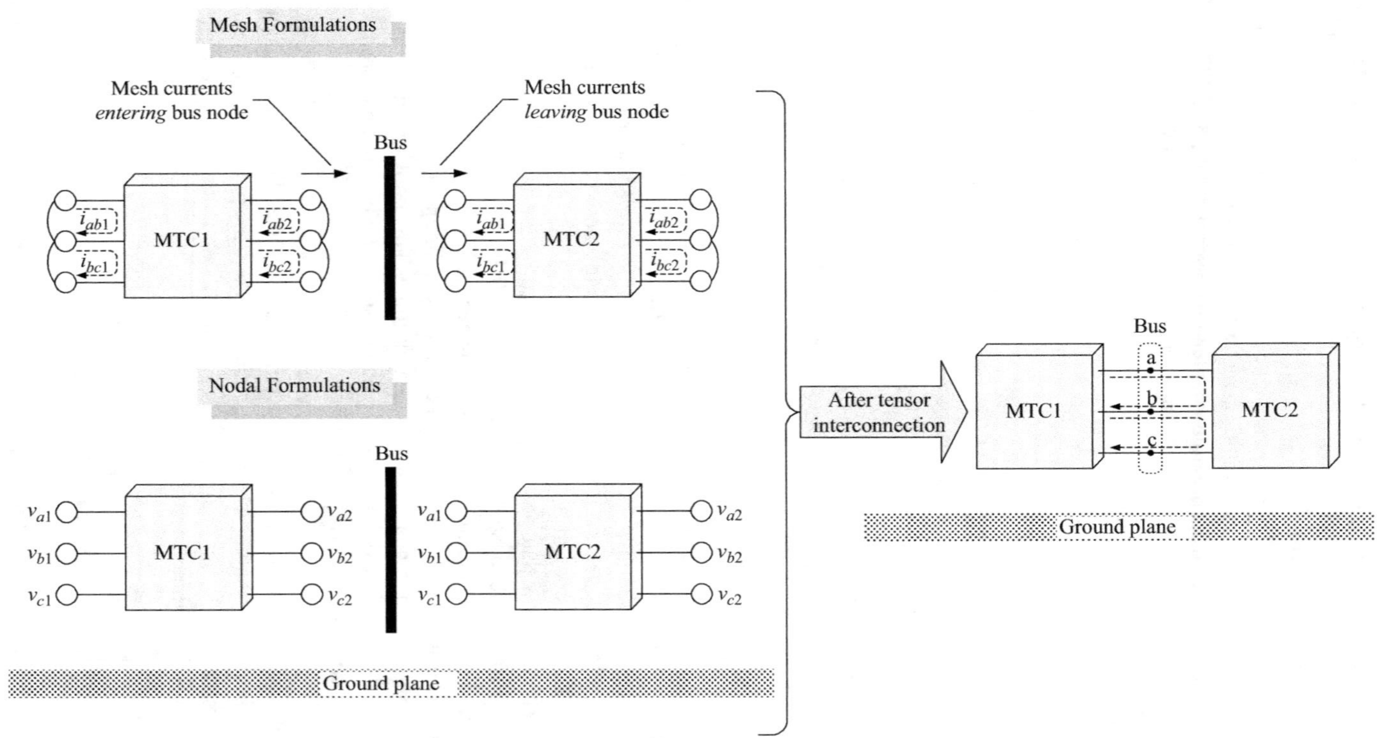

Fig. 6.2 Two isolated MTCs ready for connection (top left: mesh formulation, bottom left: nodal formulation). Right: MTCs after their interconnection at a bus

because it comprises the mesh equations of a discretized (i.e., purely resistive) power system. The traditional approach to obtain the mesh (impedance or resistance) matrix is graph theoretic in nature, where graph loops are obtained from a spanning tree to represent the network meshes. Mesh identification using graph approaches may be time consuming, may yield dense mesh matrices, involves programming overhead, may yield dense mesh matrices, and requires developing search heuristics. This section provides an alternate approach to form the mesh resistance that does not require graph theory.

The previous chapter presented the mesh equations for each power apparatus model. To obtain the mesh resistance matrix $\mathbf{R}_{mesh}$ of the entire network, the power-apparatus-level mesh equations must be interconnected first. To do so, two matrices are required to form the mesh resistance matrix $\mathbf{R}_{mesh}$: a block-diagonal matrix $\mathbf{R}_{block}$ and a connection tensor $\mathbf{C}$. Formation of $\mathbf{R}_{block}$ is straightforward, but formation of $\mathbf{C}$ is not. Formation of $\mathbf{R}_{block}$ and $\mathbf{C}$ are presented in the next two sub-sections, respectively.

6.3.1 Block-diagonal matrix

Consider a matrix $\mathbf{R}_{block}$ arranging in block-diagonal form the mesh equations of all MTCs in a power system model. For a power system of i MTCs, $\mathbf{R}_{block}$ is defined as given by (6.1), where m represents the total number of MTC meshes when all MTCs are disconnected.

$$\mathbf{R}_{block} = \begin{bmatrix} \mathbf{R}_{MTC1} & & & \\ & \mathbf{R}_{MTC2} & & \\ & & \ddots & \\ & & & \mathbf{R}_{MTCi} \end{bmatrix}_{m \times m} \tag{6.1}$$

6.3.2 Connection tensor

The power-invariant connection (or transformation) tensor $\mathbf{C}$: $\mathrm{R}^{m \times m} \rightarrow \mathrm{R}^{M \times M}$ transforms $\mathbf{R}_{block}$ into $\mathbf{R}_{mesh}$, where $M < m$ represents the number of meshes *after* all MTCs are interconnected. For clarity of the discussion, it is mentioned that other equivalent names for $\mathbf{C}$ are *connection* or *transformation* matrix. The term *tensor* is maintained due to its relevance to Kron's work on this method of analysis [123,125].

It should be pointed out, however, that the tensor $\mathbf{C}$, as used here, is *different* from the tensor $\mathbf{C}$ defined by Kron in References 123 and 125. Kron used another tensor $\mathbf{C}$ to transform primitive branch currents into the meshes of small networks. The method presented here extends this concept by transforming these small network meshes into the meshes of a large power system. The use of $\mathbf{C}$ as presented here is valid whether an MTC—or even the final power system—is planar or not. The following example demonstrates how a tensor produces the mesh resistance matrix.

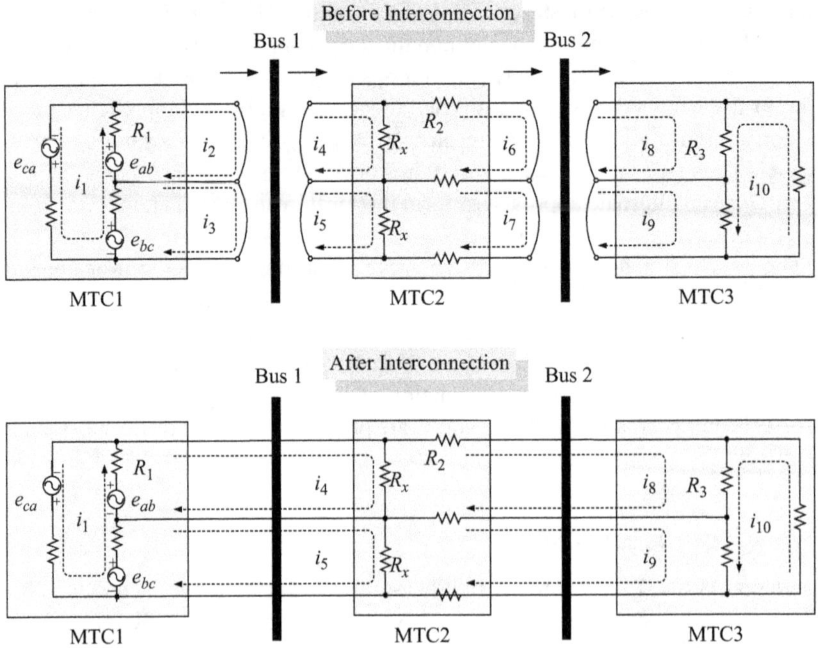

Fig. 6.3 Three MTCs before and after interconnection. Mesh currents are indicated as entering or leaving a multinode using directed arrows

Consider the three MTCs shown in Fig. 6.3 *before* (top) and *after* (bottom) their interconnection at buses 1 and 2 take place. The block-diagonal equations, having $\mathbf{R}_{block}$ as coefficient matrix, are given in (6.2), where $\mathbf{i}_{block}$ and $\mathbf{e}_{block}$ are the mesh current and EMF vectors *before* interconnection.

$$\mathbf{R}_{block}\mathbf{i}_{block} = \mathbf{e}_{block} \tag{6.2}$$

$$\left[\begin{array}{ccc|cccc|ccc}
3R_1 & R_1 & R_1 & \cdot & \cdot & \cdot & \cdot & \cdot & \cdot & \cdot \\
R_1 & R_1 & \cdot & \cdot & \cdot & \cdot & \cdot & \cdot & \cdot & \cdot \\
R_1 & \cdot & R_1 & \cdot & \cdot & \cdot & \cdot & \cdot & \cdot & \cdot \\
\hline
\cdot & \cdot & \cdot & R_x & \cdot & -R_x & \cdot & \cdot & \cdot & \cdot \\
\cdot & \cdot & \cdot & \cdot & R_x & \cdot & -R_x & \cdot & \cdot & \cdot \\
\cdot & \cdot & \cdot & -R_x & \cdot & 2R_2+R_x & -R_2 & \cdot & \cdot & \cdot \\
\cdot & \cdot & \cdot & \cdot & -R_x & -R_2 & 2R_2+R_x & \cdot & \cdot & \cdot \\
\hline
\cdot & \cdot & \cdot & \cdot & \cdot & \cdot & \cdot & R_3 & \cdot & R_3 \\
\cdot & \cdot & \cdot & \cdot & \cdot & \cdot & \cdot & \cdot & R_3 & R_3 \\
\cdot & \cdot & \cdot & \cdot & \cdot & \cdot & \cdot & R_3 & R_3 & 3R_3
\end{array}\right]
\begin{bmatrix} i_1 \\ i_2 \\ i_3 \\ i_4 \\ i_5 \\ i_6 \\ i_7 \\ i_8 \\ i_9 \\ i_{10} \end{bmatrix}
=
\begin{bmatrix} e_{ab}+e_{bc}+e_{ca} \\ e_{ab} \\ e_{bc} \\ \cdot \\ \cdot \\ \cdot \\ \cdot \\ \cdot \\ \cdot \\ \cdot \end{bmatrix}$$

A visual inspection of the disconnected (top) and connected (bottom) meshes shown in Fig. 6.3 yields the relations stated by (6.3), where $\mathbf{C}$ is the desired connection tensor

$$\mathbf{i}_{block} = \mathbf{C} \cdot \mathbf{i}_{mesh} \tag{6.3}$$

$$\begin{bmatrix} i_1 \\ i_2 \\ i_3 \\ i_4 \\ i_5 \\ i_6 \\ i_7 \\ i_8 \\ i_9 \\ i_{10} \end{bmatrix}_{10\times 1} = \begin{bmatrix} 1 & \cdot & \cdot & \cdot & \cdot & \cdot \\ \cdot & 1 & \cdot & \cdot & \cdot & \cdot \\ \cdot & \cdot & 1 & \cdot & \cdot & \cdot \\ \cdot & 1 & \cdot & \cdot & \cdot & \cdot \\ \cdot & \cdot & 1 & \cdot & \cdot & \cdot \\ \cdot & \cdot & \cdot & 1 & \cdot & \cdot \\ \cdot & \cdot & \cdot & \cdot & 1 & \cdot \\ \cdot & \cdot & \cdot & 1 & \cdot & \cdot \\ \cdot & \cdot & \cdot & \cdot & 1 & \cdot \\ \cdot & \cdot & \cdot & \cdot & \cdot & 1 \end{bmatrix}_{10\times 6} \begin{bmatrix} i_1 \\ i_4 \\ i_5 \\ i_8 \\ i_9 \\ i_{10} \end{bmatrix}_{6\times 1}$$

Substituting (6.3) into (6.2) produces an over-determined equation set. To make the equation set determined, the principle of power conservation is applied. To conserve power when transforming disconnected meshes into connected ones, the power circulating in the disconnected MTCs must equal the power circulating in the connected meshes. This principle is stated with (6.4), where $\mathbf{e}_{mesh}$ and $\mathbf{i}_{mesh}$ are, respectively, the mesh current and mesh EMF vectors *after* interconnection. Substituting (6.3) into (6.4) yields (6.5).

$$\mathbf{e}^{\mathrm{T}}_{block}\mathbf{i}_{block} = \mathbf{e}^{\mathrm{T}}_{mesh}\mathbf{i}_{mesh} \tag{6.4}$$

$$\mathbf{e}^{\mathrm{T}}_{block}(\mathbf{C} \cdot \mathbf{i}_{mesh}) = \mathbf{e}^{\mathrm{T}}_{mesh}\mathbf{i}_{mesh} \tag{6.5}$$

Comparing the left- and right-hand side of (6.5) suggests its equivalence in (6.6).

$$(\mathbf{e}^{\mathrm{T}}_{block}\mathbf{C}) = \mathbf{e}^{\mathrm{T}}_{mesh} \tag{6.6}$$

Transposing both sides of (6.6) results in (6.7). Equation (6.7) suggests that, to conserve power during a tensor transformation, both sides of (6.2) should be multiplied by $\mathbf{C}^{\mathrm{T}}$ after using the substitution of (6.3).

$$\mathbf{C}^{\mathrm{T}}\mathbf{e}_{block} = \mathbf{e}_{mesh} \tag{6.7}$$

To finish with the example, after substitution of (6.3) into (6.2), multiplication of both sides of (6.2) by $\mathbf{C}^{\mathrm{T}}$ results in (6.8), which produces the mesh resistance matrix $\mathbf{R}_{mesh}$ (or $\mathbf{A}$) for the example case shown in Fig. 6.3.

$$\left[\begin{array}{c|cc|ccc} 3R_1 & R_1 & R_1 & \cdot & \cdot & \cdot \\ \hline R_1 & R_1 + R_x & \cdot & -R_x & \cdot & \cdot \\ R_1 & \cdot & R_1 + R_x & \cdot & -R_x & \cdot \\ \hline \cdot & -R_x & \cdot & 2R_2 + R_x + R_3 & -R_2 & R_3 \\ \cdot & \cdot & -R_x & -R_2 & 2R_2 + R_x + R_3 & R_3 \\ \cdot & \cdot & \cdot & R_3 & R_3 & 3R_3 \end{array}\right] \begin{bmatrix} i_1 \\ i_4 \\ i_5 \\ i_8 \\ i_9 \\ i_{10} \end{bmatrix} = \begin{bmatrix} e_{ab} + e_{bc} + e_{ca} \\ e_{ab} \\ e_{bc} \\ \cdot \\ \cdot \\ \cdot \end{bmatrix} \tag{6.8}$$

In short, the following three steps summarize how $\mathbf{R}_{mesh}$ was obtained for the present example. First, substitution of (6.3) into (6.2) resulted in (6.9).

$$\mathbf{R}_{block}(\mathbf{C} \cdot \mathbf{i}_{mesh}) = \mathbf{e}_{block} \tag{6.9}$$

Second, multiplying both sides of (6.9) by $\mathbf{C}^{\mathrm{T}}$ resulted in (6.10).

$$\underbrace{(\mathbf{C}^{\mathrm{T}}\mathbf{R}_{block}\mathbf{C})}_{\mathbf{R}_{mesh}}\mathbf{i}_{mesh} = \underbrace{\mathbf{C}^{\mathrm{T}}\mathbf{e}_{block}}_{\mathbf{e}_{mesh}} \tag{6.10}$$

The final system of electrical network equations in the form $\mathbf{A} \cdot \mathbf{x} = \mathbf{b}$ is given by (6.11).

$$\mathbf{R}_{mesh}\mathbf{i}_{mesh} = \mathbf{e}_{mesh} \tag{6.11}$$

This example showed how the tensor **C** was used to obtain the mesh resistance matrix of the electrical network. Formation of **C** by visual inspection, however, is far more theoretical than practical. In practical scenarios, the connected meshes of power system models are not readily visible as shown here, which suggests that formation of **C** is a graph-theoretic task. However, the following subsection introduces a programmatic algorithm to form **C** *without* resorting to graph theory.

6.3.3 Algorithm to form tensor

The following algorithm to form the tensor **C** is based on Reference 149. The tensor **C** can be formed by observing that, at every bus, the net sum of mesh currents *between* conductors is zero. This observation is depicted in Fig. 6.4 and stated by (6.12).

$$\begin{array}{c}\text{Bus}\\\text{equation}\end{array} \rightarrow \begin{cases} i_{ab1} - i_{ab2} - i_{ca1} + i_{ca2} = 0 \\ i_{bc1} - i_{bc2} - i_{ca1} + i_{ca2} = 0 \end{cases} \tag{6.12}$$

In (6.12), one mesh current must be marked as *dependent* for each phase pair (*ab* or *bc*). For example, marking i_{ab1} and i_{bc1} as dependent allows their elimination and leaves a connected set of independent mesh currents at the bus. This is the central idea to forming the connection tensor **C**.

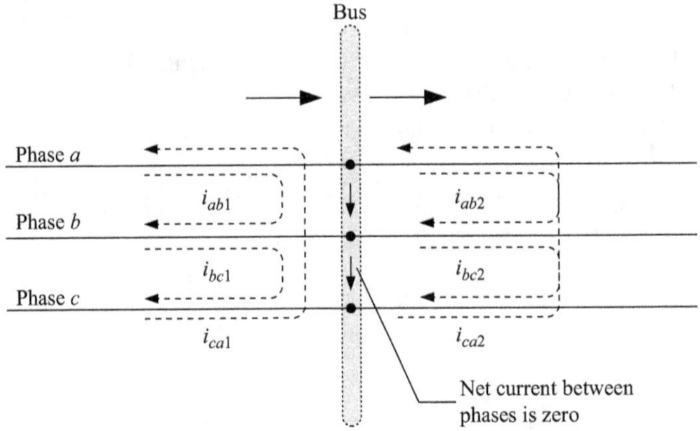

Fig. 6.4 Mesh currents incident at a multinode. The net current between phase conductors is always zero

Fig. 6.5 shows an annotated flow chart to outline the steps required to form **C**. This algorithm visits each bus in a power system and interconnects all incident meshes. By doing so, the algorithm defines a mesh current (for each phase pair) as dependent and marks it for deletion. At the end of the algorithm, the indicated column operations and column removals yield the connection tensor **C**.

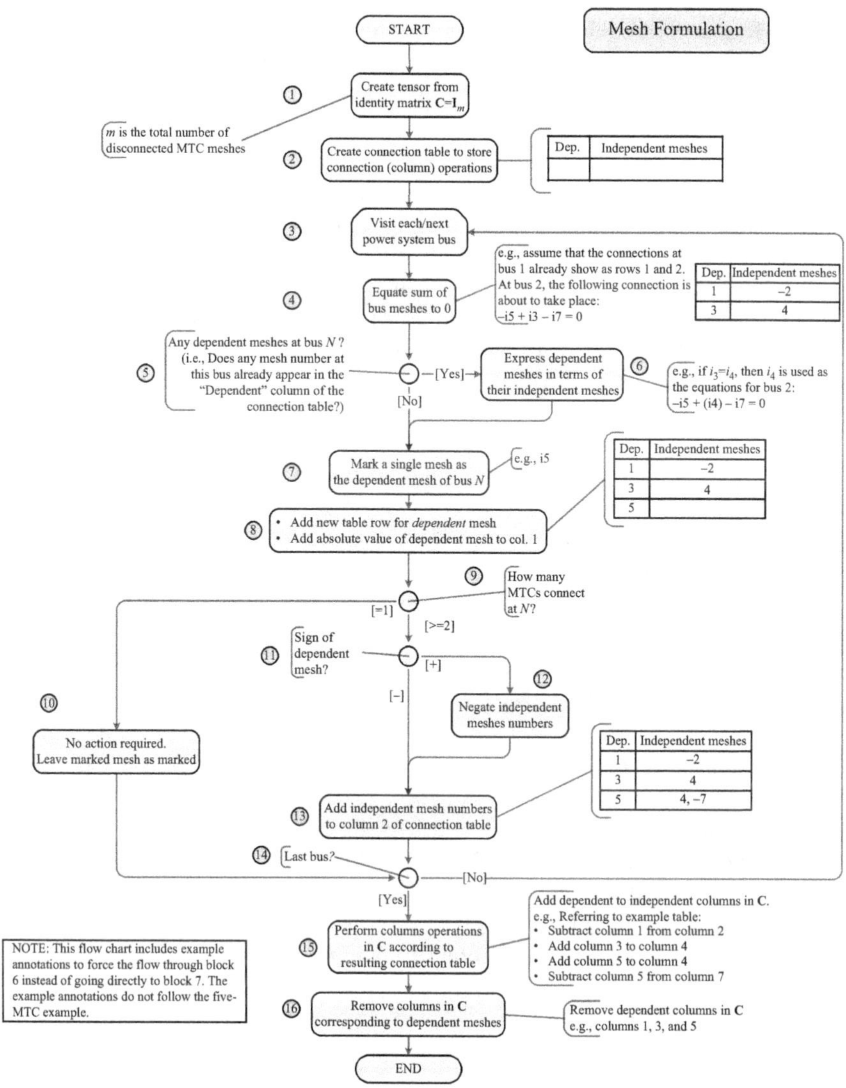

*Fig. 6.5 Flow chart illustrating steps to form connection tensor **C** for mesh formulations*

Each annotation in the flow chart of Fig. 6.5 is explained next in the following list of 16 items or steps. A context example is also included in the description to facilitate understanding of the flow chart and to force the explanations through the more complicated path of the flow chart:

1. The algorithm starts by creating $\mathbf{C} = \mathbf{I}_m$, where $\mathbf{I}_m$ is an identity matrix of order m. This matrix is not used until the connection table is completely filled.
2. Create an empty two-column connection table. Column 1 will store, in each row, the numbers of all dependent ("*Dep.*") meshes in the system. Column 2, in each row, will store the independent mesh numbers of each phase pair at each bus. Each row represents one bus connection and each bus has two connections: *ab* and *bc*.
3. The connection algorithm enters a loop that visits each power system bus in any order.
4. At each bus, the net mesh current sum of each phase pair is $\sum_{m=1}^{M} i_m = 0$, where m represents the mesh number of the MTC. Meshes entering bus N have a "+" sign and meshes leaving it have a "−" sign. This sign convention can be reversed, but the reversed convention must be consistent at each bus N.

   For explanation purposes, assume that two connections have already taken place at bus 1.[60] These connections are already included in the connection table as $i_1 = -i_2$ and $i_3 = i_4$. Assume also that the current loop iteration is visiting bus 2, where the connection described as $-i_5 + i_3 - i_7 = 0$ is about to take place. (i_3 is assumed to enter bus 2.)
5. Check whether any of the meshes at bus N (bus 2 for the context example) are dependent. Dependent meshes appear in previous rows in column 1. To identify this situation, a recursive search through the connection table is required.
6. If there are dependent meshes at bus N (i.e., meshes used previously as dependent meshes), express these previously used meshes in terms of their independent mesh numbers. Following the context example,[61] i_3 in $-i_5 + i_3 - i_7 = 0$ is expressed in terms of i_4 as $-i_5 + i_4 - i_7 = 0$
7. From the context example equation $-i_5 + i_4 - i_7 = 0$, one mesh number must be marked as *dependent*. Here, we arbitrarily choose i_5 as the dependent mesh of the connection at bus 2.
8. Create a new table row for the equation in context ($-i_5 + i_4 - i_7 = 0$). Place the dependent mesh number without its sign in column 1 (i_5). The connection table shown to the right of it shows two (previously formed) connections at bus 1: $i_1 + i_2 = 0$ for phase *ab* and $i_3 - i_4 = 0$ for phase *bc*. It is assumed that algorithm is currently visiting bus 2, and that the table is appended in step 8 by adding the number 5 (representing i_5) to column 1 (the column of dependent mesh currents).

[60] This example intends to how to account for dependent meshes previously used in the connection table.
[61] This arbitrary in-context example purposefully forces the decision through block 6 to explain the case of previously used meshes.

9. Determine how many MTCs should interconnect at bus N.
10. If there is only one MTC at N, proceed to the next node. The marked mesh (e.g., i_5) remains so. (Column 5 of **C** will be removed later to avoid a short circuit condition).
11. Check the sign of the dependent mesh.
12. In the context example, $-i_5$ has a negative sign; therefore, no negation is necessary to 4 and -7 is necessary. (If i_5 were positive, the signs of 4 and -7 would have been negated to -4 and 7, respectively.)
13. Add independent mesh currents numbers to column 2 of the connection table.
14. Check whether there are more buses to visit; if so, go back to step 3.
15. Perform the indicated column operations according to the connection table.
16. Lastly, remove all columns marked for deletion in **C** (these column numbers appear in column 1 of the connection table). This reduces the column dimensions of **C**.

The connection table tracks the column operations that should be performed on **C**, which includes column additions, subtractions, and removals. The simplicity by which **C** is formed avoids graph routines to identify meshes in connected power system models.

The following explanations now re-visit the interconnection example discussed for Fig. 6.3 this time using the flow chart of Fig. 6.5 to demonstrate how to form **C** programmatically. Each numbered step in the flow chart of Fig. 6.5 is followed in order. Referring to the schematic in Fig. 6.3:

1. Create $\mathbf{C} = \mathbf{I}_{10}$ as expressed in (6.13).
2. Create (empty) connection table:

Dependent meshes	Independent meshes

3. Visit bus 1.
4. Write the following equations for bus 1:

$$\begin{cases} -i_2 + i_4 = 0 \\ -i_3 + i_5 = 0 \end{cases}$$

5. Since the connection table is empty (so far), there are no dependent meshes.
6. (This step does not apply.)
7. Choose i_2 and i_3 as dependent meshes.
8. Add the following dependent mesh numbers to column 1:

Dependent meshes	Independent meshes
2	
3	

9. Two MTCs interconnect at bus 1.

10. (This step does not apply.)
11. i_2 and i_3 have negative signs.
12. (This step does not apply.)
13. The connection table is updated as:

Dependent meshes	Independent meshes
2	4
3	5

14. There are more buses to visit. Repeat steps 3–13 for bus 2. The resulting connection table is:

Dependent meshes	Independent meshes
2	4
3	5
6	8
7	9

15. The next to last step is to perform the following column operations according to the operations indicated by the connection table:
    - add column 2 to column 4
    - add column 3 to column 5

    These operations result in (6.14). Continuing:
    - add column 6 to column 8
    - add column 7 to column 9

    These operations result in (6.15)
16. Lastly, remove columns 2, 3, 6, and 7 from **C**. This removal leads to the final result: the connection tensor **C** shown in (6.16) ahead.

The tensor **C** in (6.16) is identical to the one obtained visually in (6.3). This procedure can be extended to interconnect a large number of MTCs as well.

$$\mathbf{C} = \overset{\textbf{Start}}{\begin{bmatrix} 1 & \cdot & \cdot & \cdot & \cdot & \cdot & \cdot & \cdot & \cdot & \cdot \\ \cdot & 1 & \cdot & \cdot & \cdot & \cdot & \cdot & \cdot & \cdot & \cdot \\ \cdot & \cdot & 1 & \cdot & \cdot & \cdot & \cdot & \cdot & \cdot & \cdot \\ \cdot & \cdot & \cdot & 1 & \cdot & \cdot & \cdot & \cdot & \cdot & \cdot \\ \cdot & \cdot & \cdot & \cdot & 1 & \cdot & \cdot & \cdot & \cdot & \cdot \\ \cdot & \cdot & \cdot & \cdot & \cdot & 1 & \cdot & \cdot & \cdot & \cdot \\ \cdot & \cdot & \cdot & \cdot & \cdot & \cdot & 1 & \cdot & \cdot & \cdot \\ \cdot & \cdot & \cdot & \cdot & \cdot & \cdot & \cdot & 1 & \cdot & \cdot \\ \cdot & \cdot & \cdot & \cdot & \cdot & \cdot & \cdot & \cdot & 1 & \cdot \\ \cdot & \cdot & \cdot & \cdot & \cdot & \cdot & \cdot & \cdot & \cdot & 1 \end{bmatrix}} \tag{6.13}$$

After connection at bus 1

$$\mathbf{C} = \begin{bmatrix} 1 & \cdot & \cdot & \cdot & \cdot & \cdot & \cdot & \cdot & \cdot & \cdot \\ \cdot & 1 & \cdot & 1 & \cdot & \cdot & \cdot & \cdot & \cdot & \cdot \\ \cdot & \cdot & 1 & \cdot & 1 & \cdot & \cdot & \cdot & \cdot & \cdot \\ \cdot & \cdot & \cdot & 1 & \cdot & \cdot & \cdot & \cdot & \cdot & \cdot \\ \cdot & \cdot & \cdot & \cdot & 1 & \cdot & \cdot & \cdot & \cdot & \cdot \\ \cdot & \cdot & \cdot & \cdot & \cdot & 1 & \cdot & \cdot & \cdot & \cdot \\ \cdot & \cdot & \cdot & \cdot & \cdot & \cdot & 1 & \cdot & \cdot & \cdot \\ \cdot & \cdot & \cdot & \cdot & \cdot & \cdot & \cdot & 1 & \cdot & \cdot \\ \cdot & \cdot & \cdot & \cdot & \cdot & \cdot & \cdot & & 1 & \cdot \\ \cdot & \cdot & \cdot & \cdot & \cdot & \cdot & \cdot & \cdot & \cdot & 1 \end{bmatrix} \qquad (6.14)$$

After connection at bus 2

$$\mathbf{C} = \begin{bmatrix} 1 & \cdot & \cdot & \cdot & \cdot & \cdot & \cdot & \cdot & \cdot & \cdot \\ \cdot & 1 & \cdot & 1 & \cdot & \cdot & \cdot & \cdot & \cdot & \cdot \\ \cdot & \cdot & 1 & \cdot & 1 & \cdot & \cdot & \cdot & \cdot & \cdot \\ \cdot & \cdot & \cdot & 1 & \cdot & \cdot & \cdot & \cdot & \cdot & \cdot \\ \cdot & \cdot & \cdot & \cdot & 1 & \cdot & \cdot & \cdot & \cdot & \cdot \\ \cdot & \cdot & \cdot & \cdot & \cdot & 1 & \cdot & 1 & \cdot & \cdot \\ \cdot & \cdot & \cdot & \cdot & \cdot & \cdot & 1 & \cdot & 1 & \cdot \\ \cdot & \cdot & \cdot & \cdot & \cdot & \cdot & \cdot & 1 & \cdot & \cdot \\ \cdot & \cdot & \cdot & \cdot & \cdot & \cdot & \cdot & & 1 & \cdot \\ \cdot & \cdot & \cdot & \cdot & \cdot & \cdot & \cdot & \cdot & \cdot & 1 \end{bmatrix} \qquad (6.15)$$

End

$$\mathbf{C} = \begin{bmatrix} 1 & \cdot & \cdot & \cdot & \cdot & \cdot \\ \cdot & 1 & \cdot & \cdot & \cdot & \cdot \\ \cdot & \cdot & 1 & \cdot & \cdot & \cdot \\ \cdot & 1 & \cdot & \cdot & \cdot & \cdot \\ \cdot & \cdot & 1 & \cdot & \cdot & \cdot \\ \cdot & \cdot & \cdot & 1 & \cdot & \cdot \\ \cdot & \cdot & \cdot & \cdot & 1 & \cdot \\ \cdot & \cdot & \cdot & 1 & \cdot & \cdot \\ \cdot & \cdot & \cdot & \cdot & 1 & \cdot \\ \cdot & \cdot & \cdot & \cdot & \cdot & 1 \end{bmatrix}_{10\times 6} \qquad (6.16)$$

It can be demonstrated that the foregoing tensor approach works even when multiple MTCs connect at a bus. Consider the five-MTC example shown in Fig. 6.6 below. Of particular interest is the connection at bus 3. At this bus, there is an MTC causing a short circuit condition. The meshes of this MTC will be removed in the example.

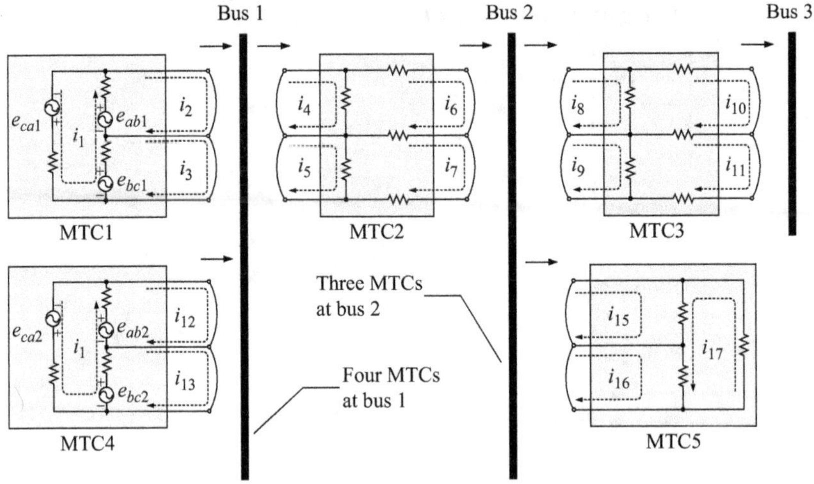

Fig. 6.6 Example of interconnection of five MTCs at three buses for a mesh formulation

Each numbered step in the flow chart of Fig. 6.5 is followed in order. Thus, referring to the schematic in Fig. 6.6:

1. Create $\mathbf{C} = \mathbf{I}_{17}$ as expressed in (6.13). (Note that $m = 17$.)
2. Create the connection table:

Dependent meshes	Independent meshes

3. Visit bus 1.
4. Write the following equations for bus 1:

$$\begin{cases} -i_2 + i_4 - i_{12} = 0 \\ -i_3 + i_5 - i_{13} = 0 \end{cases}$$

5. Since the table connection is empty, there are no dependent meshes.
6. (This step does not apply so far.)
7. Choose i_2 and i_3 as the dependent meshes (one for each phase pair).
8. Add the following table rows:

Dependent meshes	Independent meshes
2	
3	

9. Two MTCs interconnect at bus 1.
10. (This step does not apply.)
11. i_2 and i_3 have negative signs, which means 4, −12 and 5, −13 do not require negation.
12. (This connection step does not apply.)
13. The table is updated as:

Dependent meshes	Independent meshes
2	4, −12
3	5, −13

14. Since there are more buses to visit, repeat steps 3–13 for bus 2. At bus 3, mark meshes i_{10} and i_{11} as dependent and add to the table. (Since there is one MTC at bus 3, all incoming meshes must be eliminated to avoid a short circuit condition.) At this step it is observed that the resulting connection table (where N/A stands for *not applicable*) is:

Dependent meshes	Independent meshes
2	4, −12
3	5, −13
6	8, 15
7	9, 16
10	N/A
11	N/A

15. Perform the following column operations according to the connection table:
    - add column 2 to column 4; subtract column 2 from column 12
    - add column 3 to column 5; subtract column 3 from column 13
    - add column 6 to columns 8 and 15
    - add column 7 to columns 9 and 16
16. Finally, remove columns 2, 3, 6, 7, 10, and 11 from **C**. The final connection tensor is given below in (6.17).

The resulting **C** for this example is shown in (6.17). This example reinforces the observation that graph theory is not required to form **C**. Moreover, only *terminal* meshes were operated on, which leaves the meshes internal to power apparatus "as-is." It should be noticed that rows 10 and 11 in (6.17) were all zero because *columns* 10 and 11 were removed from **C**. These *rows* 10 and 11, however, must remain in **C** to constrain that $i_{10} = 0$ and $i_{11} = 0$, which removes the terminal meshes at bus 3 and avoids the aforementioned short circuit condition.

$$
\begin{bmatrix} i_1 \\ i_2 \\ i_3 \\ i_4 \\ i_5 \\ i_6 \\ i_7 \\ i_8 \\ i_9 \\ i_{10} \\ i_{11} \\ i_{12} \\ i_{13} \\ i_{14} \\ i_{15} \\ i_{16} \\ i_{17} \end{bmatrix}
=
\underbrace{\begin{bmatrix}
1 & \cdot & \cdot & \cdot & \cdot & \cdot & \cdot & \cdot & \cdot & \cdot & \cdot \\
\cdot & 1 & \cdot & \cdot & \cdot & \cdot & -1 & \cdot & \cdot & \cdot & \cdot \\
\cdot & \cdot & 1 & \cdot & \cdot & \cdot & \cdot & -1 & \cdot & \cdot & \cdot \\
\cdot & 1 & \cdot & \cdot & \cdot & \cdot & \cdot & \cdot & \cdot & \cdot & \cdot \\
\cdot & \cdot & 1 & \cdot & \cdot & \cdot & \cdot & \cdot & \cdot & \cdot & \cdot \\
\cdot & \cdot & \cdot & 1 & \cdot & \cdot & \cdot & \cdot & 1 & \cdot & \cdot \\
\cdot & \cdot & \cdot & \cdot & 1 & \cdot & \cdot & \cdot & \cdot & 1 & \cdot \\
\cdot & \cdot & \cdot & 1 & \cdot & \cdot & \cdot & \cdot & \cdot & \cdot & \cdot \\
\cdot & \cdot & \cdot & \cdot & 1 & \cdot & \cdot & \cdot & \cdot & \cdot & \cdot \\
\cdot & \cdot & \cdot & \cdot & \cdot & \cdot & \cdot & \cdot & \cdot & \cdot & \cdot \\
\cdot & \cdot & \cdot & \cdot & \cdot & \cdot & \cdot & \cdot & \cdot & \cdot & \cdot \\
\cdot & \cdot & \cdot & \cdot & \cdot & 1 & \cdot & \cdot & \cdot & \cdot & \cdot \\
\cdot & \cdot & \cdot & \cdot & \cdot & \cdot & 1 & \cdot & \cdot & \cdot & \cdot \\
\cdot & \cdot & \cdot & \cdot & \cdot & \cdot & \cdot & 1 & \cdot & \cdot & \cdot \\
\cdot & \cdot & \cdot & \cdot & \cdot & \cdot & \cdot & \cdot & 1 & \cdot & \cdot \\
\cdot & \cdot & \cdot & \cdot & \cdot & \cdot & \cdot & \cdot & \cdot & 1 & \cdot \\
\cdot & \cdot & \cdot & \cdot & \cdot & \cdot & \cdot & \cdot & \cdot & \cdot & 1
\end{bmatrix}_{17\times 11}}_{\mathbf{C}}
\begin{bmatrix} i_1 \\ i_4 \\ i_5 \\ i_8 \\ i_9 \\ i_{12} \\ i_{13} \\ i_{14} \\ i_{15} \\ i_{16} \\ i_{17} \end{bmatrix}
\tag{6.17}
$$

6.4 Forming the nodal matrix

This section shows how to use the tensor approach introduced in the foregoing section to form the nodal conductance matrix. The nodal matrix is termed nodal *admittance* matrix in networks with complex impedances and nodal *conductance* matrix in purely resistive (discrete) networks. Like in mesh formulations, where the immittance matrix $\mathbf{A}$ refers to the mesh resistance matrix $\mathbf{R}_{mesh}$, in the nodal formulations the immittance matrix $\mathbf{A}$ refers to the nodal conductance matrix $\mathbf{G}_{nodal}$.

The method introduced in the previous section to obtain $\mathbf{R}_{mesh}$ can be re-used (with minor modifications) to form $\mathbf{G}_{nodal}$. Implementing one algorithm to form both immittance matrices promotes code re-use, reduces code length, and reduces maintenance time. Readers interested in nodal formulations (and not in mesh formulations) will not find advantages in using the tensor approach to form $\mathbf{G}_{nodal}$ over using the well-known branch stamping method [49] (other than the ability to treat power apparatus models as "miniature networks"). In addition to the branch-stamping method, cutset (instead of nodal) methods can be used to obtain immittance matrices as well [87,95,150,151], but their sparsity is usually not high.

The duals of equation sets (6.9)–(6.11) for mesh formulations above are given for nodal formulations in (6.18)–(6.20) below, where $\mathbf{G}_{block}$ is the block-diagonal matrix of power apparatus conductance matrices, $\mathbf{C}$ is a connection tensor that transforms open circuit voltages to connected node voltages, $\mathbf{v}_{block}$ is the vector of disconnected power apparatus voltages, and $\mathbf{j}_{block}$ is the vector of disconnected power apparatus current injections. The tensor transformation yields $\mathbf{G}_{nodal}$ in (6.20), which is the nodal conductance matrix of the *connected* power system, $\mathbf{v}_{nodal}$ is the vector of node voltages, and $\mathbf{j}_{nodal}$ is the vector of current injections.

$$\mathbf{G}_{block}(\mathbf{C}\cdot\mathbf{v}_{block}) = \mathbf{j}_{block} \tag{6.18}$$

$$\underbrace{(\mathbf{C}^{\mathrm{T}}\mathbf{G}_{block}\mathbf{C})}_{\mathbf{G}_{nodal}}\mathbf{v}_{nodal} = \underbrace{\mathbf{C}^{\mathrm{T}}\mathbf{j}_{block}}_{\mathbf{j}_{nodal}} \tag{6.19}$$

$$\mathbf{G}_{nodal}\mathbf{v}_{nodal} = \mathbf{j}_{nodal} \tag{6.20}$$

The branch-stamping method also yields the nodal conductance matrix in (6.20) without the need for a tensor, and is the preferred way to obtain $\mathbf{G}_{nodal}$ in practice. However, readers interested in implementing mesh and nodal formulations and interested in software modularity by treating power apparatus as building blocks, may find it useful to follow the steps outlined by the flow chart in Fig. 6.7.

The example employing disconnected MTCs shown in Fig. 6.6 above is now re-visited to demonstrate the tensor approach for nodal analysis. Consistent with discretization for nodal formulations, the branches inside each MTC have current injections (Norton equivalents) in Fig. 6.8.

Because most readers will likely already be familiar with nodal formulations, the flow chart steps illustrating formation of $\mathbf{C}$ for the nodal case are brief. The numbered 14 steps in flow chart Fig. 6.7 are followed in order. Referring to the schematic in Fig. 6.8:

1. Create $\mathbf{C} = \mathbf{I}_{21}$.
2. Create the connection table:

Dependent nodes	Independent nodes

3. Visit bus 1.
4. Write the following equations for bus 1:

$$\begin{cases} v_4 = v_1; v_7 = v_1 \text{ (phase } a) \\ v_5 = v_2; v_8 = v_2 \text{ (phase } b) \\ v_6 = v_3; v_9 = v_3 \text{ (phase } c) \end{cases}$$

5. (This step does not apply.)
6. (This step does not apply.)
7. Mark v_1, v_2, and v_3 as independent voltage numbers.
8. Add the independent voltage numbers to column 2 of the connection table:

Dependent nodes	Independent nodes
	1
	2
	3

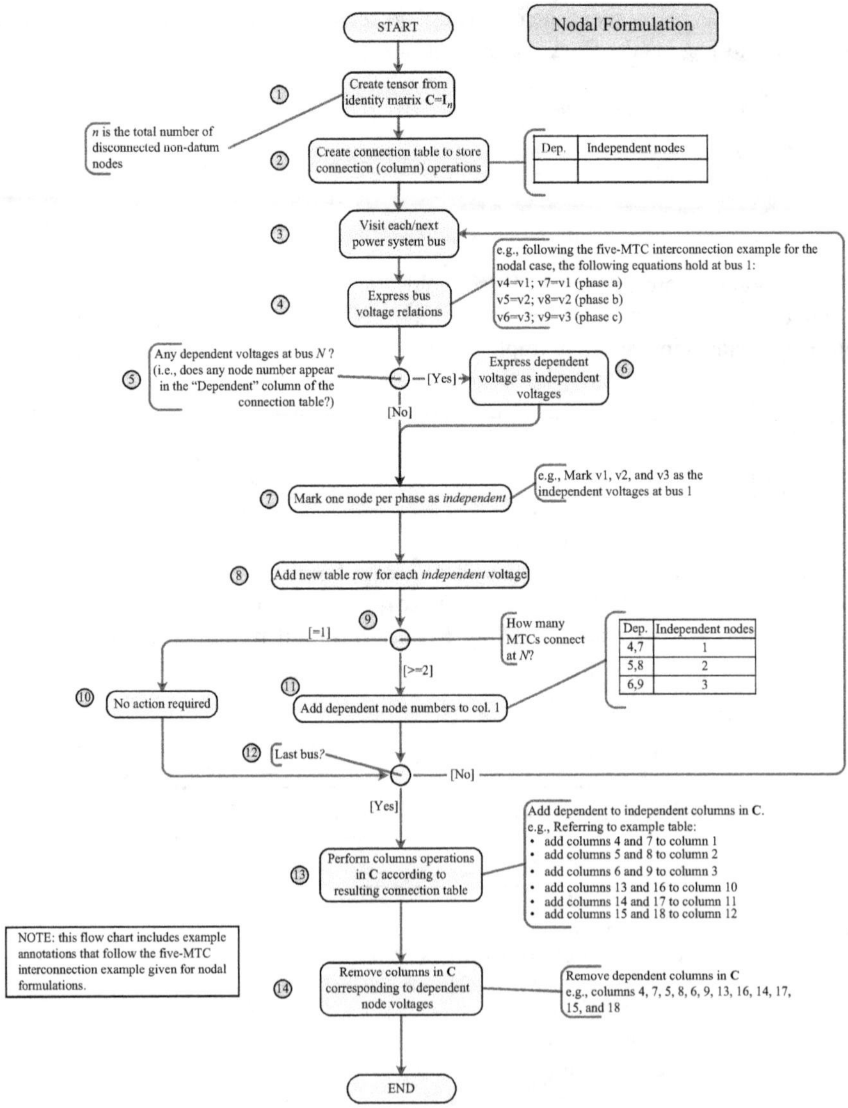

Fig. 6.7 Flow chart illustrating steps to form connection tensor **C** *for nodal formulations*

9. Two MTCs interconnect at bus 1.
10. (This step does not apply.)
11. Column 2 of the connection table is updated with the dependent node values as follows:

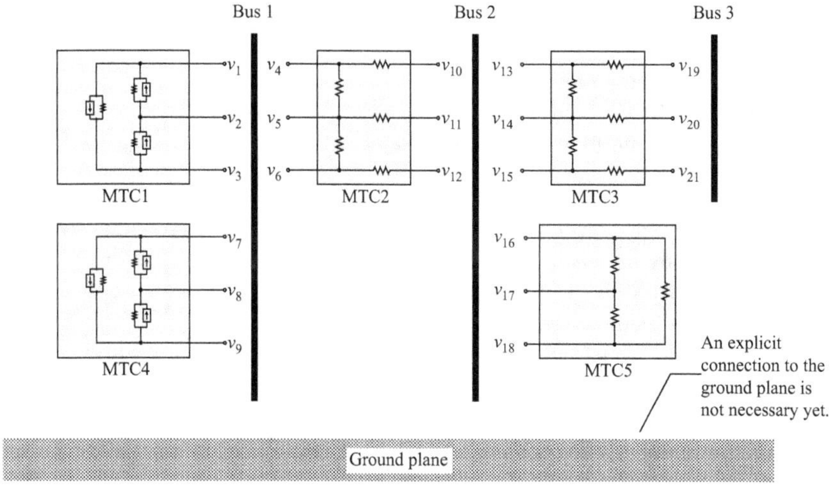

Fig. 6.8 Example of interconnection of five MTCs at three buses for nodal formulations

Dependent nodes	Independent nodes
4, 7	1
5, 8	2
6, 9	3

12. Since there are more buses to visit, repeat steps 3–11 for bus 2. At bus 3, step 10 does not require any action. This will effectively leave these nodes as open circuits. The resulting connection table then becomes:

Dependent nodes	Independent nodes
4, 7	1
5, 8	2
6, 9	3
13, 16	10
14, 17	11
15, 18	12

13. Perform the following column operations according to the foregoing columns of the connection table:
    - add columns 4 and 7 to column 1
    - add columns 5 and 8 to column 2

- add columns 6 and 9 to column 3
- add columns 13 and 16 to column 10
- add columns 14 and 17 to column 11
- add columns 15 and 18 to column 12

14. Lastly, remove columns 4, 7, 5, 8, 6, 9, 13, 16, 14, 17, 15, and 18 from **C**. The final connection tensor is given in (6.21).

In contrast to the mesh case, the nodal case requires only column additions and presents exactly one *independent* node (voltage) per phase. Additionally, (6.21) shows that 21 nodes were reduced to nine, which creates a discontinuity in the node numbering. Since node numbers normally correspond to the same row and column number in the nodal conductance matrix, to prevent zero rows and columns in $\mathbf{G}_{nodal}$ nodes 1, 2, 3, 10, 11, 12, 19, 20, 21 should be enumerated in canonical order *after* the tensor is formed.

$$
\begin{bmatrix} v_1 \\ v_2 \\ v_3 \\ v_4 \\ v_5 \\ v_6 \\ v_7 \\ v_8 \\ v_9 \\ v_{10} \\ v_{11} \\ v_{12} \\ v_{13} \\ v_{14} \\ v_{15} \\ v_{16} \\ v_{17} \\ v_{18} \\ v_{19} \\ v_{20} \\ v_{21} \end{bmatrix}
=
\underbrace{\begin{bmatrix}
1 & \cdot & \cdot & \cdot & \cdot & \cdot & \cdot & \cdot & \cdot \\
\cdot & 1 & \cdot & \cdot & \cdot & \cdot & \cdot & \cdot & \cdot \\
\cdot & \cdot & 1 & \cdot & \cdot & \cdot & \cdot & \cdot & \cdot \\
1 & \cdot & \cdot & \cdot & \cdot & \cdot & \cdot & \cdot & \cdot \\
\cdot & 1 & \cdot & \cdot & \cdot & \cdot & \cdot & \cdot & \cdot \\
\cdot & \cdot & 1 & \cdot & \cdot & \cdot & \cdot & \cdot & \cdot \\
1 & \cdot & \cdot & \cdot & \cdot & \cdot & \cdot & \cdot & \cdot \\
\cdot & 1 & \cdot & \cdot & \cdot & \cdot & \cdot & \cdot & \cdot \\
\cdot & \cdot & 1 & \cdot & \cdot & \cdot & \cdot & \cdot & \cdot \\
\cdot & \cdot & \cdot & 1 & \cdot & \cdot & \cdot & \cdot & \cdot \\
\cdot & \cdot & \cdot & \cdot & 1 & \cdot & \cdot & \cdot & \cdot \\
\cdot & \cdot & \cdot & \cdot & \cdot & 1 & \cdot & \cdot & \cdot \\
\cdot & \cdot & \cdot & 1 & \cdot & \cdot & \cdot & \cdot & \cdot \\
\cdot & \cdot & \cdot & \cdot & 1 & \cdot & \cdot & \cdot & \cdot \\
\cdot & \cdot & \cdot & \cdot & \cdot & 1 & \cdot & \cdot & \cdot \\
\cdot & \cdot & \cdot & 1 & \cdot & \cdot & \cdot & \cdot & \cdot \\
\cdot & \cdot & \cdot & \cdot & 1 & \cdot & \cdot & \cdot & \cdot \\
\cdot & \cdot & \cdot & \cdot & \cdot & 1 & \cdot & \cdot & \cdot \\
\cdot & \cdot & \cdot & \cdot & \cdot & \cdot & 1 & \cdot & \cdot \\
\cdot & \cdot & \cdot & \cdot & \cdot & \cdot & \cdot & 1 & \cdot \\
\cdot & \cdot & \cdot & \cdot & \cdot & \cdot & \cdot & \cdot & 1
\end{bmatrix}_{21\times 9}}_{\mathbf{C}}
\begin{bmatrix} v_1 \\ v_2 \\ v_3 \\ v_{10} \\ v_{11} \\ v_{12} \\ v_{19} \\ v_{20} \\ v_{21} \end{bmatrix}
\tag{6.21}
$$

6.5 Summary

This chapter presented a tensor approach to form the electrical network immittance matrices $\mathbf{R}_{mesh}$ (used in mesh formulations) and $\mathbf{G}_{nodal}$ (used in nodal formulations). The first step of this approach was to block-diagonalize the equations of all discretized power apparatus presented in Chapter 5. The second step was to form a transformation tensor by column-operating an identity matrix. The third and last step was to multiply together the block-diagonal matrix and tensor to obtain the desired electrical network immittance.

While in the nodal case there is no major advantage in using the tensor approach over the branch-stamping method, in the mesh case it avoids resorting to graph theory. A benefit of implementing the tensor method in the nodal case is that it promotes code re-use. That is, implementing the tensor method as a routine call in a solver can be used to return both $\mathbf{R}_{mesh}$ and $\mathbf{G}_{nodal}$ as needed. Returning both of these immittance matrices allows comparing their structures, their number of non-zeros, and gives the option to choose the most appropriate for the given simulation scenario. For example, Chapter 9 will show that, in a parallel context, the solution of the notional power system of Fig. 2.1 in mesh variables can be more efficient than its solution in nodal variables.

Chapter 7
Partitioning

The previous chapter explained how to form electrical network (immittance) matrices by interconnecting power apparatus together using a tensor. The concept of forming immittance matrices applies to both *unpartitioned* and *partitioned* power system models, and it is a necessary background for this chapter.

This chapter explains how to partition a power system model. Diakoptics[62] is introduced first in honor of the legacy partitioning method developed by Gabriel Kron. Following the introduction of diakoptics, the reader will find the partitioning approaches used in this book, which are node and mesh tearing.

Power system partitioning requires knowing *where* and *how many* partitions to create. The answers to these two questions are an ongoing research topic, a power system partitioning issue, and they are easy to get wrong. Erroneous assertions about these questions can lead to sub-optimal parallel simulation performance. This chapter will show how to determine where to tear by resorting to graph theory. How many partitions to create is not covered by this book, but empirical results addressing this question are provided in Chapter 9.

Readers should be aware that the partitioning approach presented here is *one* way to partition a power system for its parallel simulation. That is, in addition to what is presented in this chapter, other partitioning methods are available in the literature [40,43,100,152–159]. For instance, a well-known partitioning method is MATE [40,100,154,160,161], which has seen application in shipboard power systems [65]. This partitioning approach is led by the well-known author J. R. Martí of The University of British Columbia. He is a world-recognized expert in power system modeling and partitioning, and in parallel and real time simulation. He has done extensive work in developing partitioning techniques for the solution of large systems on multi-computer hardware [40,52]. His research group frequently publishes scholarly articles on their partitioning technique, which is a competitive method readers should consider in addition to what is presented in this chapter.

[62] The term *diakoptics* will not be found in a dictionary. *Diakoptics* is a term that was specifically *coined* for Kron (by a friend of his) to describe his piecewise solution work. The etymology of the term *diakoptics* is Latin and Greek; it is composed of two parts: *dia*, which is interpreted in English as "systems," (but this prefix in Latin means "across" or "through") and *Kopto*, which means "breaking apart, cutting, separating, or tearing apart into smaller pieces or subparts from a large, whole, system, or problem." Kron's method of tearing (i.e., his partitioning method) is well known as *diakoptics*.

Other partitioning approaches include diakoptics-based methods [124,162–164], the leap-frog method [165], state-space methods [43,166], wave-propagation delay or Bergeron's method [52,167], partitioning at locations where natural time delays occur [62,152,157,165,168–170], nested method [159], distributed load flow methods [171,172], asynchronous and multi-rate methods [53,74–76], among others. Partitioning methods that have been used *specifically* on shipboard power system models can be found in References 20, 62, 63, 158, 168, 169, and 173–178. Readers following the references above will find that the field of power system partitioning is not new [179], and that it has progressed significantly over the last two decades through its many contributors [100,153,155,157,159,160,165,166,169,180–183].

7.1 Diakoptics

Diakoptics is a term coined for Gabriel Kron's work on piecewise solutions of large-scale networks [125,184]. Starting in 1957, Kron published in London a series of papers called "*Diakoptics: The Piecewise Solution of Large-Scale Systems*" in the *Electrical Journal* (formerly the *Electrician*). These papers spanned two years, and later became available as a book under one cover [184]. Kron's original motivation was to calculate inter-area power flows from *intra*-area results [124]. Stated differently, Kron considered the use of isolated (simpler) subsystem-level solutions to calculate the flows traversing to and from neighbor subsystems. Kron succeeded at this work and developed a new partitioning theory, in which he was able to solve large network problems by piecing back together the individual solutions found from their subparts or sub-problems.

The success of diakoptics appears to have been hindered by the arrival of the personal computer. Some reasons that likely hindered the success of diakoptics at its inception were: the success of sparse-matrix ordering techniques suggested by Tinney [185], the introduction of trapezoidal discretization for computer simulation proposed by Dommel [79], modified nodal analysis as proposed by Ho [86], and the disadvantage of diakoptics presenting itself as a straightforward longhand method [40]. The combination of Tinney, Dommel, and Ho's techniques became increasingly dominant, preferred, and deemed sufficient at the time. The introduction of transient simulation using these techniques further led to the belief that computers were sufficient to solve power system problems and that no gains could be obtained from longhand partitioning approaches. Diakoptics lost popularity before it was amply accepted, but the few who used it, confirmed its advantages [124,162–164]. Although Kron himself was not a mathematician to prove or disprove his own method of tearing (diakoptics) [184], "supporters and critics alike agree that diakoptics works" [162]. (Reference 186 provides a mathematical derivation of diakoptics.)

Kron [184,162] showed that an unpartitioned electrical network characterized by linear equations in the form of (7.1) could be re-formulated as (7.2). (The details of going from (7.1) to (7.2) are presented later in this chapter.) In (7.1), $\mathbf{A}_{orig}$ is the original electrical network (coefficient) or immittance matrix of the *unpartitioned* power system, $\mathbf{x}$ is the vector of unknown variables (i.e., node voltages or mesh currents depending on the formualtion choice), $\mathbf{b}$ is the input vector, and the subscript p represents the number of subsystems or network partitions.

$$\mathbf{A}_{orig}\mathbf{x} = \mathbf{b} \tag{7.1}$$

$$\left[\begin{array}{cccc|c} \mathbf{A}_1 & \cdot & \cdot & \cdot & \mathbf{D}_1 \\ \cdot & \mathbf{A}_2 & \cdot & \cdot & \mathbf{D}_2 \\ \cdot & \cdot & \ddots & \cdot & \vdots \\ \cdot & \cdot & \cdot & \mathbf{A}_p & \mathbf{D}_p \\ \hline \mathbf{D}_1^{\mathrm{T}} & \mathbf{D}_2^{\mathrm{T}} & \cdots & \mathbf{D}_p^{\mathrm{T}} & \mathbf{Q} \end{array}\right] \left[\begin{array}{c} \mathbf{x}_1 \\ \mathbf{x}_2 \\ \vdots \\ \mathbf{x}_p \\ \hline \mathbf{u} \end{array}\right] = \left[\begin{array}{c} \mathbf{b}_1 \\ \mathbf{b}_2 \\ \vdots \\ \mathbf{b}_p \\ \hline \mathbf{0} \end{array}\right] \tag{7.2}$$

where:

$\mathbf{0}$ = zero matrix or vector (the symbol "·" is sometimes used instead)
$\mathbf{A}_i$ = subsystem coefficient matrix for subsystem i
$\mathbf{b}_i$ = excitation vector for subsystem i
$\mathbf{D}_i$ = connection matrix coupling variables in subsystem i to the boundary network
p = number of partitions (specified manually by the user)
$\mathbf{Q}$ = boundary network immittance matrix
$\mathbf{u}$ = vector of boundary network variables (torn branch voltages or currents)
$\mathbf{x}_i$ = vector of unknown variables for subsystem i

The compound-matrix form of (7.2) is given by (7.3), where the constituent matrices are expanded in (7.4) and (7.5). In (7.5), the subscript r represents the number of boundary variables formed after partitioning a network into p partitions.

$$\left[\begin{array}{c|c} \mathbf{A}_{block} & \mathbf{D} \\ \hline \mathbf{D}^{\mathrm{T}} & \mathbf{Q} \end{array}\right] \left[\begin{array}{c} \mathbf{x} \\ \hline \mathbf{u} \end{array}\right] = \left[\begin{array}{c} \mathbf{b} \\ \hline \mathbf{0} \end{array}\right] \tag{7.3}$$

$$\mathbf{A}_{block} = \begin{bmatrix} \mathbf{A}_1 & & & \\ & \mathbf{A}_2 & & \\ & & \ddots & \\ & & & \mathbf{A}_p \end{bmatrix}; \quad \mathbf{x} = \begin{bmatrix} \mathbf{x}_1 \\ \mathbf{x}_2 \\ \vdots \\ \mathbf{x}_p \end{bmatrix}; \quad \mathbf{b} = \begin{bmatrix} \mathbf{b}_1 \\ \mathbf{b}_2 \\ \vdots \\ \mathbf{b}_p \end{bmatrix}; \quad \mathbf{D} = \begin{bmatrix} \mathbf{D}_1 \\ \mathbf{D}_2 \\ \vdots \\ \mathbf{D}_p \end{bmatrix} \tag{7.4}$$

$$\mathbf{u} = \begin{bmatrix} u_1 \\ u_2 \\ \vdots \\ u_r \end{bmatrix}; \qquad \mathbf{Q} = \begin{bmatrix} Q_1 & & & \\ & Q_2 & & \\ & & \ddots & \\ & & & Q_r \end{bmatrix} \tag{7.5}$$

$$\mathbf{D}(i,j) \rightarrow \begin{cases} = 1, & \text{if } x_i \text{ is positively coupled to } \mathrm{u}_j \\ = -1, & \text{if } x_i \text{ is negatively coupled to } \mathrm{u}_j \\ = 0, & \text{if } x_i \text{ is not coupled to } \mathrm{u}_j \end{cases}$$

Expanding the compound matrix in (7.3) as two matrix equations:

$$\mathbf{x} = \mathbf{A}_{block}^{-1}\mathbf{b} - \mathbf{A}_{block}^{-1}\mathbf{D}\cdot\mathbf{u} \tag{7.6}$$

$$\mathbf{Q}\cdot\mathbf{u} = -\mathbf{D}^{\mathrm{T}}\mathbf{x}. \tag{7.7}$$

Substituting (7.6) into (7.7):

$$\mathbf{Q}\cdot\mathbf{u} = -\mathbf{D}^{\mathrm{T}}\overbrace{(\mathbf{A}_{block}^{-1}\mathbf{b} - \mathbf{A}_{block}^{-1}\mathbf{D}\cdot\mathbf{u})}^{\mathbf{x}} \tag{7.8}$$

$$\mathbf{Q}\cdot\mathbf{u} = -\mathbf{D}^{\mathrm{T}}\mathbf{A}_{block}^{-1}\mathbf{b} + \mathbf{D}^{\mathrm{T}}\mathbf{A}_{block}^{-1}\mathbf{D}\cdot\mathbf{u} \tag{7.9}$$

Grouping terms **u** on the left side:

$$\mathbf{Q}\cdot\mathbf{u} - \mathbf{D}^{\mathrm{T}}\mathbf{A}_{block}^{-1}\mathbf{D}\cdot\mathbf{u} = -\mathbf{D}^{\mathrm{T}}\mathbf{A}_{block}^{-1}\mathbf{b} \tag{7.10}$$

Factoring **u**:

$$(\mathbf{Q} - \mathbf{D}^{\mathrm{T}}\mathbf{A}_{block}^{-1}\mathbf{D})\mathbf{u} = -\mathbf{D}^{\mathrm{T}}\mathbf{A}_{block}^{-1}\mathbf{b} \tag{7.11}$$

Negating both sides:

$$\underbrace{(\mathbf{D}^{\mathrm{T}}\mathbf{A}_{block}^{-1}\mathbf{D} - \mathbf{Q})}_{\boldsymbol{\Lambda}}\mathbf{u} = \underbrace{\mathbf{D}^{\mathrm{T}}\mathbf{A}_{block}^{-1}\mathbf{b}}_{\boldsymbol{\beta}} \tag{7.12}$$

Solving for **u**:

$$\mathbf{u} = (\mathbf{D}^{\mathrm{T}}\mathbf{A}_{block}^{-1}\mathbf{D} - \mathbf{Q})^{-1}(\mathbf{D}^{\mathrm{T}}\mathbf{A}_{block}^{-1}\mathbf{b}) = \boldsymbol{\Lambda}^{-1}\boldsymbol{\beta} \tag{7.13}$$

Substituting (7.13) back into (7.6) results in (7.14). (Equation (7.14) is also known in mathematics as Woodbury's method for inverting modified matrices [186–188].)

$$\mathbf{x} = \mathbf{A}_{block}^{-1}\mathbf{b} - \mathbf{A}_{block}^{-1}\mathbf{D}\overbrace{(\mathbf{D}^{\mathrm{T}}\mathbf{A}_{block}^{-1}\mathbf{D} - \mathbf{Q})^{-1}(\mathbf{D}^{\mathrm{T}}\mathbf{A}_{block}^{-1}\mathbf{b})}^{\mathbf{u}} \tag{7.14}$$

Substituting the expanded-form matrices in (7.4) and (7.5) into (7.14) produces (7.15) and (7.16), which is the diakoptics solution to (7.1) proposed by Kron.

$$\mathbf{x} = \begin{bmatrix} \mathbf{x}_1 \\ \mathbf{x}_2 \\ \vdots \\ \mathbf{x}_p \end{bmatrix} = \begin{bmatrix} \mathbf{A}_1^{-1}\mathbf{b}_1 \\ \mathbf{A}_2^{-1}\mathbf{b}_2 \\ \vdots \\ \mathbf{A}_p^{-1}\mathbf{b}_p \end{bmatrix} - \begin{bmatrix} \mathbf{A}_1^{-1}\mathbf{D}_1 \\ \mathbf{A}_2^{-1}\mathbf{D}_2 \\ \vdots \\ \mathbf{A}_p^{-1}\mathbf{D}_p \end{bmatrix}\mathbf{u} \tag{7.15}$$

$$\mathbf{u} = \left(\sum_{i=1}^{p}(\mathbf{D}_i^T\mathbf{A}_i^{-1}\mathbf{D}_i - \mathbf{Q})^{-1}\right)\left(\sum_{i=1}^{p}(\mathbf{D}_i^T\mathbf{A}_i^{-1}\mathbf{b}_i)\right) = \boldsymbol{\Lambda}^{-1}\boldsymbol{\beta} \tag{7.16}$$

The suggested steps to solve (7.15) and (7.16) are the following [187]:

1. Compute the first term on the right-hand side of (7.15). The solution of each $\mathbf{A}_i^{-1}\mathbf{x}_i = \mathbf{b}_i$ term can be executed in parallel.
2. Using the $\mathbf{x}_i$ vectors found from step 1, compute (7.16). This step requires synchronizing (i.e., pausing) the parallel execution to compute **u.**

3. After computing $\mathbf{u}$, compute the second term on the right-hand side of (7.15). Similar to step 1, this step is also executed in parallel.
4. Subtract the second term on the right-hand side of (7.15) from the first term. This step is also executed in parallel.

From the four steps outlined above, it may now be gathered that this is a "fork/join" [189] or parallel-sequential algorithm type [190]. While some steps are parallelizable (i.e., can be forked), the solutions must be *joined* before advancing to the next timestep. Although solving (7.16) is the bottleneck of diakoptics, the advantages of parallelizing the solution given by expression (7.1) can in many cases (but not always) outweigh the overhead of said bottleneck. For example, small power system models may not benefit from partitioning as much as large models do as will be shown in Chapter 9.

7.2 Accuracy

A note on the accuracy of partitioning a model is in order. It is interesting to note that factoring $\mathbf{b}$ from (7.14) results in (7.17) or equivalently in (7.18).

$$\mathbf{x} = \underbrace{\left(\mathbf{A}_{block}^{-1} - \mathbf{A}_{block}^{-1}\mathbf{D}\left(\mathbf{S}^{-1} + \mathbf{D}^{\mathrm{T}}\mathbf{A}_{block}^{-1}\mathbf{D}\right)^{-1}\mathbf{D}^{\mathrm{T}}\mathbf{A}_{block}^{-1}\right)}_{\mathbf{A}_{equiv}^{-1}}\mathbf{b} \tag{7.17}$$

$$\mathbf{x} = \mathbf{A}_{equiv}^{-1}\mathbf{b} \tag{7.18}$$

Comparing $\mathbf{A}_{orig}$ from (7.1) to $\mathbf{A}_{equiv}$ in (7.18) reveals that $\mathbf{A}_{orig}$ and $\mathbf{A}_{equiv}$ *are equivalent* as stated by (7.19). This equivalency makes a strong case for validation concerns such as whether or not partitioned simulations return the same results as unpartitioned simulations do. Equation (7.19) shows that they do. Moreover, it has been suggested that partitioning may very well improve accuracy [162].

$$\mathbf{A}_{orig} = \mathbf{A}_{equiv} = \left(\mathbf{A}_{block}^{-1} - \mathbf{A}_{block}^{-1}\mathbf{D}\left(\mathbf{S}^{-1} + \mathbf{D}^{\mathrm{T}}\mathbf{A}_{block}^{-1}\mathbf{D}\right)^{-1}\mathbf{D}^{\mathrm{T}}\mathbf{A}_{block}^{-1}\right)^{-1} \tag{7.19}$$

Comparing the results of unpartitioned and partitioned simulations of the same solver, one could take into account the possible computational (finite) differences. These differences are usually too small to be noticeable and too small to make a case. The residual $\mathbf{A}_{orig} - \mathbf{A}_{equiv} \neq \mathbf{0}$ is purely computational and introduces negligible differences in diakoptics-based solutions. As this residual is small it can be safely ignored.

7.3 Zero-immittance tearing

A property related to diakoptics that is often overlooked is that boundary branches are *not* required to tear networks. This section introduces how to tear meshes and nodes – instead of branches – which is a more convenient approach than tearing branches (Kron's diakoptics).

The advantage of tearing meshes and nodes is that there are more places to tear from. For instance, in diakoptics, branches must exist at an intended disconnection point. Stated differently, disconnection points can only be defined where there are boundary branches.

The differences between diakoptics and mesh or node tearing are illustrated by referring to Figs. 7.1 and 7.2. Fig. 7.1 shows that, in mesh tearing, a disconnection point can be defined at places where shunt branches do not exist. Similarly, Fig. 7.2 illustrates that, in node tearing, a disconnection point can be defined where series branches do not exist.

Tearing networks as illustrated by Fig. 7.2, however, introduces boundary variables that did not exist before partitioning (as shown later in this chapter). Furthermore, depending on how many times a system is torn, the number of boundary variables can grow rapidly. (This set of boundary variables is often termed a *boundary network*, which in the diakoptics derivations presented earlier in this chapter was described by the square matrix **Q**.) It will be shown later that the *rate* at which the boundary network grows depends on the formulation choice. That is, it is possible for one formulation method to outperform the other in parallel scenarios, as the size of each boundary network can differ.

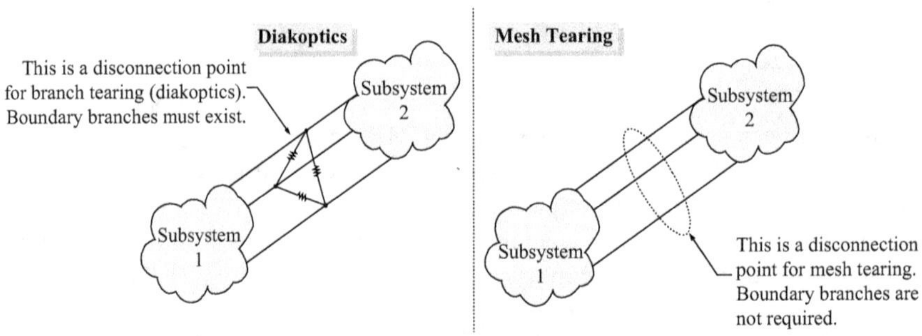

Fig. 7.1 Difference between branch tearing (diakoptics) and mesh tearing

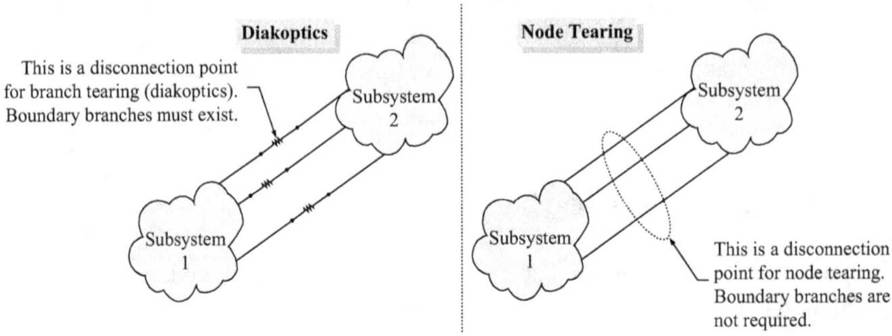

Fig. 7.2 Difference between branch tearing (diakoptics) and node tearing

The term zero-immittance stems from the following abstractions. With regard to the mesh tearing situation shown on the right of Fig. 7.1, it can be abstracted that across open space between conductor pairs there are branches of zero admittance. Along the same lines, regarding the node tearing situation on the right of Fig. 7.2, the galvanic continuity in the lines crossing from side 1 to side 2 can be abstracted as branches of zero impedance. In consequence, the term *zero-immittance tearing* stems from tearing branches of zero-immittance. Because the term *zero-immittance tearing* is somewhat unfriendly, the remainder of the book will continue the use of *mesh* and *node* tearing terminology instead.

The following explanations (based on Reference 191) elucidate the differences between diakoptics and mesh and node tearing. In relation to (7.3), it is not necessary for the boundary branch-immittance matrix $\mathbf{Q}$ to exist, that is the matrix can be $\mathbf{Q} = \mathbf{0}$. The matrix $\mathbf{Q}$ (in diakoptics) represents the immittance matrix of all torn branches, which is a densely coupled square matrix.

Referring to mesh tearing in Fig. 7.1 and node tearing in Fig. 7.2, because boundary branches are *not* torn to produce subsystems (or partitions), the matrix $\mathbf{Q}$ equals $\mathbf{0}$. In the mesh case, meshes crossing from one side to the other (e.g., from subsystem 1 to subsystem 2) can be thought of as meshes circulating around an open space of zero admittance. In the nodal case, the galvanic continuity throughout the lines can be thought of as a junction of zero impedance. If said meshes or nodes are bisected, it constitutes a condition of tearing branches of zero-immittance, which effectively makes $\mathbf{Q} = \mathbf{0}$.

Substituting $\mathbf{Q} = \mathbf{0}$ in (7.3) results in (7.20). Following the derivations from expressions (7.8) to (7.14) shown above produces expressions (7.21) and (7.22). Notice that $\mathbf{Q}$ no longer appears in the computation of $\mathbf{u}$ as it did in (7.16).

$$\left[\begin{array}{c|c} \mathbf{A}_{block} & \mathbf{D} \\ \hline \mathbf{D}^{\mathrm{T}} & \mathbf{0} \end{array}\right] \left[\begin{array}{c} \mathbf{x} \\ \hline \mathbf{u} \end{array}\right] = \left[\begin{array}{c} \mathbf{b} \\ \hline \mathbf{0} \end{array}\right] \tag{7.20}$$

$$\mathbf{x} = \begin{bmatrix} \mathbf{x}_1 \\ \mathbf{x}_2 \\ \vdots \\ \mathbf{x}_p \end{bmatrix} = \begin{bmatrix} \mathbf{A}_1^{-1}\mathbf{b}_1 \\ \mathbf{A}_2^{-1}\mathbf{b}_2 \\ \vdots \\ \mathbf{A}_p^{-1}\mathbf{b}_p \end{bmatrix} - \begin{bmatrix} \mathbf{A}_1^{-1}\mathbf{D}_1 \\ \mathbf{A}_2^{-1}\mathbf{D}_2 \\ \vdots \\ \mathbf{A}_p^{-1}\mathbf{D}_p \end{bmatrix} \mathbf{u} \tag{7.21}$$

$$\mathbf{u} = \left(\sum_{i=1}^{p} (\mathbf{D}_i^T \mathbf{A}_i^{-1} \mathbf{D}_i)^{-1} \right) \left(\sum_{i=1}^{p} (\mathbf{D}_i^T \mathbf{A}_i^{-1} \mathbf{b}_i) \right) = \mathbf{\Lambda}^{-1}\boldsymbol{\beta} \tag{7.22}$$

It is worthy to note that the zero-immittance tearing method has some advantages. Some of these are:

1. In diakoptics, matrix $\mathbf{Q}$ exists if a user introduces branches at the intended partitioning points. This is not always possible and is a limitation that does not exist in zero-immittance tearing.
2. In zero-immittance tearing, networks can be partitioned inside or outside power apparatus, or at single- or three-phase buses. This flexibility increases the number of available disconnection points [192].

The zero-immittance tearing method is not free of drawbacks; some of its *disadvantages* can also be pointed out:

1. Zero-immittance tearing retains the non-zero structure of $\mathbf{A}_{orig}$ (something of importance to sparse solvers [193]). Diakoptics, on the other hand, can reduce the non-zero structure by tearing mutual inductances (in nodal formulations [18]) or tearing densely coupled shunt-impedances (in mesh formulations) [149]. Tearing mutual inductances not only requires users to include them in a model, but it also is a counter-intuitive approach to selecting disconnection points. Tearing shunt-impedances is more intuitive: for example, meshes commonly intersect at bus cable capacitances [177] and produce dense regions in $\mathbf{A}$. The removal of these branches with diakoptics favorably affects the non-zero count of $\mathbf{A}_{orig}$.
2. The system order in $\mathbf{A}_{orig}$ increases for every single-phase zero-immittance tear. This occurs in mesh formulations by the introduction of bisecting voltage sources, and in nodal formulations by the introduction of series current sources. These sources create bisections and new variables on all sides of the tearing point, which does not occur in diakoptics.

However, it is not difficult to show that the advantages of zero-immittance tearing outweigh its disadvantages. The following subsections illustrate the mesh and node tearing partitioning approaches pictorially.

7.4 Mesh tearing

Consider a three-phase power system showing a possible disconnection point as shown in Fig. 7.3. Furthermore, assume that this disconnection point has been identified as a good candidate location to disconnect the network.[63]

The first step in tearing this disconnection point is to insert a voltage source across each conductor pair as shown in Fig. 7.4. The introduction of these voltage

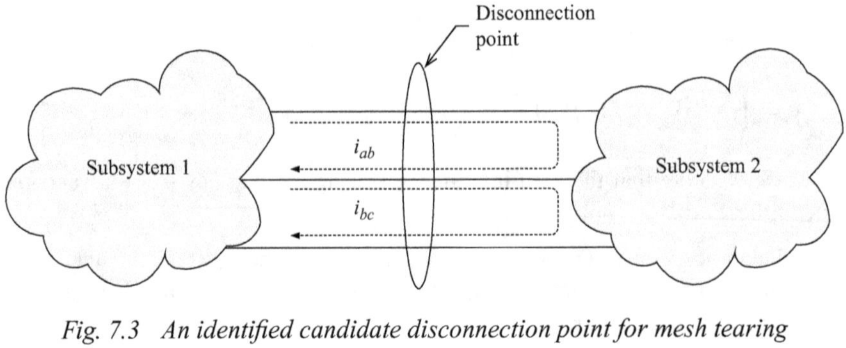

Fig. 7.3 An identified candidate disconnection point for mesh tearing

[63] How to determine the disconnection points for a large power system model is covered in section 7.8.

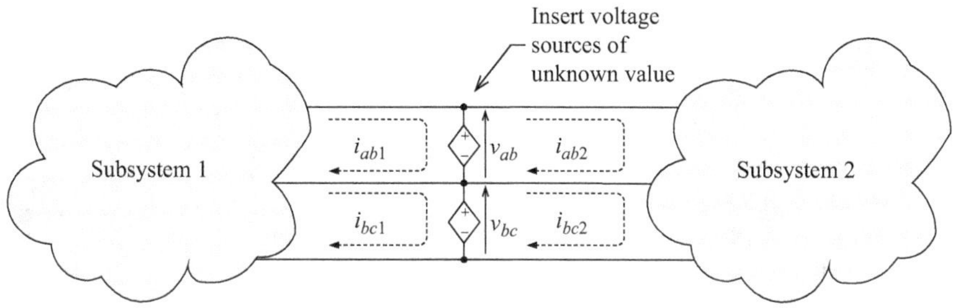

Fig. 7.4 Addition of unknown-valued voltage sources bisects the disconnection point and forms new boundary variables

sources bisects existing mesh currents $i_{ab}$and i_{bc} into four newly formed meshes i_{ab1}, i_{ab2}, i_{bc1}, and i_{bc2}, respectively. There are two things to note about the boundary variables v_{ab} and v_{bc}: (1) they constitute new boundary variables which did not exist before; and (2) they create new mesh currents which did not exist before either. Considering that such new boundary variables are added at *each* disconnection point in a power system model, there is a permissible number of tearing that results in simulation performance gains. Beyond this critical value, it is common to experience diminishing simulation performance.

From electric circuit theory, voltage sources can be torn apart without violating the physical properties of the network as shown in Fig. 7.5. Insertion and tearing of these voltage sources at *all* disconnection points completes the mesh tearing process.

A few useful observations are in order. In practice, but depending on the power system model, it may be required to tear many disconnection points to produce only two partitions. For example, *Systems* 3 and 4 (Figs. 2.6 and 2.1) are power system models tightly coupled by their cables running across switchboards and load centers. Even though several cables are placed on normally open (alternate) paths, these cables always form part of the model. This type of bus coupling suggests that it may be required to tear many disconnection points before subsystems can be formed.

Experience suggests, however, that the disconnection points that should be considered first are at buses that connect (and can potentially disconnect) the most number

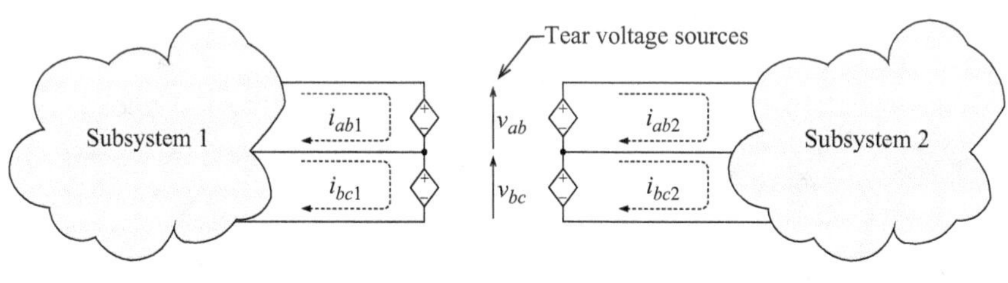

Fig. 7.5 Tearing of voltage sources at disconnection point produces subsystems

of power apparatus in one place. For the power system models shown in Chapter 2, these buses correspond to switchboards. Tearing switchboards (or buses) can produce many partitions without increasing the number of boundary variables. This choice reduces the number of needed disconnection points, but may not be obvious from the radial depiction in Fig. 7.5.

To illustrate the concept of tearing buses to mitigate introducing too many voltage sources, consider the bus disconnection point shown in Fig. 7.6. Inserting the unknown voltage sources at this location results in the system shown in Fig. 7.7, where tearing the voltage sources results in the partitioned system shown in Fig. 7.8.

It is interesting to note that to produce three subsystems from the bus disconnection point, only two boundary variables were introduced in Fig. 7.8. This is the

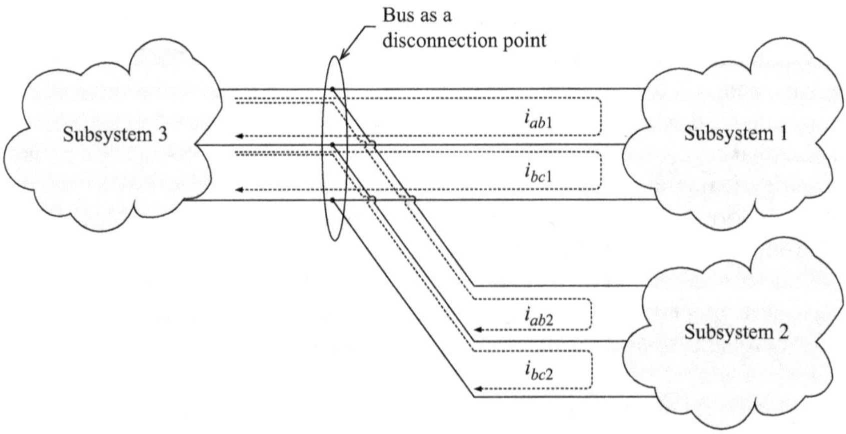

Fig. 7.6 Identification of a bus disconnection point to produce three subsystems

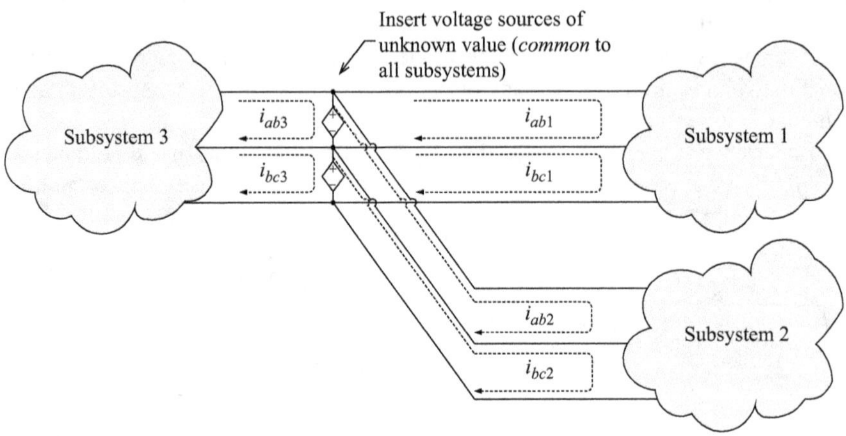

Fig. 7.7 Addition of voltage sources at disconnection point to produce three subsystems

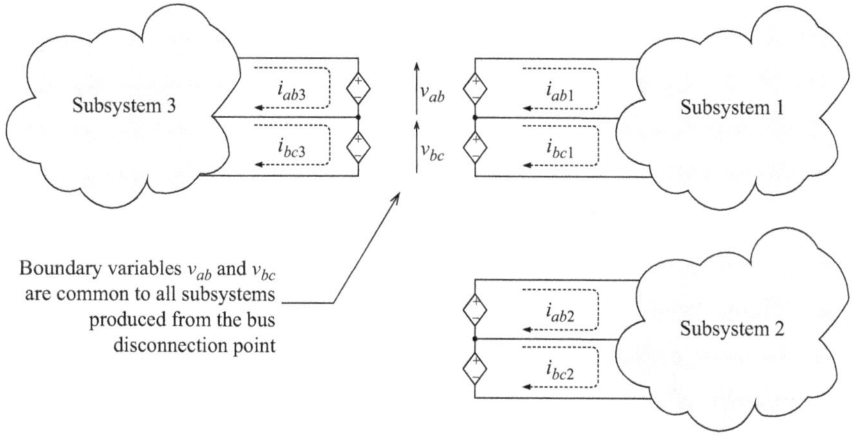

Fig. 7.8 Three subsystems are created from tearing two boundary (voltage) sources. The boundary variables v_{ab} and v_{bc} are common to the three subsystems

same number of boundary variables that was required to produce two subsystems in Fig. 7.5. The property of tearing buses while *not* increasing the number of boundary variables is natural to mesh tearing (i.e., it does not apply to nodal formulations) and extends to any number of partitions.

The electrical network equations of each subsystem in Fig. 7.5 or 7.8 can now be formulated in a parallelizable, doubly bordered block-diagonal form as given in (7.23), where the values of $\mathbf{D}_p(i,j)$ are given in (7.24). In this equation, the unknown voltage source values are arranged in the vector $\mathbf{u}^{k+1}$. The advantage of a formulation like the one given in (7.23) is manifested through its solution, which is given as (7.25) and (7.26). (Details on this solution method can be found in References 162, 181, and 187.)

$$\left[\begin{array}{cccc|c} \mathbf{R}_{mesh1}^{k+1} & & & & \mathbf{D}_1 \\ & \mathbf{R}_{mesh2}^{k+1} & & & \mathbf{D}_2 \\ & & \ddots & & \vdots \\ & & & \mathbf{R}_{mesh_p}^{k+1} & \mathbf{D}_p \\ \hline \mathbf{D}_1^{\mathrm{T}} & \mathbf{D}_2^{\mathrm{T}} & \cdots & \mathbf{D}_p^{\mathrm{T}} & \mathbf{0} \end{array}\right] \left[\begin{array}{c} \mathbf{i}_{mesh1}^{k+1} \\ \mathbf{i}_{mesh2}^{k+1} \\ \vdots \\ \mathbf{i}_{mesh_p}^{k+1} \\ \hline \mathbf{u}^{k+1} \end{array}\right] = \left[\begin{array}{c} \mathbf{e}_{mesh1}^{k+1} \\ \mathbf{e}_{mesh2}^{k+1} \\ \vdots \\ \mathbf{e}_{mesh_p}^{k+1} \\ \hline \mathbf{0} \end{array}\right] \tag{7.23}$$

$$\mathbf{D}_p(i,j) \rightarrow \begin{cases} = 1, & \text{if the } i\text{th mesh current in subsystem } p \text{ enters the } \textit{positive} \\ & \text{terminal of unknown boundary voltage source } j \\ = -1, & \text{if the } i\text{th mesh current in subsystem } p \text{ enters the } \textit{negative} \\ & \text{terminal of unknown boundary voltage source } j \\ = 0, & \text{otherwise} \end{cases} \tag{7.24}$$

where:

$\mathbf{e}_{mesh_p}^{k+1}$ = EMF vector of subsystem p
$\mathbf{i}_{mesh_p}^{k+1}$ = mesh current vector of subsystem p
$\mathbf{R}_{mesh_p}^{k+1}$ = mesh resistance matrix of subsystem p
$\mathbf{D}_p$ = tensor linking the mesh currents of subsystem p to its boundary voltages
$\mathbf{u}^{k+1}$ = vector of system-wide unknown boundary voltage sources
$\mathbf{0}$ = zero vector or matrix
p = number of subsystems or partitions.

$$\begin{bmatrix} \mathbf{i}_{mesh1}^{k+1} \\ \mathbf{i}_{mesh2}^{k+1} \\ \vdots \\ \mathbf{i}_{mesh_p}^{k+1} \end{bmatrix} = \underbrace{\begin{bmatrix} \left(\mathbf{R}_{mesh1}^{k+1}\right)^{-1} \mathbf{e}_{mesh1}^{k+1} \\ \left(\mathbf{R}_{mesh2}^{k+1}\right)^{-1} \mathbf{e}_{mesh2}^{k+1} \\ \vdots \\ \left(\mathbf{R}_{mesh_p}^{k+1}\right)^{-1} \mathbf{e}_{mesh_p}^{k+1} \end{bmatrix}}_{\mathbf{x}_1^{k+1}} - \begin{bmatrix} \left(\mathbf{R}_{mesh1}^{k+1}\right)^{-1} \mathbf{D}_1 \\ \left(\mathbf{R}_{mesh2}^{k+1}\right)^{-1} \mathbf{D}_2 \\ \vdots \\ \left(\mathbf{R}_{mesh_p}^{k+1}\right)^{-1} \mathbf{D}_p \end{bmatrix} \mathbf{u}^{k+1} \tag{7.25}$$

$$\mathbf{u}^{k+1} = \left(\sum_{i=1}^{p} \overbrace{\left(\mathbf{D}_i^{\mathrm{T}} \left(\mathbf{R}_{mesh_i}^{k+1}\right)^{-1} \mathbf{D}_i\right)}^{\mathbf{\Lambda}_i} \right)^{-1} \left(\sum_{i=1}^{p} \overbrace{\left(\mathbf{D}_i^{\mathrm{T}} \mathbf{x}_1^{k+1}\right)}^{\boldsymbol{\beta}_i} \right) = \mathbf{\Lambda}^{-1} \boldsymbol{\beta} \tag{7.26}$$

Equation (7.23) comprises two rows of matrix equations denoted with dashed lines. The top row of this equation corresponds to the mesh equations of each subsystem and their relation to the boundary variable vector $\mathbf{u}^{k+1}$. The bottom row states that the net (mesh) current through the unknown voltage sources equals zero, as was seen from the unpartitioned networks in Figs. 7.3 and 7.6.

Readers should note that (7.25) and (7.26) have the same form as (7.21) and (7.22) derived earlier. The difference is that (7.21) and (7.22) are given in the general ($\mathbf{A} \cdot \mathbf{x} = \mathbf{b}$) notation, whereas (7.25) and (7.26) are given in a notation familiar to mesh analysts—that is, using $\mathbf{R} \cdot \mathbf{i} = \mathbf{e}$ notation, instead of an $\mathbf{A} \cdot \mathbf{x} = \mathbf{b}$ notation.

It is opportune to show what the matrix structure of (7.23) looks like for the notional shipboard power system model introduced in Chapter 2 (Fig. 2.1, *System* 4). The structure plot of the doubly bordered coefficient matrix of (7.23) is shown in Fig. 7.9. This structure plot shows the case of $p = 4$ partitions, which is the number of partitions that will result in the fastest simulation runtime for the mesh formulation (explained in Chapter 9).

The structure plot in Fig. 7.9 represents the coefficient matrix in (7.23) for *System* 4. This plot shows the immittance matrix of each subsystem arranged along the main diagonal. These subsystem immittance matrices are the electrical network coefficient matrices of subsystems 1 through 4, respectively, and are termed mesh resistance matrices ($\mathbf{R}_{mesh}$) since they are based on a mesh formulation. Each mesh resistance matrix was formed using the tensor approach presented in section 6.3. The $\mathbf{D}_i$ matrices are shown along the right side of Fig. 7.9 and are termed disconnection matrices. (Formation of the disconnection matrices is the focus of section 7.6 presented later.) The column dimension of each $\mathbf{D}_i$ matrix equals the number of

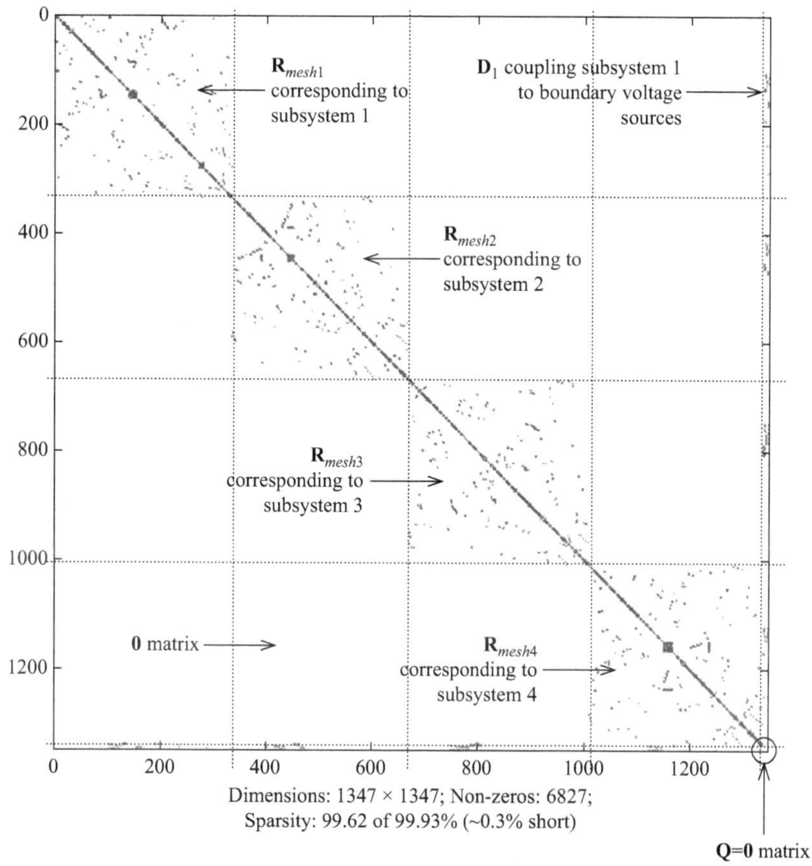

Fig. 7.9 Mesh structure plot of coefficient matrix of (7.23) for System 4 model ($p = 4$ partitions, $r = 10$ boundary variables)

boundary variables and is denoted r. That is, partitioning *System* 4 (Fig. 2.1) four times ($p = 4$) introduced ten ($r = 10$) boundary voltage sources. As will be seen in the next section, when using node tearing, it is required to introduce $r = 33$ boundary current sources to partition the same system $p = 4$ times as well (i.e., 3.3s times as many boundary variables r).

The bottom-right corner in Fig. 7.9 shows the boundary-branch immittance matrix $\mathbf{Q} = \mathbf{0}$. This matrix is all zeros in mesh and node tearing, but not in diakoptics. The caption below the structure plot in Fig. 7.9 shows the matrix dimensions, the number of non-zeros, and the sparsity of the doubly bordered block-diagonal matrix.

Another representation of (7.23) can be obtained by combining the concepts of modeling power apparatus as MTCs (Chapter 5) and forming the mesh resistance matrix using a tensor (Chapter 6) as shown in (7.27) (compound matrix form) and (7.28) (expanded matrix form). These views highlight that $\mathbf{Q} = \mathbf{0}$ and show where the connection tensor $\mathbf{C}_i$ of each subsystem is used. The parallel solution to (7.27) was given in (7.25) and (7.26).

$$\left[\begin{array}{c|c} \mathbf{R}_{mesh}^{k+1} & \mathbf{D} \\ \hline \mathbf{D}^{\mathrm{T}} & \cdot \end{array}\right] \begin{bmatrix} \mathbf{i}_{mesh}^{k+1} \\ \mathbf{u}^{k+1} \end{bmatrix} = \left[\begin{array}{c} \mathbf{e}_{mesh}^{k+1} \\ \hline \cdot \end{array}\right] \tag{7.27}$$

$$\left[\begin{array}{cccc|c} \overbrace{\mathbf{C}_1^{\mathrm{T}} \underbrace{\begin{bmatrix} \mathbf{R}_{MTC11} & & \\ & \mathbf{R}_{MTC12} & \\ & & \ddots \end{bmatrix}}_{\mathbf{R}_{block1}} \mathbf{C}_1}^{\mathbf{R}_{mesh1}^{k+1}} & \cdot & \cdot & \cdot & \mathbf{D}_1 \\ \cdot & \overbrace{\mathbf{C}_2^{\mathrm{T}} \underbrace{\begin{bmatrix} \mathbf{R}_{MTC21} & & \\ & \mathbf{R}_{MTC22} & \\ & & \ddots \end{bmatrix}}_{\mathbf{R}_{block2}} \mathbf{C}_2}^{\mathbf{R}_{mesh2}^{k+1}} & \cdot & \cdot & \mathbf{D}_2 \\ \cdot & \cdot & \ddots & \cdot & \vdots \\ \cdot & \cdot & & \overbrace{\mathbf{C}_p^{\mathrm{T}} \underbrace{\begin{bmatrix} \mathbf{R}_{MTC31} & & \\ & \mathbf{R}_{MTC32} & \\ & & \ddots \end{bmatrix}}_{\mathbf{R}_{block_p}} \mathbf{C}_p}^{\mathbf{R}_{mesh_p}^{k+1}} & \mathbf{D}_p \\ \hline \mathbf{D}_1^{\mathrm{T}} & \mathbf{D}_2^{\mathrm{T}} & \cdots & \mathbf{D}_p^{\mathrm{T}} & \cdot \end{array}\right] \left[\begin{array}{c} \mathbf{i}_{mesh1}^{k+1} \\ \mathbf{i}_{mesh2}^{k+1} \\ \vdots \\ \mathbf{i}_{mesh_p}^{k+1} \\ \hline \mathbf{u}^{k+1} \end{array}\right] = \left[\begin{array}{c} \mathbf{e}_{mesh1}^{k+1} \\ \mathbf{e}_{mesh2}^{k+1} \\ \vdots \\ \mathbf{e}_{mesh_p}^{k+1} \\ \hline \cdot \end{array}\right] \tag{7.28}$$

7.5 Node tearing

This section presents the node tearing partitioning approach. Both mesh and node tearing approaches will be used on the power system models introduced in Chapter 2 (*Systems* 1, 2, 3, and 4) to measure their speedups in Chapter 9.

To illustrate node tearing, consider the disconnection point shown in Fig. 7.10. To tear this disconnection point, two current sources of unknown value are inserted in line with the conductors as shown by Fig. 7.11.

The introduction of these unknown-valued current sources bisects each node voltage v_a, v_b, and v_c into two pairs each: v_{a1} and v_{a2}, v_{b1} and v_{b2}, and v_{c1} and v_{c2}, respectively. The new boundary variables i_a, i_b, and i_c just created produced two effects. One is that they form a new boundary network that did not exist before, and the other is that they created node voltages, which did not exist before either. Considering that, in node tearing, a new current source is introduced in each phase of each subsystem created from a disconnection point, there will be a

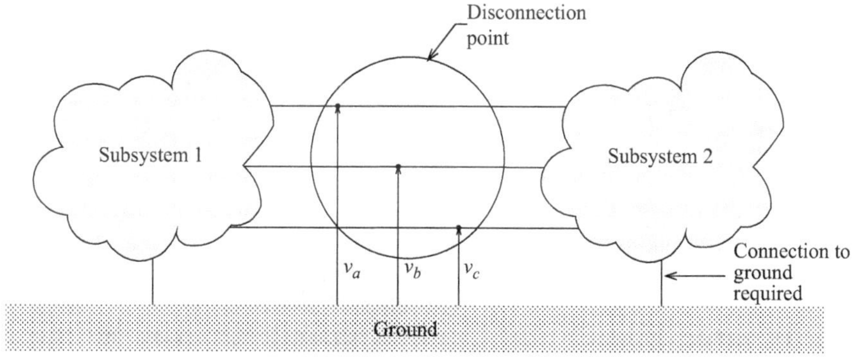

Fig. 7.10 A candidate disconnection point for node tearing

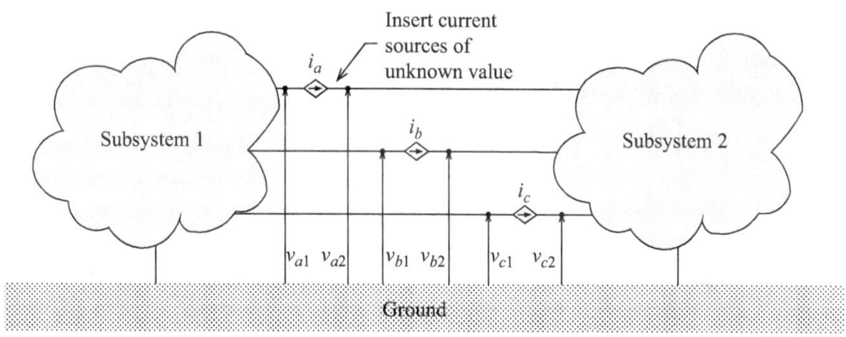

Fig. 7.11 Addition of unknown-valued current sources bisects the disconnection point and forms new boundary variables

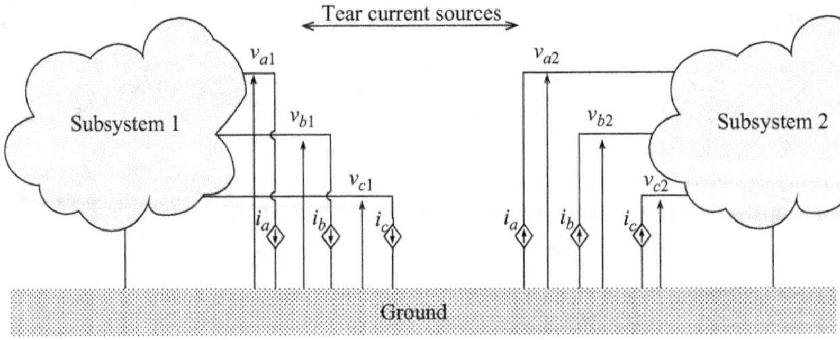

Fig. 7.12 Tearing of current sources at disconnection point produces subsystems

critical number of network tears that can be made within which the performance of the solver will be ameliorated. Beyond this critical value, diminished speedups in the performance of the solver are likely to occur. This diminishing benefit effect will be shown in Chapter 9.

From circuit theory, the unknown-valued current sources can be torn apart as seen in Fig. 7.12 without violating electrical properties. Inserting and tearing these current sources at a disconnection point completes node tearing.

Like mesh tearing, and depending on the power system model being partitioned, it may be required to tear more than one disconnection point before two subsystems are formed. In contrast to mesh tearing, the boundary network produced from node tearing increases rapidly when multiple subsystems are formed from the same disconnection point. This is an important (and perhaps not apparent) contrast between the mesh and node tearing approaches. The reason for this difference in boundary network growth is that in mesh tearing, the boundary voltage sources impress voltages common to all subsystems produced from the same disconnection point. In node tearing, on the other hand, a current source of different value is injected into each phase of each subsystem produced from the disconnection point. This difference should not be overlooked.

To exemplify the preceding point, consider the bus disconnection point shown in Fig. 7.13, where the creation of $p = 3$ subsystems is intended. Insertion of unknown-valued current sources at this bus results in the in-line current placements shown in Fig. 7.14. It is important to note that the current source values in each subsystem are different. (In mesh analysis, the boundary voltage sources were common to all subsystems produced from the same disconnection point.) Tearing the current sources shown in (7.14) results in the partitioned network shown in Fig. 7.15.

Readers may have noticed by inspecting Fig. 7.15 that the current sinks on subsystem 1 can be expressed as the sum of the current injections in subsystems 2 and 3. This dependency of boundary variables, which should be observed during implementation in software code, reduces the number of boundary variables from nine to six.

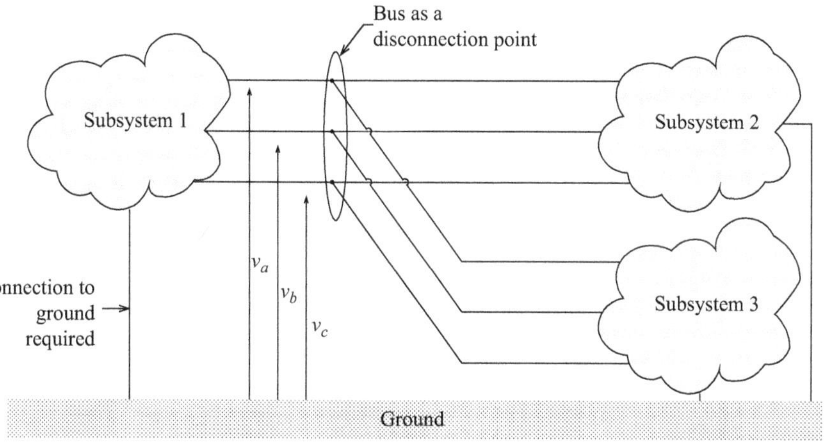

Fig. 7.13 Tearing a bus to produce three subsystems

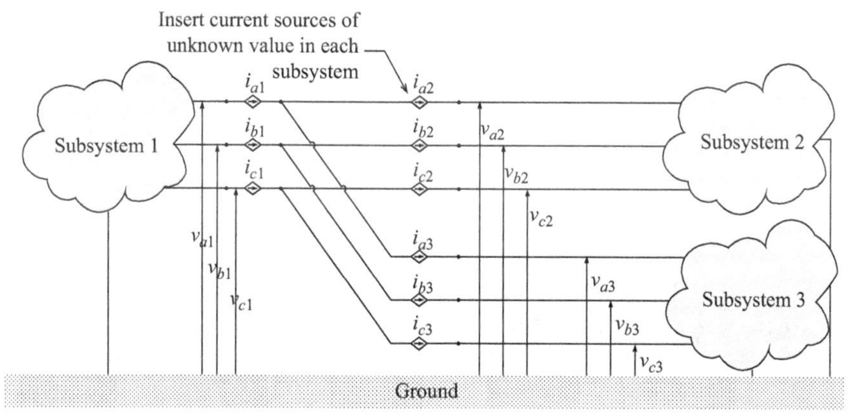

Fig. 7.14 Addition of unknown-valued current sources at disconnection point

As may be already apparent, tearing nodes creates $N - 1$ unknown currents for each torn conductor (or electrical phase), where N represents the number of subsystems formed by tearing the conductor. This is more than the number of voltage sources required to tear buses in mesh tearing. Therefore, node tearing can lead to larger boundary networks for the same power system model when compared to mesh tearing.

Similar to what was illustrated for mesh tearing, the partitioned networks depicted in either Fig. 7.12 or 7.15 lend themselves for the parallelizable, doubly bordered block-diagonal form given in (7.29) below, where the unknown current source values are arranged in the vector $\mathbf{u}^{k+1}$ (the values of $\mathbf{D}_p(i,j)$ are given in (7.30). The advantage of a formulation like that presented in (7.29) is manifested through its solution, which is given as (7.31) and (7.32) [162,181].

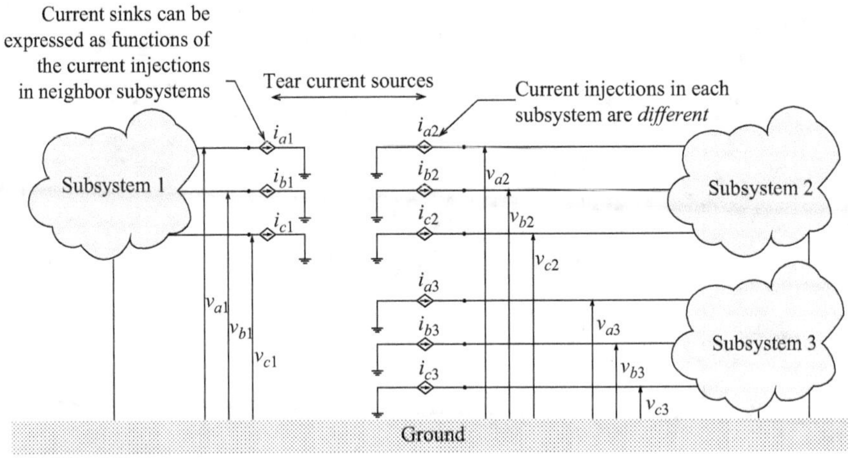

Fig. 7.15 Creation of three subsystems from a bus disconnection point creates six boundary variables instead of nine

$$\left[\begin{array}{cccc|c} \mathbf{G}_{nodal1}^{k+1} & & & & \mathbf{D}_1 \\ & \mathbf{G}_{nodal2}^{k+1} & & & \mathbf{D}_2 \\ & & \ddots & & \vdots \\ & & & \mathbf{G}_{nodal_p}^{k+1} & \mathbf{D}_p \\ \hline \mathbf{D}_1^{\mathrm{T}} & \mathbf{D}_2^{\mathrm{T}} & \cdots & \mathbf{D}_p^{\mathrm{T}} & \mathbf{0} \end{array}\right] \left[\begin{array}{c} \mathbf{v}_{nodal1}^{k+1} \\ \mathbf{v}_{nodal2}^{k+1} \\ \vdots \\ \mathbf{v}_{nodal_p}^{k+1} \\ \hline \mathbf{u}^{k+1} \end{array}\right] = \left[\begin{array}{c} \mathbf{j}_{nodal1}^{k+1} \\ \mathbf{j}_{nodal2}^{k+1} \\ \vdots \\ \mathbf{j}_{nodal_p}^{k+1} \\ \mathbf{0} \end{array}\right] \tag{7.29}$$

$$\mathbf{D}_p(i,j) \rightarrow \begin{cases} = 1, & \text{if current source } j \text{ is a sink at node } i \\ = -1, & \text{if current source } j \text{ is an injection at node } i \\ = 0, & \text{otherwise} \end{cases} \tag{7.30}$$

where:

$\mathbf{j}_{mesh_p}^{k+1}$ = current injection vector of subsystem p
$\mathbf{v}_{mesh_p}^{k+1}$ = node voltage vector of subsystem p
$\mathbf{G}_{mesh_p}^{k+1}$ = nodal conductance matrix of subsystem p
$\mathbf{D}_p$ = tensor linking node voltages of subsystem p to boundary injections
$\mathbf{u}^{k+1}$ = vector of system-wide unknown boundary current sources
$\mathbf{0}$ = zero vector or matrix
p = number of subsystems or partitions.

$$\begin{bmatrix} \mathbf{v}_{nodal1}^{k+1} \\ \mathbf{v}_{nodal2}^{k+1} \\ \vdots \\ \mathbf{v}_{nodal_p}^{k+1} \end{bmatrix} = \underbrace{\begin{bmatrix} \left(\mathbf{G}_{nodal1}^{k+1}\right)^{-1} \mathbf{j}_{nodal1}^{k+1} \\ \left(\mathbf{G}_{nodal2}^{k+1}\right)^{-1} \mathbf{j}_{nodal2}^{k+1} \\ \vdots \\ \left(\mathbf{G}_{nodal_p}^{k+1}\right)^{-1} \mathbf{j}_{nodal_p}^{k+1} \end{bmatrix}}_{\mathbf{x}_1^{k+1}} - \begin{bmatrix} \left(\mathbf{G}_{nodal1}^{k+1}\right)^{-1} \mathbf{D}_1 \\ \left(\mathbf{G}_{nodal2}^{k+1}\right)^{-1} \mathbf{D}_2 \\ \vdots \\ \left(\mathbf{G}_{nodal_p}^{k+1}\right)^{-1} \mathbf{D}_p \end{bmatrix} \mathbf{u}^{k+1} \tag{7.31}$$

$$\mathbf{u}^{k+1} = \left(\sum_{i=1}^{p} \overbrace{\left(\mathbf{D}_i^{\mathrm{T}} \left(\mathbf{G}_{nodal_i}^{k+1} \right)^{-1} \mathbf{D}_i \right)}^{\mathbf{\Lambda}_i} \right)^{-1} \left(\sum_{i=1}^{p} \overbrace{\left(\mathbf{D}_i^{\mathrm{T}} \mathbf{x}_1^{k+1} \right)}^{\boldsymbol{\beta}_i} \right) = \mathbf{\Lambda}^{-1} \boldsymbol{\beta} \tag{7.32}$$

Equation (7.29) comprises two rows of matrix equations denoted with dashed lines. The top row corresponds to the node voltage equations of each subsystem and their relation to the boundary variable vector $\mathbf{u}$. The bottom row states that the phase voltages at all sides of a disconnection are equal.

Readers should note that (7.31) and (7.32) have the same form as (7.21) and (7.22) derived earlier. The difference is that (7.21) and (7.22) are given in general ($\mathbf{A} \cdot \mathbf{x} = \mathbf{b}$) notation, whereas (7.31) and (7.32) are in a notation familiar to nodal analysts—that is, using $\mathbf{G} \cdot \mathbf{v} = \mathbf{j}$ notation instead of $\mathbf{A} \cdot \mathbf{x} = \mathbf{b}$ notation.

Similar to the mesh case, it is timely to show what the matrix structure of (7.29) looks like for *System* 4 (Fig. 2.1). The structure plot of the doubly bordered coefficient matrix of (7.29) is shown in Fig. 7.16. This structure plot shows the case of $p = 4$ partitions, which is the number of partitions that resulted in the fastest simulation runtime for the nodal formulation (explained in Chapter 9).

The immittance matrix of each subsystem is arranged along the main diagonal. These subsystem immittance matrices are the electrical network coefficient matrices of subsystems 1 through 4, respectively, and are termed nodal conductance matrices ($\mathbf{G}_{nodal}$) since they are based on a nodal formulation. Each nodal conductance matrix was formed using the tensor approach presented in section 6.3. The disconnection ($\mathbf{D}_i$) matrices are shown along the right side of Fig. 7.16. In contrast to the mesh formulation, partitioning the notional shipboard power system (*System* 4, Fig. 2.1) four times using node tearing required the introduction of $r = 33$ boundary current sources. When compared to the mesh case shown in Fig. 7.16, node tearing produces a boundary network that is $33/10 = 3.3$ times larger. The size of the boundary network (the value of r) negatively affects the performance of the solver as it increases the solution time of $\mathbf{u}$ by having a larger, denser $\mathbf{\Lambda}$ matrix in (7.32).

The bottom-right corner in Fig. 7.16 shows the boundary-branch immittance matrix $\mathbf{Q} = \mathbf{0}$. This matrix is all zeros in mesh and node tearing, but not in diakoptics. The bottom caption of the structure plot in Fig. 7.16 shows the matrix dimensions, the number of non-zeros, and the sparsity of the doubly bordered block-diagonal matrix. Comparing Figs. 7.9 and 7.16, it is noticed that there are significantly more meshes than nodes in the notional power system (*System* 4, Fig. 2.1). This difference between them is a direct result of having cable capacitances (Fig. 5.1), which increases the model's meshing, and of having shunt resistances in some protective devices to make voltage measurements (Figs. 5.4 and 5.5). Cables and protective devices are the dominant power apparatus (Table 2.2), and their shunt branches significantly increase the mesh count of the system.

Like it was shown for the mesh case, another representation of (7.29) can be obtained by combining the concepts of modeling power apparatus as MTCs (Chapter 5) and forming the nodal conductance matrix using a tensor (Chapter 6) as shown

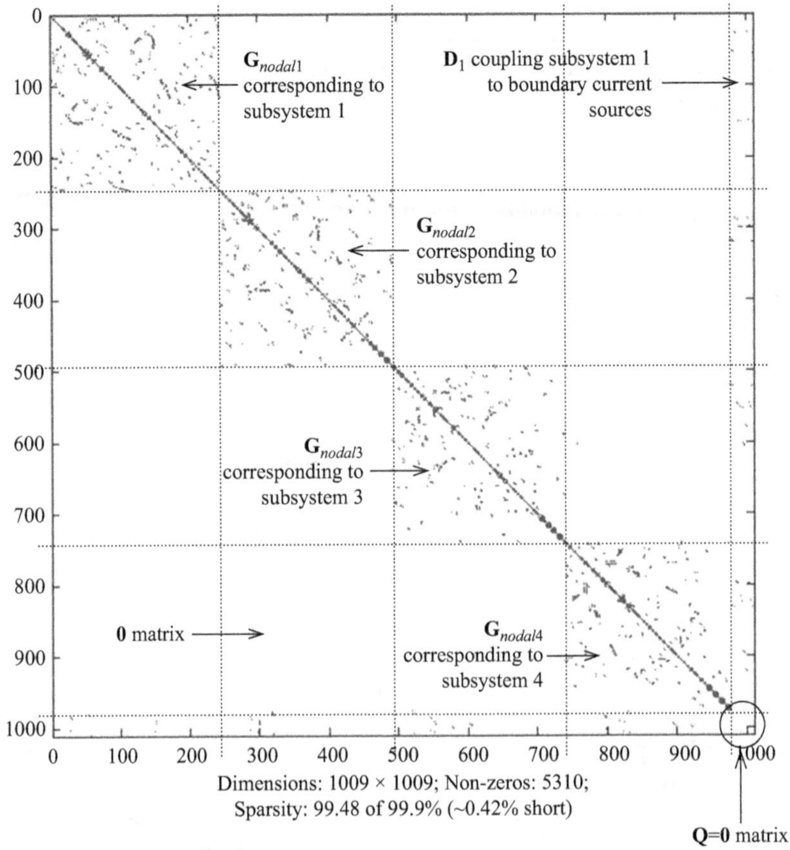

Fig. 7.16 Nodal structure plot of coefficient matrix in (7.29) for System 4 model ($p = 4$ partitions, $r = 33$ boundary variables)

in (7.33) (compound matrix form) and (7.34) (expanded matrix form). These views highlight that $\mathbf{Q} = \mathbf{0}$ and also show where the connection tensor $\mathbf{C}_i$ of each subsystem fits. The parallel solution to (7.34) was given in (7.31) and (7.32).

To clarify possible ambiguities in the notations introduced so far, Fig. 7.17 shows the generalized, mesh, and node tearing notations side by side. The annotations show the meaning of each term, and they are enumerated in their suggested order of reading. Readers should be able to relate the general annotations (top) to the notations shown for each tearing formulation below it. The superscript $k + 1$ explicitly means a discrete notation.

7.6 Tearing examples

The previous sections in this chapter introduced partitioning theory starting from diakoptics and leading toward mesh and node tearing. The solution equations

$$\left[\begin{array}{c:c} \mathbf{G}_{nodal}^{k+1} & \mathbf{D} \\ \hdashline \mathbf{D}^{\mathrm{T}} & \cdot \end{array}\right] \begin{bmatrix} \mathbf{v}_{nodal}^{k+1} \\ \mathbf{u}^{k+1} \end{bmatrix} = \left[\begin{array}{c} \mathbf{j}_{mesh}^{k+1} \\ \hdashline \cdot \end{array}\right] \tag{7.33}$$

$$\left[\begin{array}{cccc:c} \overbrace{\mathbf{C}_1^{\mathrm{T}} \underbrace{\begin{bmatrix} \mathbf{G}_{MTC11} & & \\ & \mathbf{G}_{MTC12} & \\ & & \ddots \end{bmatrix}}_{\mathbf{C}_{block1}} \mathbf{C}_1}^{\mathbf{G}_{nodal1}^{k+1}} & \cdot & \cdot & \cdot & \mathbf{D}_1 \\ \cdot & \overbrace{\mathbf{C}_2^{\mathrm{T}} \underbrace{\begin{bmatrix} \mathbf{G}_{MTC21} & & \\ & \mathbf{G}_{MTC22} & \\ & & \ddots \end{bmatrix}}_{\mathbf{G}_{block2}} \mathbf{C}_2}^{\mathbf{G}_{nodal2}^{k+1}} & \cdot & \cdot & \mathbf{D}_2 \\ \cdot & \cdot & \ddots & \cdot & \vdots \\ \cdot & \cdot & & \overbrace{\mathbf{C}_p^{\mathrm{T}} \underbrace{\begin{bmatrix} \mathbf{G}_{MTC31} & & \\ & \mathbf{G}_{MTC32} & \\ & & \ddots \end{bmatrix}}_{\mathbf{G}_{block_p}} \mathbf{C}_p}^{\mathbf{G}_{nodal_p}^{k+1}} & \mathbf{D}_p \\ \hdashline \mathbf{D}_1^{\mathrm{T}} & \mathbf{D}_2^{\mathrm{T}} & \cdots & \mathbf{D}_p^{\mathrm{T}} & \cdot \end{array}\right] \left[\begin{array}{c} \mathbf{v}_{nodal1}^{k+1} \\ \mathbf{v}_{nodal2}^{k+1} \\ \vdots \\ \mathbf{v}_{nodal_p}^{k+1} \\ \hdashline \mathbf{u}^{k+1} \end{array}\right] = \left[\begin{array}{c} \mathbf{j}_{nodal1}^{k+1} \\ \mathbf{j}_{nodal2}^{k+1} \\ \vdots \\ \mathbf{j}_{nodal_p}^{k+1} \\ \hdashline \cdot \end{array}\right] \tag{7.34}$$

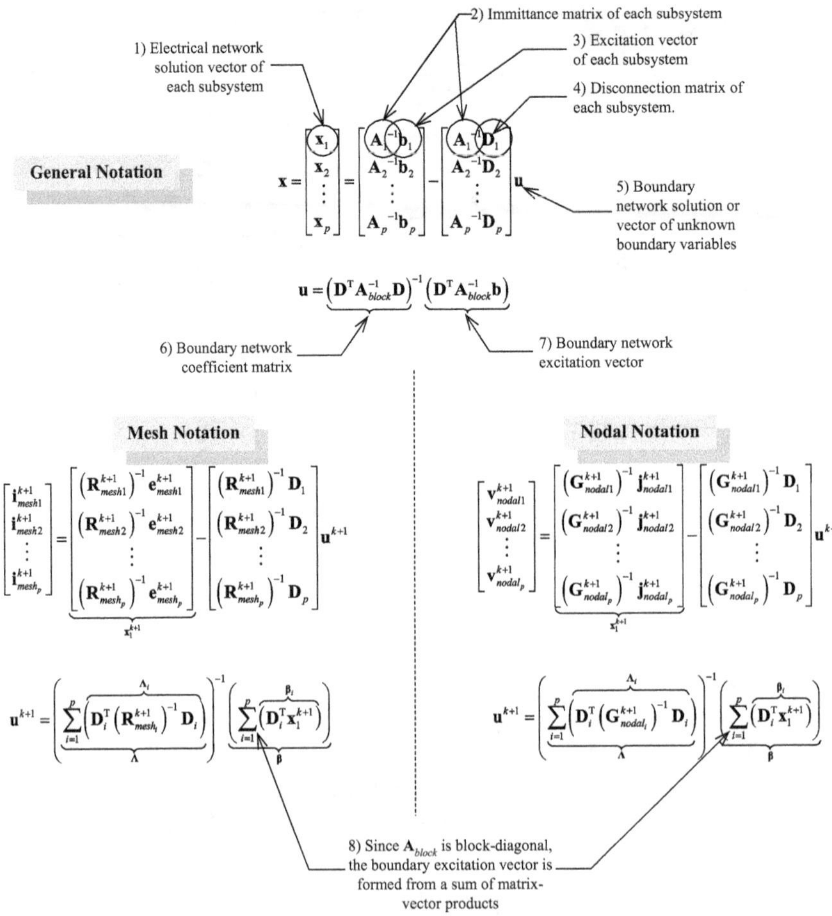

Fig. 7.17 Summary of tearing notations. Top: general notation irrespective of formulation type. Bottom: notations used in mesh and node tearing

(summarized by Fig. 7.17) showed disconnection matrices noted as $\mathbf{D}_i$ and their transpose as $\mathbf{D}_i^{\mathrm{T}}$ but their formation was not discussed. These disconnection matrices are as important as the immittance matrices are as they are required to solve the parallel equations summarized in Fig. 7.17. The formation of these disconnection matrices is best explained by *graphical* examples as it is done in this section.

Furthermore, this section will present several examples to elucidate node and mesh tearing, and to show how to form the disconnection $\mathbf{D}_i$ matrices. For pedagogical and clarity reasons, the illustrative examples presented do *not* use the earlier power system models *Systems* 1, 2, 3, and 4 (Figs. 2.4–2.6, and 2.1, respectively) presented in Chapter 2 as they are too large and unwieldy. Instead, this section introduces now a new, smaller, and simpler power system model, which in itself is very convenient to explain how said disconnection matrices are formed.[64] Undoubtedly, the use of a simpler and smaller power system model as an example facilitates more succinct elucidations, allows for clearer diagrams, and promotes comprehension rather than complexity.

In addition to exemplifying how to form the $\mathbf{D}_i$ matrices, the following examples will show the immittance matrices for node and mesh tearing directly. The procedure to form the immittance matrices was the subject of Chapter 6. Therefore, formation of the immittance matrices (i.e., nodal conductance and mesh resistance matrices) is not covered in the next section. However, because the example simple power system model used next is small, the immittance matrices can be formed by visual inspection and do not require the use of a tensor.

The problems analyzed next are purposefully kept basic; there is no complexity that may enshroud the partitioning methodology presented by this book. That is, the principles and techniques that follow can be applied equally well to *Systems* 1–4 presented in Chapter 2. Moreover, the reader should keep in mind that the principles introduced in this chapter will be used on *Systems* 1–4 to run the performance metrics presented later in Chapter 9.

7.6.1 Node tearing

To begin with, consider the simple power system model represented by its one-line diagram view shown on the left-hand of Fig. 7.18; the corresponding computer model created in *Simulink* is shown on the right, which shows blocks from the *SimPowerSystems* library. One-line diagrams, computer models, and impedance diagrams [140] often look different, but all represent the same power system. The one-line diagram is a simplified view of the power system it represents.

In Fig. 7.18, *GEN1* and *GEN2* represent three-phase sources, *RCT1* represents a three-phase rectifier, *Lod1* represents a DC load, and *LOD1* represents a three-phase load series-RC load. The power apparatus models for the three-phase sources, rectifier, and three-phase loads are the same as those introduced in sections 5.6, 5.4.1, and 5.2, respectively. The three-phase series RC load conforms with the three-phase static load model shown in Fig. 7.18, where the load capacitance can be calculated in a similar way as done for the inductance in expression (5.5).

The parameters of these power apparatus are neglected as they do not affect the partitioning method. Furthermore, during the partitioning of this simple power system the network parameters will be shown only symbolically to illustrate where the matrix entries stem from. As before, to resemble a realistic simulation scenario, a three-phase voltage and current measurement block (*VIM1*) was inserted between *GEN1* and the AC bus.

The *circuit*-level view of the simple power system model under analysis (Fig. 7.18) is shown in Fig. 7.19, where its nodes are numbered arbitrarily. This view

[64] The explanations given for the new, small power system model introduced this chapter apply to *System* 1–4 of Chapter 2. The three-phase power apparatus of the simple power system discussed in this chapter are the same as those introduced in Chapter 5.

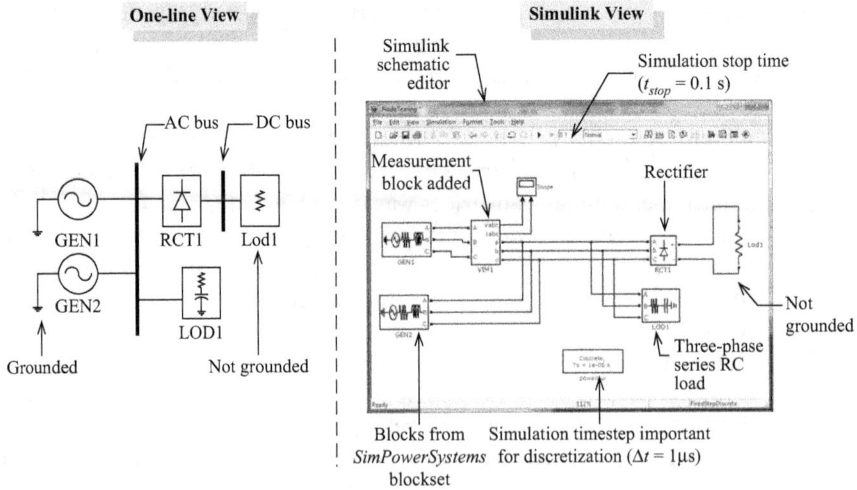

Fig. 7.18 Simple power system to demonstrate node and mesh tearing and how to form the disconnection matrices. Left: one-line diagram representation. Right: actual computer model in Simulink

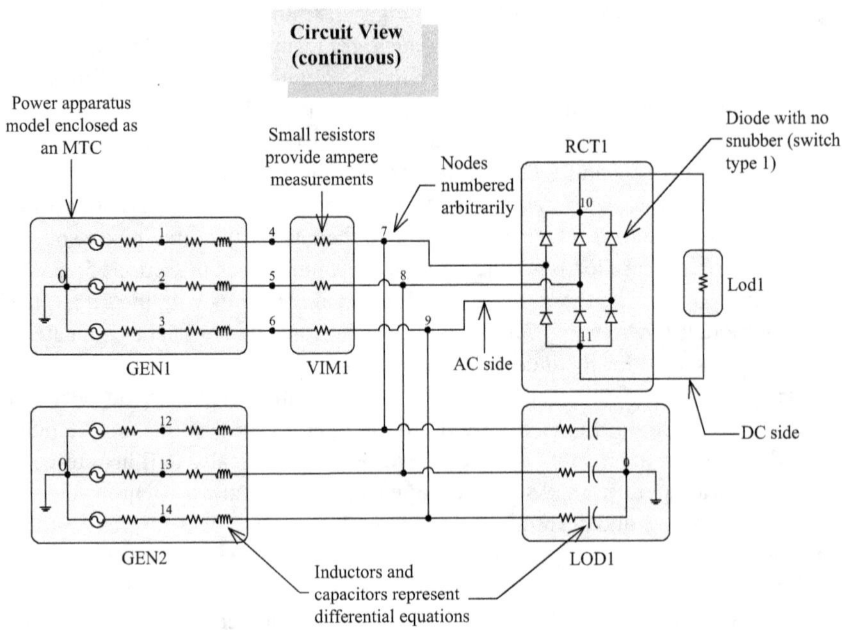

Fig. 7.19 The simple power system shown at the circuit level

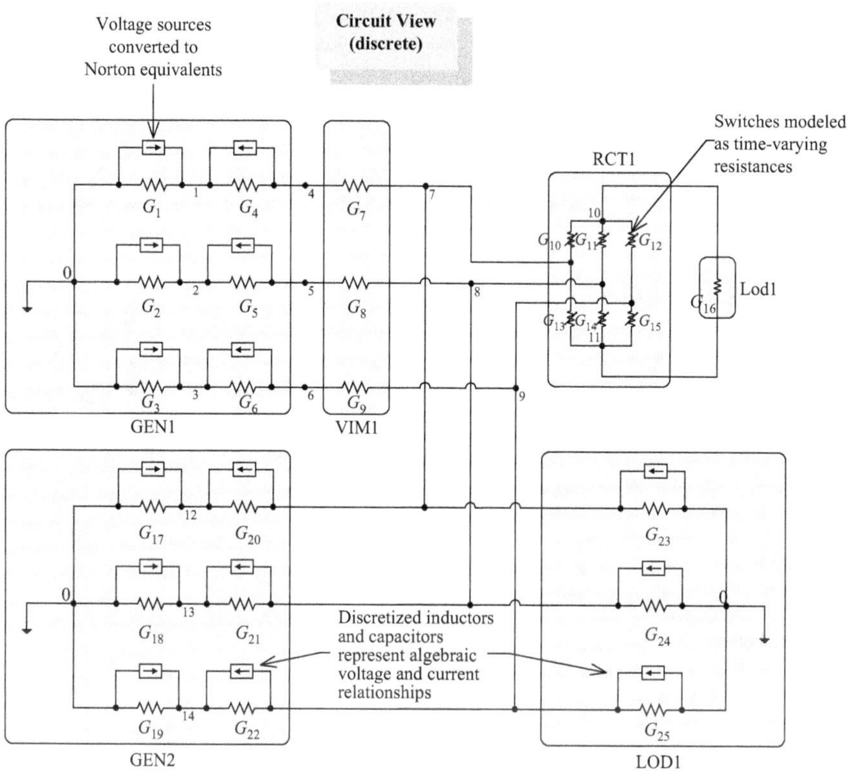

Fig. 7.20 Discretized power system. Resistance is shown as conductance

is a *continuous* representation of the power system where the inductor and capacitor symbols [194] imply this simple power system model includes differential voltage and current relations.

With respect to *GEN1*, *GEN2*, and *LOD1*, after discretizing the simple power system, the continuous relationships of the inductors and capacitors become algebraic relationships as shown by the discrete view of Fig. 7.20.[65] Turning now to the measurement block *VIM1*, this block is treated as a power apparatus for consistency and ease of implementation in software code. The parasitic inline resistors make available current measurements inside the solver.

With respect to *GEN1* and *GEN2*, insofar as classical nodal analysis does not readily handle voltage sources, the voltage sources use their series resistance to form Norton equivalents instead. As for *RCT1*, the diodes are modeled in the simple power system as time-varying resistances. These resistances will be included in immittance matrix $\mathbf{A}_i$ for each subsystem as shown later. Finally, regarding *Lod1*, this

[65] In nodal analysis, it is common to express resistance (R) as conductance ($G = 1/R$) to avoid fractions.

component is not grounded and, therefore, it will require caution when defining the boundary variables.

Since the simple power system shown in Fig. 7.20 is already discretized, its immittance matrix (nodal conductance matrix) can be formed. The nodal conductance matrix **A** (or $\mathbf{G}_{nodal}$) matrix is shown in (7.35). It should be noticed that the matrix in (7.35) is in its unpartitioned form, which is why a corresponding disconnection matrix **D** does not exist. The disconnection matrices arise when a power system is partitioned as explained next.

7.6.1.1 Two partitions

The previous subsection introduced a simple power system model to exemplify its possible several views and to show its immittance matrix in unpartitioned ($p = 1$) form. This same model is now partitioned into two partitions ($p = 2$) to show the immittance matrix of each partition and their corresponding disconnection matrices $\mathbf{D}_i$. Following the four views of the unpartitioned case shown before, the partitioned[66] power system is first shown as a one-line diagram in Fig. 7.21.

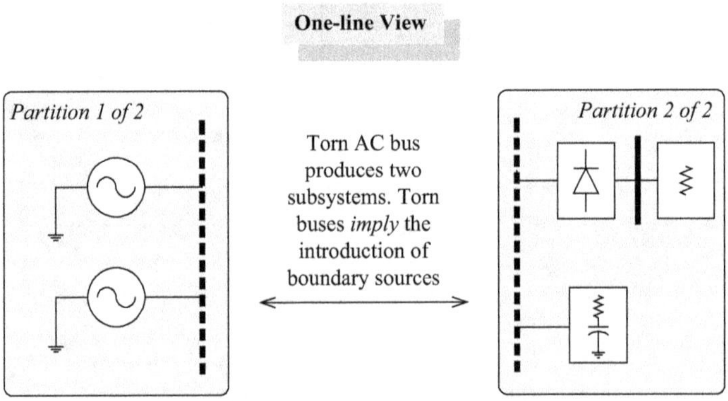

Fig. 7.21 Partitioned power system showing p = 2 partitions. The partitions were chosen manually for illustration purposes

The circuit view (continuous) of the two-partition system (i.e., $p = 2$) is shown in Fig. 7.22, where the boundary (or disconnection) matrices $\mathbf{D}_i$ are shown for each partition (or subsystem) $i = 1$ and $i = 2$. Matrix $\mathbf{D}_1$ couples subsystem 1 to the boundary variables u_1, u_2, and u_3 using "+" signs as explained by (7.30).[67] Equivalently, matrix $\mathbf{D}_2$ negatively couples subsystem 2 to the boundary variables u_1, u_2, and u_3.

The signs (± 1) indicate the direction in which the boundary current sources sink/inject current from/to each subsystem. The disconnection matrices $\mathbf{D}_i$ are

[66] As pointed out early, the terms *partition* and *subsystem* have the same meaning and are used interchangeably in this book.

[67] $\mathbf{D}_p$ or $\mathbf{D}_i$ mean the same, and refer to the submatrix of **D** corresponding to subsystem p or i. $\mathbf{D}_p$ is used over $\mathbf{D}_i$ to prevent confusion when using (i,j) indices as given in (7.24) and (7.30). That is, $\mathbf{D}_p(i,j)$ is an unambiguous way to write $\mathbf{D}_i(i,j)$.

$$
\mathbf{A} =
\begin{array}{c|cccccccccccccc|}
 & 1 & 2 & 3 & 4 & 5 & 6 & 7 & 8 & 9 & 10 & 11 & 12 & 13 & 14 \\
\hline
1 & G_1 + G_4 & \cdot & \cdot & -G_4 & \cdot & \cdot & \cdot & \cdot & \cdot & \cdot & \cdot & \cdot & \cdot & \cdot \\
2 & & G_2 + G_5 & \cdot & \cdot & -G_5 & \cdot & \cdot & \cdot & \cdot & \cdot & \cdot & \cdot & \cdot & \cdot \\
3 & & & G_3 + G_6 & \cdot & \cdot & -G_6 & \cdot & \cdot & \cdot & \cdot & \cdot & \cdot & \cdot & \cdot \\
4 & & & & G_4 + G_7 & \cdot & \cdot & -G_7 & \cdot & \cdot & \cdot & \cdot & \cdot & \cdot & \cdot \\
5 & & & & & G_5 + G_8 & \cdot & \cdot & -G_8 & \cdot & \cdot & \cdot & \cdot & \cdot & \cdot \\
6 & & & & & & G_6 + G_9 & \cdot & \cdot & -G_9 & \cdot & \cdot & \cdot & \cdot & \cdot \\
7 & & & & & & & G_{7,7} & \cdot & \cdot & -G_{10} & -G_{13} & -G_{20} & \cdot & \cdot \\
8 & & & & & & & & G_{8,8} & \cdot & -G_{11} & -G_{14} & \cdot & -G_{21} & \cdot \\
9 & & & & & & & & & G_{9,9} & -G_{12} & -G_{15} & \cdot & \cdot & -G_{22} \\
10 & & & & & & & & & & G_{10,10} & -G_{16} & \cdot & \cdot & \cdot \\
11 & & & & & & & & & & & G_{11,11} & \cdot & \cdot & \cdot \\
12 & & & & & & & & & & & & G_{17} + G_{20} & \cdot & \cdot \\
13 & & & & & & & & & & & & & G_{18} + G_{21} & \cdot \\
14 & & & & & & & & & & & & & & G_{19} + G_{22}
\end{array}
\qquad (7.35)
$$

Symmetric matrix. Only upper triangular matrix is shown. Each "." represents a zero.

where:

$$
\begin{aligned}
G_{7,7} &= G_7 + G_{10} + G_{13} + G_{20} + G_{23} \\
G_{8,8} &= G_8 + G_{11} + G_{14} + G_{21} + G_{24} \\
G_{9,9} &= G_9 + G_{12} + G_{15} + G_{22} + G_{25} \\
G_{10,10} &= G_{10} + G_{11} + G_{12} + G_{16} \\
G_{11,11} &= G_{13} + G_{14} + G_{15} + G_{16}
\end{aligned}
$$

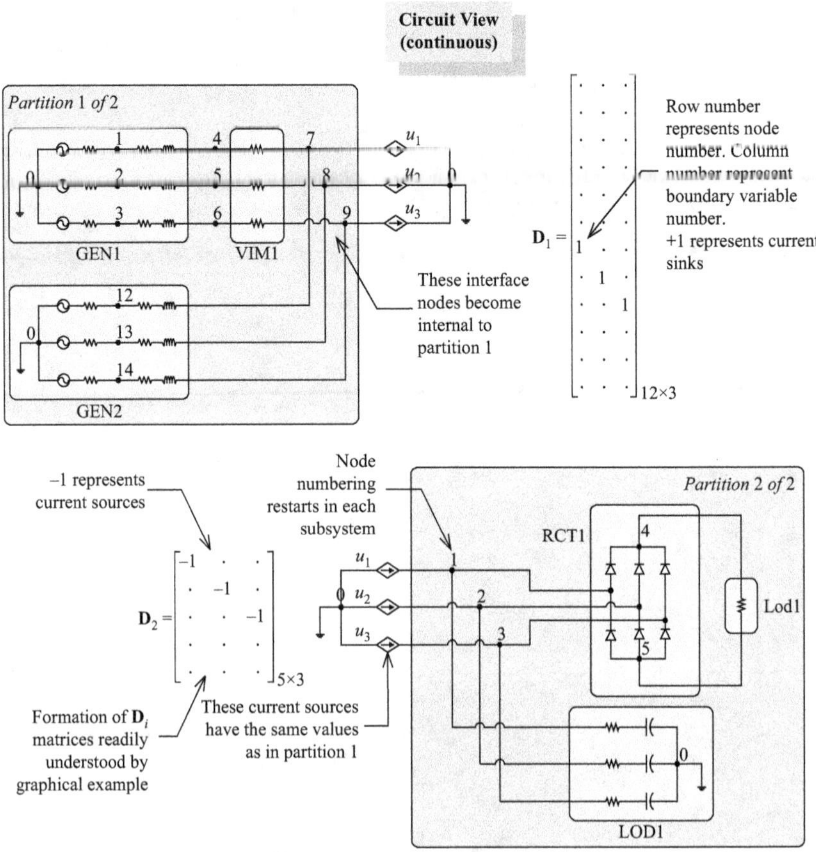

Fig. 7.22 Partitioned simple power system showing p = 2 partitions, boundary variables u_i, and disconnection matrices $\mathbf{D}_i$

shown alongside each subsystem in Fig. 7.22, for convenience. These matrices express how each subsystem relates to its boundary variables; the dimensions of these $\mathbf{D}_i$ matrices are equal to the number of non-*datum* nodes in each subsystem by the *total* number of boundary variables of the network (in this case, $r = 3$ boundary variables were introduced across the entire network). It is also noticed that the node numbering restarts in each subsystem. Restarting the node numbering prevents from having zero rows and columns in the immittance matrices.

The newly introduced boundary variables are currents u_1, u_2, and u_3, where their values (in amps) are the same in both partitions. These boundary variables are formed by partitioning the system and solved via the boundary variable vector $\mathbf{u}$.

After the two partitions in Fig. 7.22 are discretized, the simple power system appears as shown in Fig. 7.23. (Notice that the $\mathbf{D}_i$ matrices do not change.) The immittance matrices $\mathbf{A}_i$ for each subsystem are given in (7.36) and (7.37), respectively.

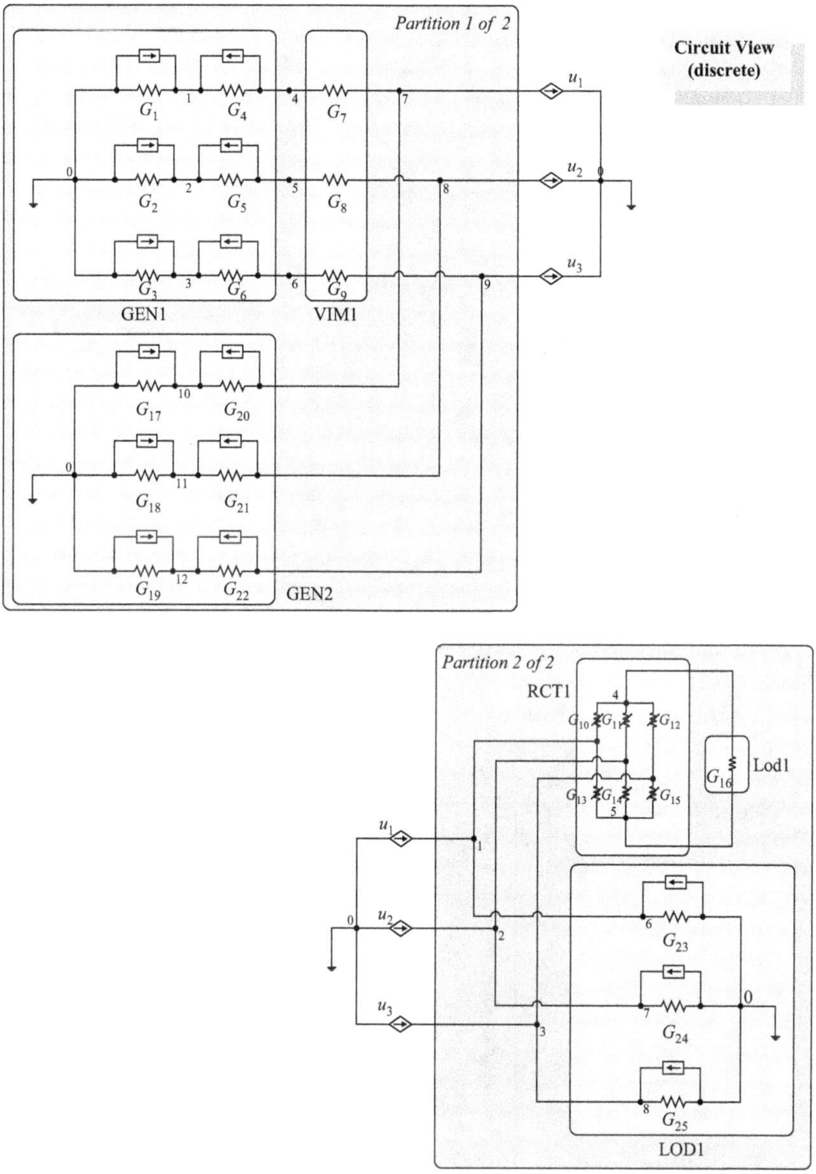

Fig. 7.23 Discretized simple power system model. Boundary variables and matrices do not change

7.6.1.2 Three partitions

In accordance with the same principles and guidelines established for the $p = 2$ case (*supra*), the three-partition $p = 3$ simple power system is first represented under the view of a one-line diagram as in Fig. 7.24. To skip repetition of the explanations

$$\mathbf{A}_1 = \begin{array}{c} \\ 1\\2\\3\\4\\5\\6\\7\\8\\9\\10\\11\\12 \end{array}\begin{array}{c} \begin{array}{cccccccccccc} 1 & 2 & 3 & 4 & 5 & 6 & 7 & 8 & 9 & 10 & 11 & 12 \end{array} \\ \left[\begin{array}{cccccccccccc}
G_1+G_4 & \cdot & \cdot & -G_4 & \cdot & \cdot & \cdot & \cdot & \cdot & \cdot & \cdot & \cdot \\
 & G_2+G_5 & \cdot & \cdot & -G_5 & \cdot & \cdot & \cdot & \cdot & \cdot & \cdot & \cdot \\
 & & G_3+G_6 & \cdot & \cdot & -G_6 & \cdot & \cdot & \cdot & \cdot & \cdot & \cdot \\
 & & & G_4+G_7 & \cdot & \cdot & -G_7 & \cdot & \cdot & \cdot & \cdot & \cdot \\
 & & & & G_5+G_8 & \cdot & \cdot & -G_8 & \cdot & \cdot & \cdot & \cdot \\
 & & & & & G_6+G_9 & \cdot & \cdot & -G_9 & \cdot & \cdot & \cdot \\
 & & & & & & G_{7,7} & \cdot & \cdot & -G_{20} & \cdot & \cdot \\
 & & & & & & & G_{8,8} & \cdot & \cdot & -G_{21} & \cdot \\
 & & & & & & & & G_{9,9} & \cdot & \cdot & -G_{22} \\
 & & & & & & & & & G_{10,10} & \cdot & \cdot \\
 & & & & & & & & & & G_{11,11} & \cdot \\
 & & & & & & & & & & & G_{12,12}
\end{array}\right] \end{array} \tag{7.36}$$

Symmetric matrix. Only upper triangular matrix is shown. Each "." represents a zero.

where:

$$G_{7,7} = G_7 + G_{20}$$

$$G_{8,8} = G_8 + G_{21}$$

$$G_{9,9} = G_9 + G_{22}$$

$$G_{10,10} = G_{17} + G_{20}$$

$$G_{11,11} = G_{18} + G_{21}$$

$$G_{12,12} = G_{19} + G_{22}$$

$$
\mathbf{A}_2 = \begin{array}{c} \\ 1 \\ 2 \\ 3 \\ 4 \\ 5 \end{array}
\begin{array}{c}
\begin{array}{ccccc} 1 & 2 & 3 & 4 & 5 \end{array} \\
\begin{bmatrix}
G_{10} + G_{13} + G_{23} & \cdot & \cdot & -G_{10} & -G_{13} \\
 & G_{11} + G_{14} + G_{24} & \cdot & -G_{11} & -G_{14} \\
 & & G_{12} + G_{15} + G_{25} & -G_{12} & -G_{15} \\
 & & & \begin{pmatrix} G_{10} + G_{11} \\ + G_{12} + G_{16} \end{pmatrix} & -G_{16} \\
 & & & & \begin{pmatrix} G_{13} + G_{14} \\ + G_{15} + G_{16} \end{pmatrix}
\end{bmatrix}
\end{array}
\tag{7.37}
$$

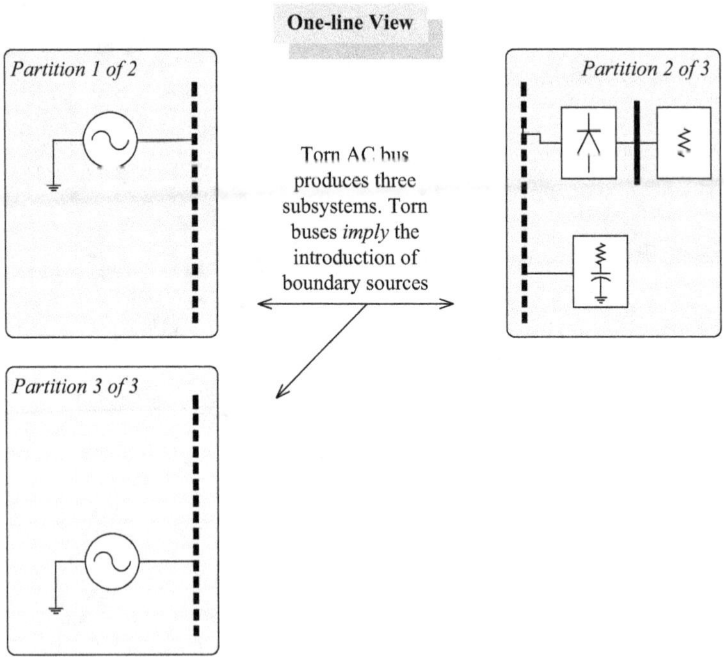

Fig. 7.24 Partitioned power system showing $p = 3$ partitions. The partitions were chosen manually for illustration purposes

already given for the $p = 1$ and $p = 2$ cases, the circuit views and matrices are now presented without detailing descriptions.

The continuous circuit view for $p = 3$ is shown in Fig. 7.25. It is important to note that the boundary variables at partition 2 of 3 are *dependent* and can be expressed in terms of the boundary variables defined for partitions 1 and 3. All boundary variables must be independent, but there is freedom in choosing which variables are considered dependent and independent. The discrete circuit view for the three-partition case is shown in Fig. 7.26 including its corresponding $\mathbf{D}_i$ matrices. The immittance matrices for the three subsystems are given in (7.38)–(7.40).

$$\mathbf{A}_3 = \begin{array}{c} \\ 1 \\ 2 \\ 3 \\ 4 \\ 5 \\ 6 \end{array} \begin{array}{c} \begin{array}{cccccc} 1 & 2 & 3 & 4 & 5 & 6 \end{array} \\ \begin{bmatrix} G_{17}+G_{20} & \cdot & \cdot & -G_{20} & \cdot & \cdot \\ & G_{18}+G_{21} & \cdot & \cdot & -G_{21} & \cdot \\ & & G_{19}+G_{22} & \cdot & \cdot & -G_{22} \\ & & & G_{20} & \cdot & \cdot \\ & & & & G_{21} & \cdot \\ & & & & & G_{22} \end{bmatrix} \end{array} \tag{7.40}$$

$$
\mathbf{A}_1 = \begin{array}{c} \\ 1\\2\\3\\4\\5\\6\\7\\8\\9 \end{array}
\begin{array}{c}
\begin{array}{ccccccccc} 1 & 2 & 3 & 4 & 5 & 6 & 7 & 8 & 9 \end{array} \\
\left[\begin{array}{ccccccccc}
G_1 + G_4 & \cdot & \cdot & -G_4 & \cdot & \cdot & \cdot & \cdot & \cdot \\
 & G_2 + G_5 & \cdot & \cdot & -G_5 & \cdot & \cdot & \cdot & \cdot \\
 & & G_3 + G_6 & \cdot & \cdot & -G_6 & \cdot & \cdot & \cdot \\
 & & & G_4 + G_7 & \cdot & \cdot & -G_7 & \cdot & \cdot \\
 & & & & G_5 + G_8 & \cdot & \cdot & -G_8 & \cdot \\
 & & & & & G_6 + G_9 & \cdot & \cdot & -G_9 \\
 & & & & & & G_7 & \cdot & \cdot \\
 & & & & & & & G_8 & \cdot \\
 & & & & & & & & G_9
\end{array}\right]_{9\times 9}
\end{array}
\tag{7.38}
$$

$$
\mathbf{A}_2 = \begin{array}{c} \\ 1\\2\\3\\4\\5 \end{array}
\begin{array}{c}
\begin{array}{ccccc} 1 & 2 & 3 & 4 & 5 \end{array} \\
\left[\begin{array}{ccccc}
G_{10} + G_{13} + G_{23} & \cdot & \cdot & -G_{10} & -G_{13} \\
 & G_{11} + G_{14} + G_{24} & \cdot & -G_{11} & -G_{14} \\
 & & G_{12} + G_{15} + G_{25} & -G_{12} & -G_{15} \\
 & & & \begin{pmatrix} G_{10} + G_{11} \\ + G_{12} + G_{16} \end{pmatrix} & -G_{16} \\
 & & & & \begin{pmatrix} G_{13} + G_{14} \\ + G_{15} + G_{16} \end{pmatrix}
\end{array}\right]
\end{array}
\tag{7.39}
$$

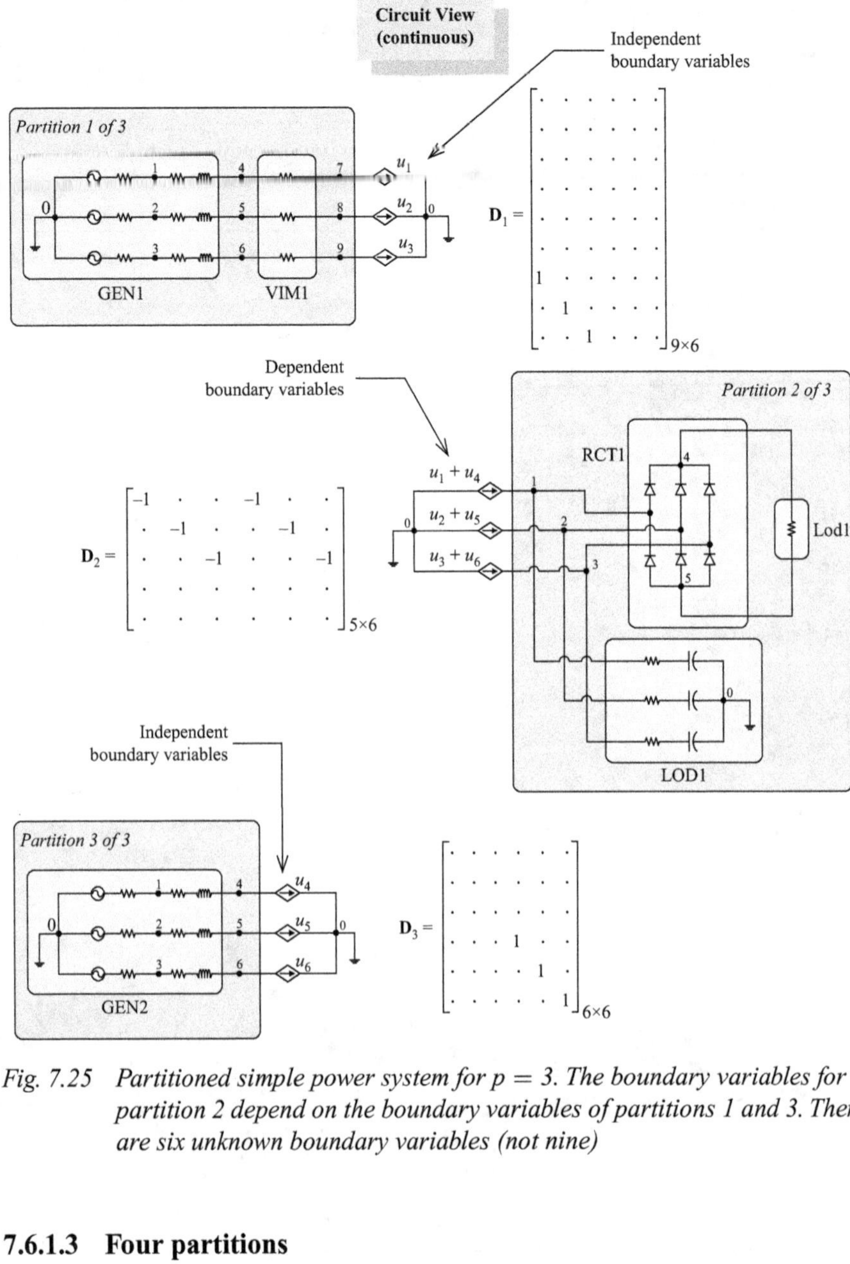

Fig. 7.25 Partitioned simple power system for p = 3. The boundary variables for partition 2 depend on the boundary variables of partitions 1 and 3. There are six unknown boundary variables (not nine)

7.6.1.3 Four partitions

This subsection partitions the simple power system model to create four partitions ($p = 4$), where special attention is paid to the boundary at *Lod1*. As usual, the one-line view diagram is given by Fig. 7.27 below, where it exhibits the partitioned power system. In this scenario, two changes are made. First, the subsystems are deliberately numbered *differently* (to show that the subsystem number does not affect the results),

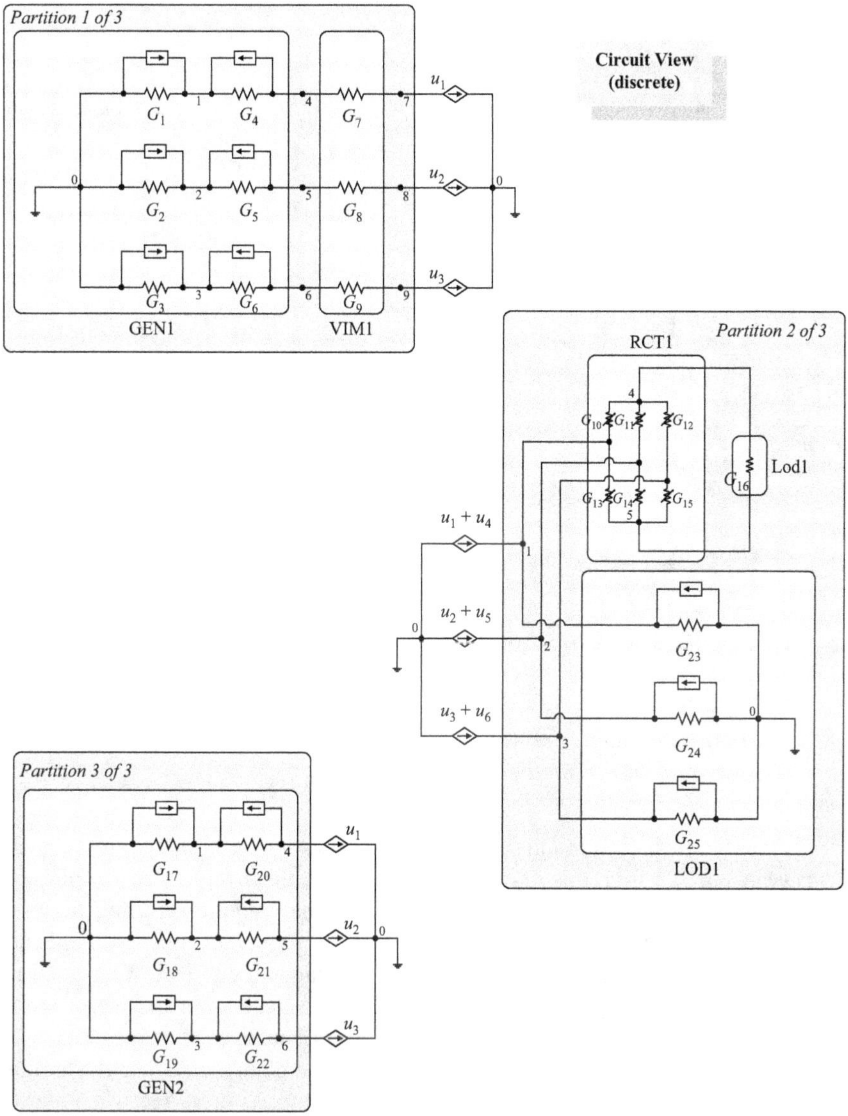

Fig. 7.26 Partitioned power system for p = 3 shown as discretized. The boundary variables for partition 2 depend on the boundary variables of partitions 1 and 3

and, second, a fourth partition is created from the DC bus. These changes demonstrate that the:

1. Partitions or subsystems can be numbered arbitrarily.
2. Partitions can be created from *any* bus.
3. Partitions can be as small as one power apparatus per subsystem.

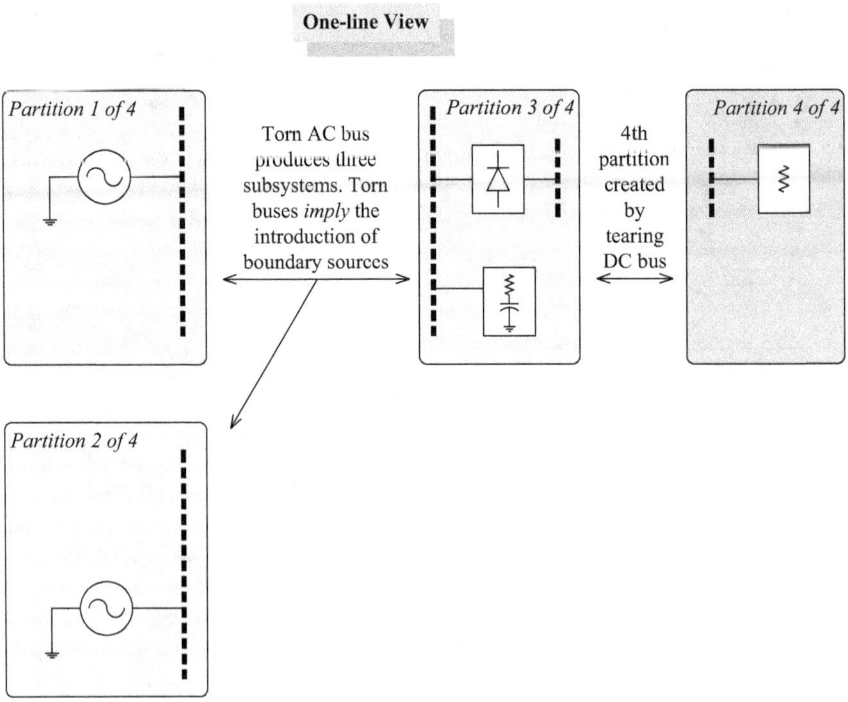

Fig. 7.27 Partitioned simple power system showing p = 4 partitions. The fourth partition was created by tearing the DC bus

The circuit view (continuous) for the four-partition case is shown in Fig. 7.28. Two caveats emerge from this $p = 4$ case. The first is that every electrical phase in every subsystem must see a path to ground. This may require inserting fictitious high-resistance grounds to force the situation. (This is numerically accomplished by adding parasitic values such as 10^{-6} to select diagonal entries in $\mathbf{G}_{nodali}$ corresponding to the floating phases. The second caveat is that tearing DC buses can be accomplished in two ways: using one planar current source as shown in Fig. 7.28, or using two current source injections from the ground plane. To reduce the number of boundary variables, the use of a single current source is preferred when tearing DC buses. Software developers, however, must catch this condition and take the correct action when identifying the disconnection points (e.g., AC vs. DC)).

The discrete view of the $p = 4$ case shown in Fig. 7.29 suggests that the number of boundary variables u_i can increase rapidly as the number of partitions increases. This fact is more pronounced in nodal formulations than it is in mesh formulations. The immittance matrices corresponding to each subsystem are given in (7.41)–(7.44). Finally, it is highlighted that automatic identification of good disconnection points is an ongoing power system partitioning issue and one that calls for good search heuristics. An approach to obtain a good set of disconnection points is presented later in section 7.8.

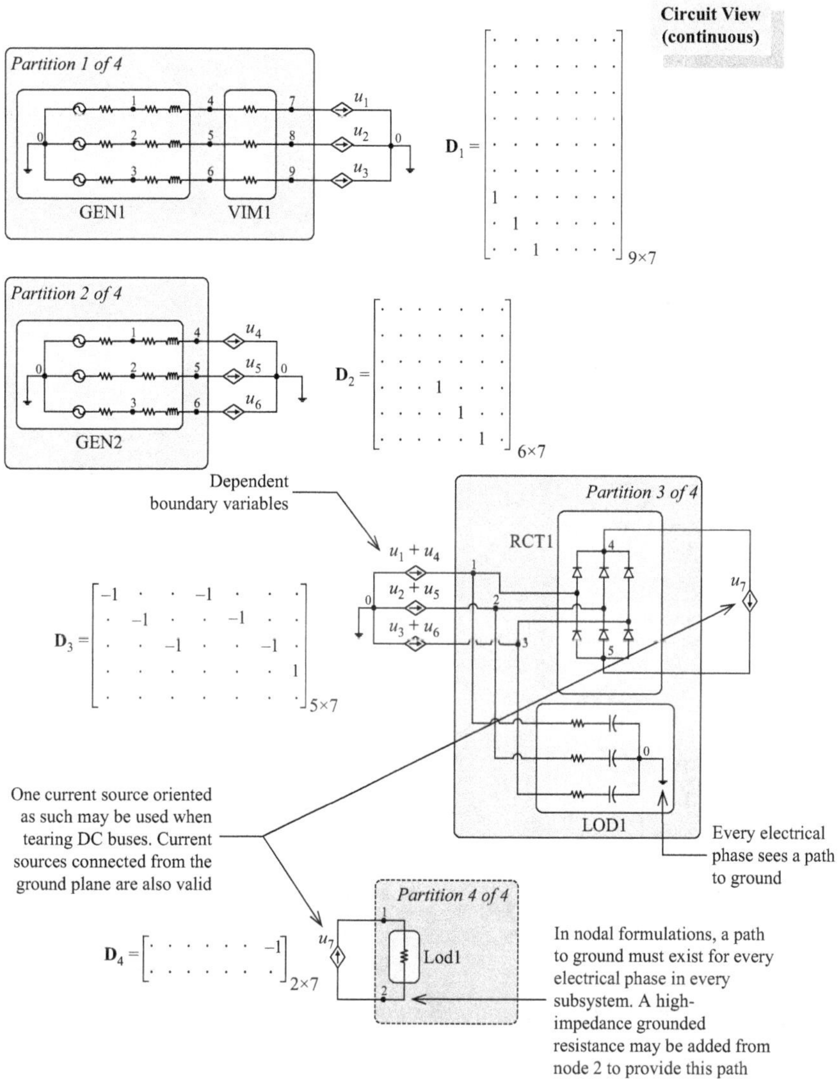

Fig. 7.28 Continuous view of simple partitioned power system for p = 4 showing $\mathbf{D}_i$ *matrices*

7.6.1.4 Observations

As demonstrated by using a simple power system model, to parallelize the simulation of a power system model using node tearing, matrices $\mathbf{A}_i$ (or $\mathbf{G}_{nodal i}$) and $\mathbf{D}_i$ for each subsystem must be formed *a priori*. As the reader should realize, Chapter 6 showed how to obtain the immittance matrices, while this section showed by way of graphical examples how to form the $\mathbf{D}_i$ matrices.

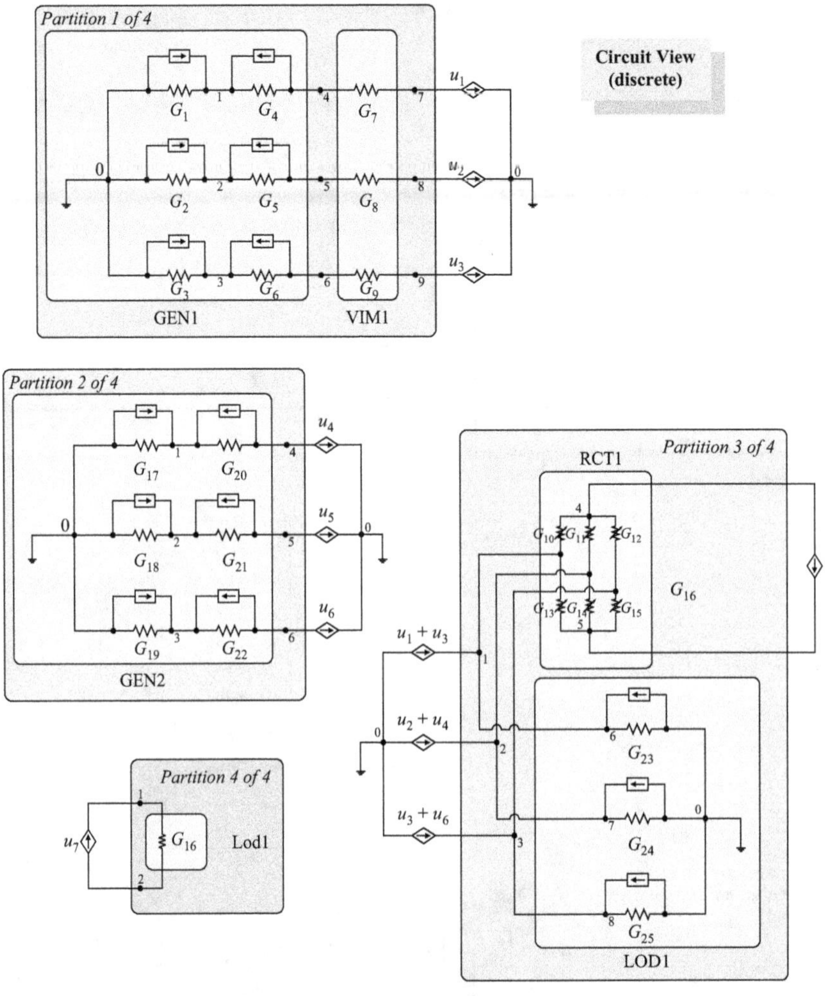

Fig. 7.29 Discrete view of partitioned simple power system for the p = 4 case

The structure of the $\mathbf{A}_i$ matrices depends on the node-numbering scheme. The preferred numbering schemes are those that reduce the fill-ins of the $\mathbf{A}_i$'s factors (e.g., LU and Cholesky factors). While numbering schemes are not covered in this book, they should be taken into consideration when numbering nodes inside each subsystem. Examples of matrix ordering schemes or node numbering schemes can be found in Reference 46.

When producing more than $p = 2$ partitions from the same bus in node tearing, boundary variables can be expressed in terms of the boundary variables defined elsewhere. It is important to notice when these dependencies occur to avoid redundancy and warrant a solvable partitioned power system model.

$$
\mathbf{A}_1 = \begin{array}{c} \\ 1 \\ 2 \\ 3 \\ 4 \\ 5 \\ 6 \\ 7 \\ 8 \\ 9 \end{array}
\begin{array}{c}
\begin{array}{ccccccccc} 1 & 2 & 3 & 4 & 5 & 6 & 7 & 8 & 9 \end{array} \\
\left[\begin{array}{ccccccccc}
G_1 + G_4 & \cdot & \cdot & -G_4 & \cdot & \cdot & \cdot & \cdot & \cdot \\
 & G_2 + G_5 & \cdot & \cdot & -G_5 & \cdot & \cdot & \cdot & \cdot \\
 & & G_3 + G_6 & \cdot & \cdot & -G_6 & \cdot & \cdot & \cdot \\
 & & & G_4 + G_7 & \cdot & \cdot & -G_7 & \cdot & \cdot \\
 & & & & G_5 + G_8 & \cdot & \cdot & -G_8 & \cdot \\
 & & & & & G_6 + G_9 & \cdot & \cdot & -G_9 \\
 & & & & & & G_7 & \cdot & \cdot \\
 & & & & & & & G_8 & \cdot \\
 & & & & & & & & G_9
\end{array}\right]_{9\times 9}
\end{array}
\tag{7.41}
$$

$$
\mathbf{A}_2 = \begin{array}{c} \\ 1 \\ 2 \\ 3 \\ 4 \\ 5 \\ 6 \end{array}
\begin{array}{c}
\begin{array}{cccccc} 1 & 2 & 3 & 4 & 5 & 6 \end{array} \\
\left[\begin{array}{cccccc}
G_{17} + G_{20} & \cdot & \cdot & -G_{20} & \cdot & \cdot \\
 & G_{18} + G_{21} & \cdot & \cdot & -G_{21} & \cdot \\
 & & G_{19} + G_{22} & \cdot & \cdot & -G_{22} \\
 & & & G_{20} & \cdot & \cdot \\
 & & & & G_{21} & \cdot \\
 & & & & & G_{22}
\end{array}\right]_{6\times 6}
\end{array}
\tag{7.42}
$$

$$\mathbf{A}_3 = \begin{array}{c} \\ 1 \\ 2 \\ 3 \\ 4 \\ 5 \end{array} \begin{array}{c} \begin{array}{ccccc} 1 & 2 & 3 & 4 & 5 \end{array} \\ \left[\begin{array}{ccccc} G_{10} + G_{13} + G_{23} & \cdot & \cdot & -G_{10} & -G_{13} \\ & G_{11} + G_{14} + G_{24} & \cdot & -G_{11} & -G_{14} \\ & & G_{12} + G_{15} + G_{25} & -G_{12} & -G_{15} \\ & & & \begin{pmatrix} G_{10} + G_{11} \\ + G_{12} \end{pmatrix} & \cdot \\ & & & & \begin{pmatrix} G_{13} + G_{14} \\ + G_{15} \end{pmatrix} \end{array}\right]_{5\times 5} \end{array} \tag{7.43}$$

$$\mathbf{A}_4 = \begin{array}{c} \\ 1 \\ 2 \end{array} \begin{array}{c} \begin{array}{cc} 1 & 2 \end{array} \\ \left[\begin{array}{cc} G_{16} + 10^{-6} & -G_{16} \\ -G_{16} & G_{16} + 10^{-6} \end{array}\right]_{2\times 2} \end{array} \tag{7.44}$$

There are situations where subsystem immittance matrices may become singular as a result of partitioning. When this occurs, in most conditions the $\mathbf{A}_i$ matrix's diagonal entries can be incremented by parasitic values to remove the singularity. This is electrically equivalent to adding high-resistance grounds to the floating phase in each subsystem to ensure that each electrical phase has a path to ground.

An important issue in power system partitioning is how the order of the boundary network (denoted by r to represent the column dimension of any $\mathbf{D}_i$ matrix or the rank of vector $\mathbf{u}$) increases with p. This issue sets a pragmatic, upper limit on the number of partitions that can be created for parallel simulations, which prevents fine-grained partitioning from performing all too well. However, this issue does not prevent all cores in a multicore desktop computer from being utilized.

Another issue, as mentioned earlier, is to determine the best number of partitions p before starting a parallel simulation. This is a difficult problem to address as the right value of p varies with the characteristics of each problem. It will be shown in Chapter 9 that a good value for p can be determined empirically by sweeping p and running enough simulations repeatedly until "rules-of-thumb" form. It is also likely that after enough experience in parallel power system simulation is gained, said rules of thumb can be embedded in software code to commence a simulation from the best number of partitions for the problem at hand.

7.6.2 Mesh tearing

The previous subsection presented different views of a basic (simple) power system model in order to exemplify the formation of $\mathbf{D}_i$ matrices. This subsection makes reference to what was just presented. The one-line and continuous views presented above are the same for mesh tearing, so they are not repeated. Furthermore, the explanations to exemplify mesh tearing start from the $p = 2$ case because the unpartitioned mesh formulation is irrelevant to both partitioning and to the formation of $\mathbf{D}_i$ matrices, and because elucidation about the formation of mesh matrices was already the focus of Chapter 6.

7.6.2.1 Two partitions

The discrete view of the two-partition case (i.e., $p = 2$) is shown in Fig. 7.30, where the boundary (or disconnection) matrices $\mathbf{D}_i$ are shown for each partition (or subsystem): $i = 1$ and $i = 2$. One peculiarity that can be noticed from this figure is that there are no node numbers since node numbers are only relevant to nodal formulations. As in the nodal case, the way the meshes are chosen (i.e., the mesh set) affects the resulting structure of $\mathbf{R}_{mesh}$ in each subsystem.

The values of ± 1 in the $\mathbf{D}_i$ matrices indicate how the boundary voltages are oriented with respect to meshes belonging to each subsystem. These $\mathbf{D}_i$ matrices have dimensions equal to the number of meshes in each subsystem (rows) by the *total* number of boundary variables of the power system problem. In this $p = 2$ case, for example, two boundary variables ($r = 2$) were introduced. Using the mesh numbering shown in Fig. 7.30 yields the mesh resistance, or immittance, matrices shown in (7.45) and (7.46) for subsystems 1 and 2, respectively.

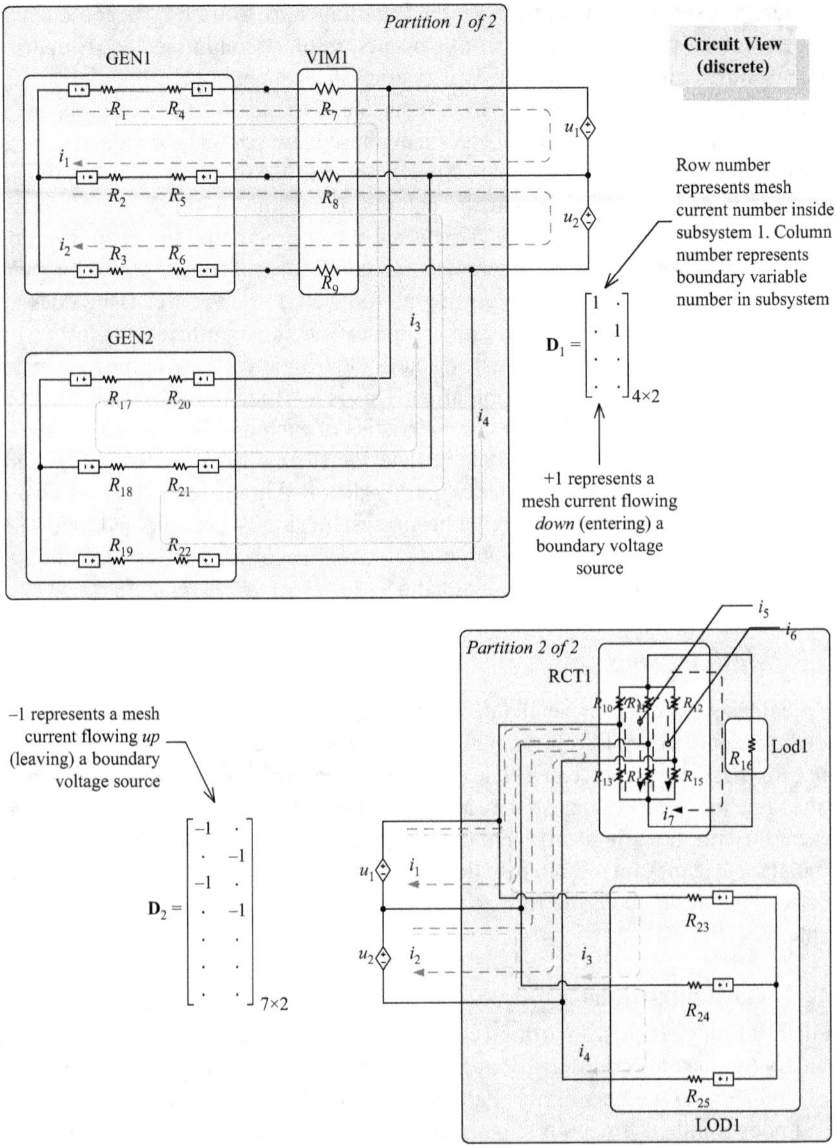

Fig. 7.30 Discrete view of partitioned simple power system showing p = 2 partitions, boundary variables, and $\mathbf{D}_i$ *matrices*

7.6.2.2 Three partitions

The $p = 3$ case is presented directly as discretized in Fig. 7.31. It should be noticed that the number of boundary variables ($r = 2$) has not changed. This particular characteristic of mesh tearing (mentioned for Fig. 7.8) maintains the partitioning overhead

$$\mathbf{A}_1 = \begin{bmatrix} \begin{pmatrix} R_1 + R_2 + R_4 \\ + R_5 + R_7 + R_8 \end{pmatrix} & -(R_2 + R_5 + R_8) & \begin{pmatrix} R_1 + R_4 + R_7 \\ + R_2 + R_5 + R_8 \end{pmatrix} & -(R_2 + R_5 + R_8) \\ \cdot & \begin{pmatrix} R_2 + R_3 + R_5 \\ + R_6 + R_8 + R_9 \end{pmatrix} & -(R_2 + R_5 + R_8) & \begin{pmatrix} R_2 + R_3 + R_5 \\ + R_6 + R_8 + R_9 \end{pmatrix} \\ \cdot & \cdot & \begin{pmatrix} R_1 + R_4 + R_7 \\ + R_{20} + R_{17} + R_{18} \\ + R_{21} + R_8 + R_5 + R_2 \end{pmatrix} & -\begin{pmatrix} R_2 + R_5 + R_8 \\ + R_{18} + R_{21} \end{pmatrix} \\ \cdot & \cdot & \cdot & \begin{pmatrix} R_2 + R_5 + R_8 \\ + R_{21} + R_{18} + R_{19} \\ + R_{22} + R_9 + R_6 + R_3 \end{pmatrix} \end{bmatrix} \tag{7.45}$$

Symmetric matrix. Only upper triangular matrix is shown. Each "." represents a zero.

$$\mathbf{A}_2 = \begin{bmatrix} (R_{10} + R_{11}) & -R_{11} & \cdot & \cdot & R_{10} + R_{11} & -R_{11} & \\ \cdot & (R_{11} + R_{12}) & \cdot & \cdot & -R_{11} & R_{11} + R_{12} & -R_{12} \\ \cdot & \cdot & (R_{23} + R_{24}) & -R_{24} & \cdot & \cdot & \cdot \\ \cdot & \cdot & \cdot & (R_{24} + R_{25}) & \cdot & \cdot & \cdot \\ & & & & \begin{pmatrix} R_{10} + R_{11} \\ + R_{13} + R_{14} \end{pmatrix} & -R_{11} - R_{14} & \cdot \\ & & & & & \begin{pmatrix} R_{11} + R_{12} \\ + R_{14} + R_{15} \end{pmatrix} & -R_{12} + R_{15} \\ & & & & & & \begin{pmatrix} R_{12} + R_{15} \\ + R_{16} \end{pmatrix} \end{bmatrix} \tag{7.46}$$

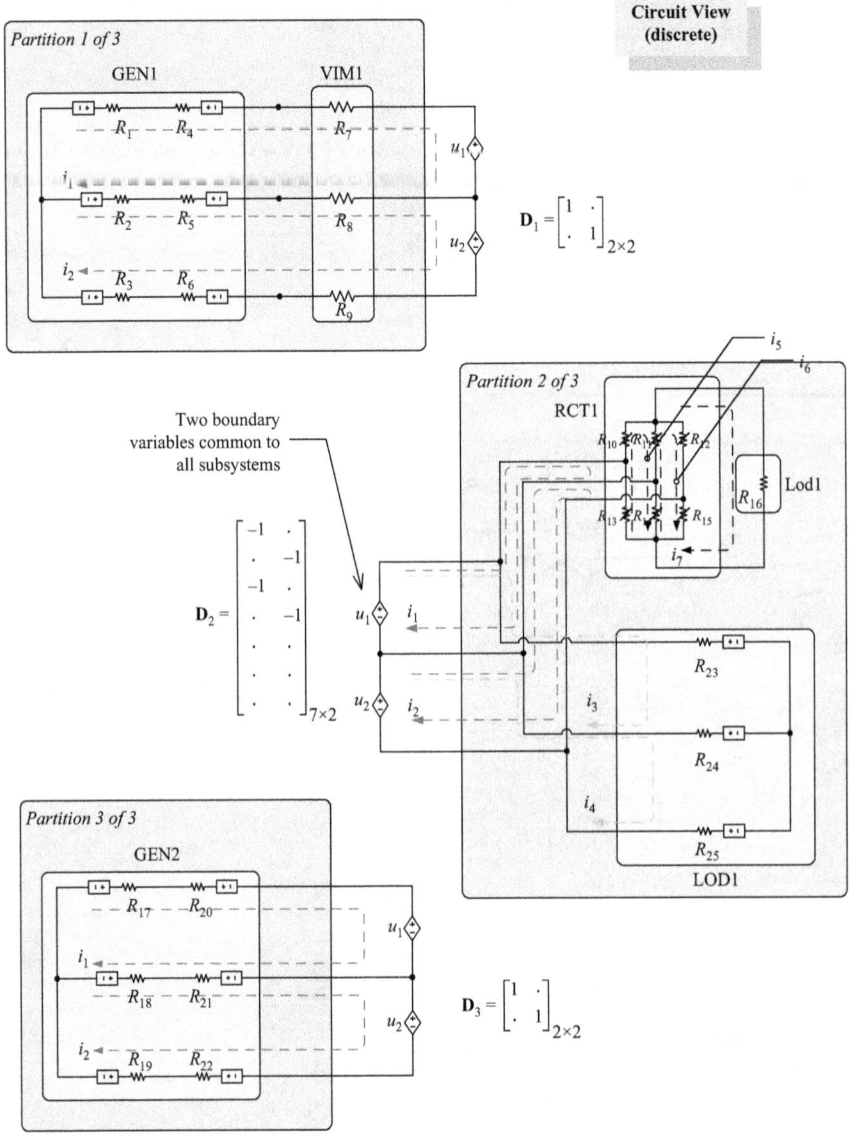

Fig. 7.31 Partitioned simple power system for $p = 3$ shown as discretized

low: an important issue in parallel power system simulation known as communication overhead. The adverse impact of this overhead is the increased solution time of the boundary network (or vector **u**), which is highly undesirable. The boundary matrices $\mathbf{D}_i$ for each subsystem are also shown in Fig. 7.31; their corresponding subsystem mesh-resistance matrices are laid out in (7.47)–(7.49), respectively.

$$\mathbf{A}_1 = \begin{bmatrix} \begin{pmatrix} R_1 + R_2 + R_4 \\ + R_5 + R_7 + R_8 \end{pmatrix} & -(R_2 + R_5 + R_8) \\ \cdot & \begin{pmatrix} R_2 + R_3 + R_5 \\ + R_6 + R_8 + R_9 \end{pmatrix} \end{bmatrix} \tag{7.47}$$

$$\mathbf{A}_2 = \begin{bmatrix} (R_{10} + R_{11}) & -R_{11} & \cdot & \cdot & R_{10} + R_{11} & -R_{11} & \\ \cdot & (R_{11} + R_{12}) & \cdot & \cdot & -R_{11} & R_{11} + R_{12} & -R_{12} \\ \cdot & \cdot & (R_{23} + R_{24}) & -R_{24} & \cdot & \cdot & \cdot \\ \cdot & \cdot & \cdot & (R_{24} + R_{25}) & \cdot & \cdot & \cdot \\ & & & & \begin{pmatrix} R_{10} + R_{11} \\ + R_{13} + R_{14} \end{pmatrix} & -R_{11} - R_{14} & \cdot \\ & & & & & \begin{pmatrix} R_{11} + R_{12} \\ + R_{14} + R_{15} \end{pmatrix} & -R_{12} + R_{15} \\ & & & & & & \begin{pmatrix} R_{12} + R_{15} \\ + R_{16} \end{pmatrix} \end{bmatrix} \tag{7.48}$$

$$\mathbf{A}_3 = \begin{bmatrix} \begin{pmatrix} R_{17} + R_{20} \\ + R_{18} + R_{21} \end{pmatrix} & -(R_{18} + R_{20}) \\ \cdot & \begin{pmatrix} R_{18} + R_{21} \\ + R_{19} + R_{22} \end{pmatrix} \end{bmatrix} \tag{7.49}$$

7.6.2.3 Four partitions

The discrete circuit view for this case is also presented directly in Fig. 7.32. The boundary variables of subsystems 1, 2, and 3 are the same as before; they remain unchanged. However, a new boundary variable was created by tearing the DC bus to

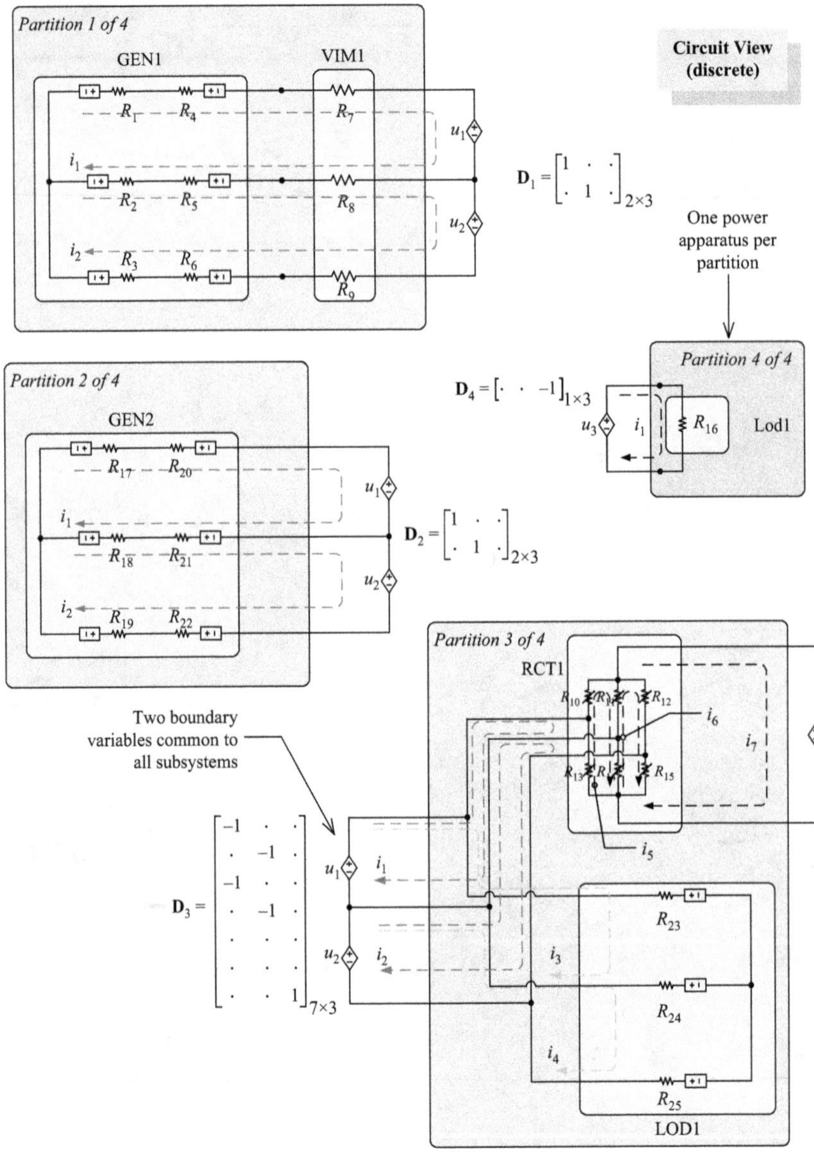

Fig. 7.32 Discrete view of partitioned simple power system for p = 4 showing $\mathbf{D}_i$ *matrices*

create the fourth partition. The $\mathbf{D}_i$ matrix of each partition is also shown in Fig. 7.32. The corresponding immittance matrices for each partition are given by (7.50)–(7.53) below.

The fourth partition contains only one power apparatus: *Lod1*. While this example is somewhat simple, the take away from here is that complex power apparatus models can be placed in their own stand-alone partitions if necessary. This segregation option is convenient when compared to other partitioning approaches as there is no need to have (or to introduce) boundary branches, and there is no requirement to have a timestep (Δt) delay to produce numerical decoupling between subsystems.

It should be highlighted that the number of boundary variables at the torn AC bus would not change even if *LOD1* were placed in its own partition. That is, in contrast to node tearing, increasing the number of subsystems produced from the *same* bus does *not* increase the number of boundary variables. This property is particular of mesh tearing and does not apply to node tearing.

While it is highly desirable to keep the size of the boundary network small to reduce the solution time of $\mathbf{u}$, another power system partitioning issue is the access time penalty of each subsystem (or thread as explained in Chapter 8) attempting to fetch the vector $\mathbf{u}$. In shared-memory computers, such as multicore computers, producing the results of $\mathbf{u}$ by one thread and, then, fetching them simultaneously from several other threads can have negative impacts on simulation performance. This is a well-known performance issue in multicore computing known as "false sharing" [189,195], which is a bottleneck that can easily (and unknowingly) be introduced by the programmer.

$$\mathbf{A}_1 = \begin{bmatrix} \begin{pmatrix} R_1 + R_2 + R_4 \\ + R_5 + R_7 + R_8 \end{pmatrix} & -(R_2 + R_5 + R_8) \\ \cdot & \begin{pmatrix} R_2 + R_3 + R_5 \\ + R_6 + R_8 + R_9 \end{pmatrix} \end{bmatrix} \tag{7.50}$$

$$\mathbf{A}_2 = \begin{bmatrix} \begin{pmatrix} R_{17} + R_{20} \\ + R_{18} + R_{21} \end{pmatrix} & -(R_{18} + R_{20}) \\ \cdot & \begin{pmatrix} R_{18} + R_{21} \\ + R_{19} + R_{22} \end{pmatrix} \end{bmatrix} \tag{7.51}$$

7.6.2.4 Observations

Mesh tearing has been exemplified by partitioning a simple power system into $p = 2$, $p = 3$, and $p = 4$ partitions. The matrices needed to produce parallel simulations are the $\mathbf{A}_i$ (same as $\mathbf{R}_{meshi}$) and $\mathbf{D}_i$ matrices for *each* subsystem. Once these matrices are obtained, they feed into (7.21) and (7.22) to produce a parallel simulation. The excitation (right-hand side) vectors of each subsystem are formed dynamically during runtime and change at each simulation timestep. In mesh analysis, they are formed from the contour sum of voltage sources in each mesh.

Analogous to the nodal formulation case, the structure of the $\mathbf{A}_i$ matrices depends on the mesh numbering scheme. The preferred numbering schemes are

$$\mathbf{A}_3 = \begin{bmatrix} (R_{10}+R_{11}) & -R_{11} & \cdot & \cdot & R_{10}+R_{11} & -R_{11} & \\ \cdot & (R_{11}+R_{12}) & \cdot & \cdot & -R_{11} & R_{11}+R_{12} & -R_{12} \\ \cdot & \cdot & (R_{23}+R_{24}) & -R_{24} & \cdot & \cdot & \cdot \\ \cdot & \cdot & \cdot & (R_{24}+R_{25}) & \cdot & \cdot & \cdot \\ & & & & \begin{pmatrix} R_{10}+R_{11} \\ +R_{13}+R_{14} \end{pmatrix} & -R_{11}-R_{14} & \cdot \\ & & & & & \begin{pmatrix} R_{11}+R_{12} \\ +R_{14}+R_{15} \end{pmatrix} & -R_{12}+R_{15} \\ & & & & & & \begin{pmatrix} R_{12}+R_{15} \\ +R_{16} \end{pmatrix} \end{bmatrix} \tag{7.52}$$

$$\mathbf{A}_4 = [R_{16}] \tag{7.53}$$

those that reduce fill-ins in the matrix factors. While these numbering schemes were not covered here, they should be considered when enumerating meshes. To reduce fill-ins in mesh formulations, the rules of thumb is to keep the branch-traversal length of each mesh as short as possible, and to avoid intersecting other meshes as well [196].

In the nodal case, it was seen that some subsystem boundary variables can be expressed in terms of boundary variables defined for other subsystems. While this flexibility does not apply in the mesh case, the boundary variables (voltage sources) defined at buses are common to all subsystems interfaced at the same boundary.

7.7 Validation

Mathematically speaking, (7.19) showed that partitioned and unpartitioned simulations by the same solver give the same result. Computationally speaking, however, finite, yet negligible differences, may emerge due to the differences in numerical precision that are observed when working with matrices of different sizes and condition number [46].

Differences in results are even more pronounced when comparing results produced with *different* solvers. It is useful to compare the results of a commercial simulator against those of a custom-made solver performing the same task; possible differences between them arise from different reasons as discussed below. This section considers caveats of validating the results of a commercial simulator versus those produced by a custom-made solver. Some of these caveats include:

1. Δt in both programs must be exactly equal (e.g., 42.296 μs is not the same as 42.2963 μs).
2. The *integration* method in both programs must be the same (e.g., the root matching method produces different results than the Tustin and backward Euler methods).
3. The use of internal parasitic impedances to facilitate programming must be included in both programs (e.g., a 1-MΩ resistance in one program is not the same as not having it in the other program).
4. Branch models must be the same (e.g., a stand-alone resistance is not the same as having a series RL branch and setting $L = 10^{-12}$ H).
5. The switch models should be equivalent (e.g., the opening of a switch equipped with a snubber produces a different response than the opening action of one without it).
6. The interpolation methods employed should be the same (e.g., some programs produce a new solution at t_{z1} as shown in Fig. 3.4 while other programs do not).
7. The way power apparatus are modeled should be exactly the same (e.g., a delta-connected source (or load) is not the same as a wye-connected one; a cable model with stray capacitances is not the same as a cable model without them).
8. Matrix ill-conditioning can cause finite simulation differences (e.g., the results of an ill-conditioned mesh formulation may disagree with the results of a better-conditioned nodal formulation).

9. Different equation formulation types can lead to differences in results (e.g., nodal formulations, specially partitioned ones, require that each subsystem be grounded; grounds are not required at all in mesh or state-variable formulations).
10. The most frustrating reasons for differences in results are those that cannot be traced. Software developers (understandably) often do not document every implementation detail about their models or solvers. When important detail is concealed from the user's knowledge, it is difficult to explain why results disagree (e.g., if the carrier signal of PWM controllers in different programs are 180° out of phase, the firing instants will then differ).

In addition to the reasons enumerated above, there are also internal optimizations in code that can cause differences. These optimizations can enhance performance, reduce code length, and reduce code maintenance time, and can also change the level of fidelity. For example, some programs may model open switches as 1-MΩ resistors; others, may treat these as 0 A current sources. These differences are usually not known to the end user.

Despite the expected differences, it is important to contrast the simulation results obtained with *Simulink* against those obtained with the multicore solver developed for this book. Taking the example of the simple power system model introduced in Fig. 7.18 and adding to it an LC filter[68] at the output terminals of the rectifier results in the side-by-side comparison shown in Fig. 7.33. On the left-hand side of this figure are the results provided by *Simulink* when using the backward Euler integration method. On the right are the results from using the multicore ($p = 2$ nodal case at $\Delta t = 46.296$ μs). The top row shows the three-phase line voltages at the terminals of GEN1. The bottom row shows the three-phase line currents leaving GEN1. Overall, these two results are in reasonable agreement. It should be noted that no claim can be made on which result set is correct; what can be said about these results, however, is that they are similar.

Moreover, the reader should be aware that the multicore solver developed for this book bases its discretization on the root-matching method solely, while discretization with *SimPowerSystems* is more flexible as it can be changed to either the backward Euler or the Tustin method. Changing the discretization method to the Tustin method results in the comparison shown in Fig. 7.34 ($p = 4$, mesh case at $\Delta t = 46.296$ μs), which shows the potential impact of changing discretization methods. Similarly, no claim is made on which result set is correct; only that the results are different.

Furthermore, it is not always possible to assert whether or not a simulation result set is correct; therefore, the word *error* is not used in this book–and should not be used when comparing results unless one result set is clearly accepted as being correct. A few alternative ways to increase confidence [88] in simulation results are:

1. Validate the simulation results against more than one other commercial tool [89,118].

[68] A large series LC filter was introduced to force a charging transient at $t = 0$. The $L = 1$ mH inductance $C = 1$ mF correspond to the filter shown in Fig. 5.11.

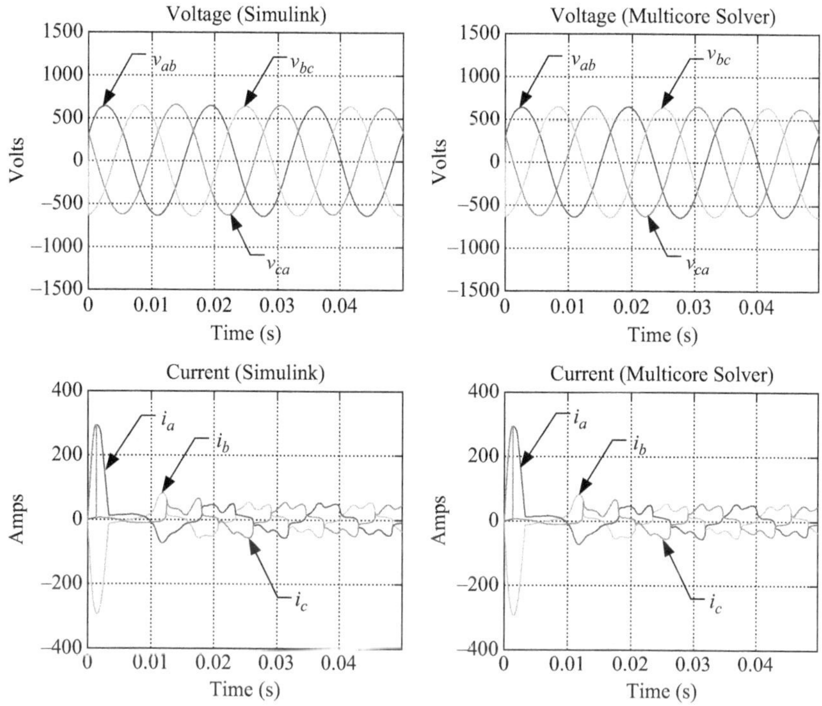

Fig. 7.33 Simulation results of the node tearing example using Simulink's backward Euler integration method (given $p = 2$ and a nodal formulation using $\Delta t = 46.296\ \mu s$)

2. Conduct experiments in a controlled setting to record data or collect primary data from an energized network [197].
3. Validate simulation results against simple RLC circuits for which analytical solutions exist.

The first approach may be trivial for a few, small and simple power circuits. But for large shipboard power system models such as the one shown in Fig. 2.1 (*System* 4), building a model of this size on different platforms requires significant research resources. And even if such resources existed, readers should expect several of the items listed (and not listed) at the beginning of this section to arise.

The second approach of conducting experiments in a controlled setting is the most desirable one, but many times unrealistic. Rarely will electric networks comprising expensive kW- or MW-level power apparatus be available for exercising disturbance contingencies. Power apparatus are too expensive to use for experimental testing.[69] Not only must the power apparatus be available for

[69] In passing, and as a reference of electrical equipment costs, real shipboard power apparatus can cost $0.50 per watt. On the other hand, residential solar photovoltaic arrays can cost $3 per watt.

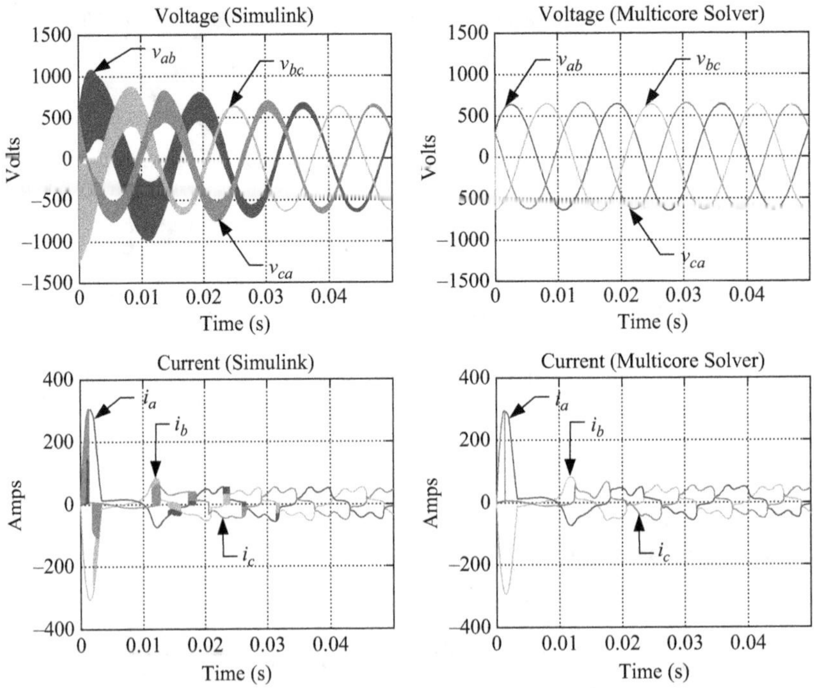

Fig. 7.34 Simulation results of the mesh tearing example using Simulink's Tustin transformation method (given p = 4 and mesh case at $\Delta t = 46.296\ \mu s$*)*

experimentation (and properly connected while meeting safety guidelines), but proper instrumentation must also be in place, too, which adds to experiment costs and preparation time. Some laboratories [62,73,197–205], however, *do* have experimental equipment for testing. In the author's experience comparing simulation results against *experimental* data (if available) is the preferred approach to validate a solver's accuracy (as exemplified in Reference 202 by a series DC arc model).

The third approach listed above is highly credible, but it is limited to small RLC circuits. Closed-form solutions can be derived for such circuits, which allow the validation of results against analytical expressions. But the use of simple RLC circuits is limited in practice. Power system networks are highly complex networks which, to say the least, include many power electronic converters and machines whose solutions cannot be reached longhand. There is, however, value in using RLC circuits for validation. The value lies in asserting correctness in the chosen discretization (or integration) method for simple circuits whose confidence can—arguably—be transferred to larger networks. It is important to keep an open mind concerning the benefits (and drawbacks) of these three validation approaches.

7.8 Graph partitioning

The motivation behind power system partitioning (in general) is to reduce simulation runtime. This can be accomplished by first partitioning a power system model and then simulating it in parallel as subsystems of less computational burden than the original (unpartitioned) model.

Parallel simulation of power systems requires addressing two issues: *where* to tear and *how many* partitions to create. The selection of good disconnection points and knowing how many partitions to create (based on one's hardware) are hard choices to get right. Judgment is required to pick the disconnection points and number of partitions that achieve the best result. Judgment is aided by experience and knowledge of the characteristics (e.g., topology and power apparatus types) of the power system model involved. Providing the right answers to these questions is not easy; it is a very laborious and repetitive task; one, however, can learn from experience (and advice). Nonetheless, the answers can be determined empirically rather quickly. The following discussion attempts to furnish answers to such strenuous questions:

Where to tear: The problem of locating adequate disconnection points for a power system is an NP-complete problem and cannot be expressed in closed form [206]. This task, however, can be outsourced to an established graph partitioning tool such as *hMetis* [192,207]. In the solution of a myriad of empirical problems in science and mathematics the aid of graphs tools and methods are important to solving topology-based problems [97,150,151,208,209]. Invoking *hMetis* by passing-in a representative graph of the power system at hand returns p different vertex sets, where p represents the number of partitions specified by the user, and each vertex represents one power apparatus (or an MTC). The design and validity of a reliable representative graph, however, is the responsibility of the team developing the parallel solver.

How many partitions to create: The question raised for determining the best number of partitions for a given situation is also not easy to answer. Matching the number of partitions p to the number of cores c on a desktop multicore computer is "simplistic in its analysis" [210]—but at times it *is* the answer and a good place to start. This correspondence between the number of partitions and cores can be learned empirically by sweeping the number of partitions and measuring the performance of the solver. Over time, patterns that allow the development of useful "rules of thumb" are formed.

To illustrate how *hMetis* can be invoked to partition a power system model, consider *System* 1 (Fig. 2.4) shown in Fig. 7.35 using MTC blocks.[70] In Fig. 7.35, the power apparatus are shown as MTCs [148] and are interconnected at the buses. To "send" this model to *hMetis*, the model must be first *mapped* to a representative

[70] Graph partitioning is demonstrated herein by referring to *System* 1. To avoid confusion, the reader should bear in mind that the simple power system model introduced earlier in Fig. 7.18 (subsection 7.6.1) to demonstrate the formation of the $\mathbf{D}_i$ matrices is no longer used in this book.

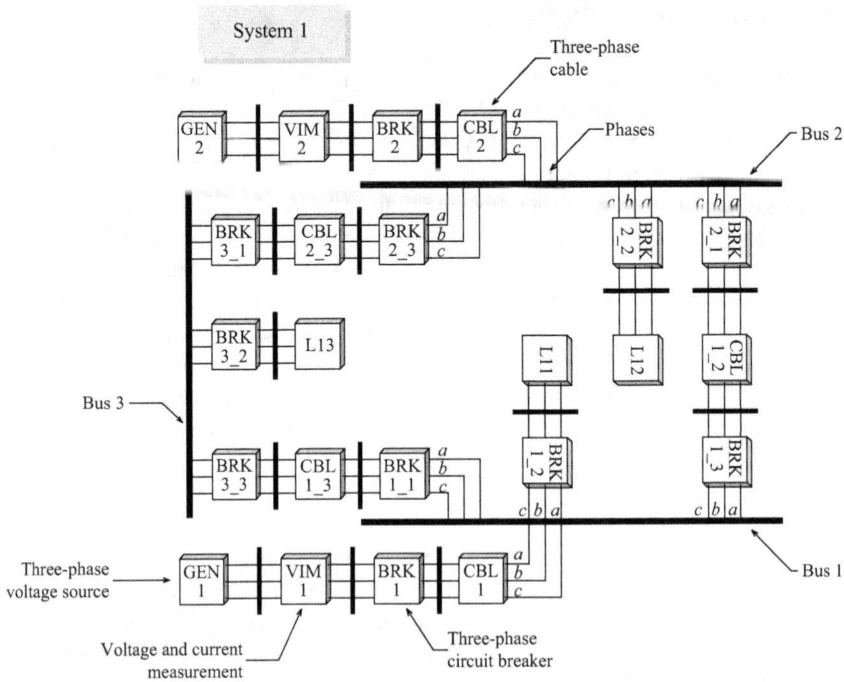

Fig. 7.35 System 1 shown with MTC blocks to illustrate graph partitioning. The one-line diagram of this model was shown in Fig. 2.4

graph. One possible mapping scheme of such graph is to represent each power apparatus (MTCs) as a graph vertex and each bus as a graph edge.[71]

Following this scheme, the graph (or hyper-graph) corresponding to the power system in Fig. 7.35 is shown in Fig. 7.36. After formation of the representative graph, the graph must be saved as an edge list[72] and used as the graph input for *hMetis* [192]. The utility of *hMetis* is that it has readily-implemented partitioning and balancing heuristics [207,211] suitable for locating the disconnection points of this type of graph [177]. The output produced by *hMetis* is a text file containing the partitioned vertex sets. A vertex set is a group of vertices that belong to the same subsystem. Knowing which power apparatus belong to each partition *implicitly* reveals the disconnection points.

Depending on the weight assigned to each vertex and edge, there are many possible outputs that *hMetis* can produce. The weights assigned to each vertex help

[71] A graph edge connects two graph vertices together. A hyper-edge connects three or more vertices together. The prefix "h" in *hMetis* suggests that *hMetis* is a hyper-graph partitioning package. This tool is suitable to partition hyper-graphs arising from electric circuits such as power systems.
[72] An edge list is a plain text file listing the graph vertices connected by each graph edge.

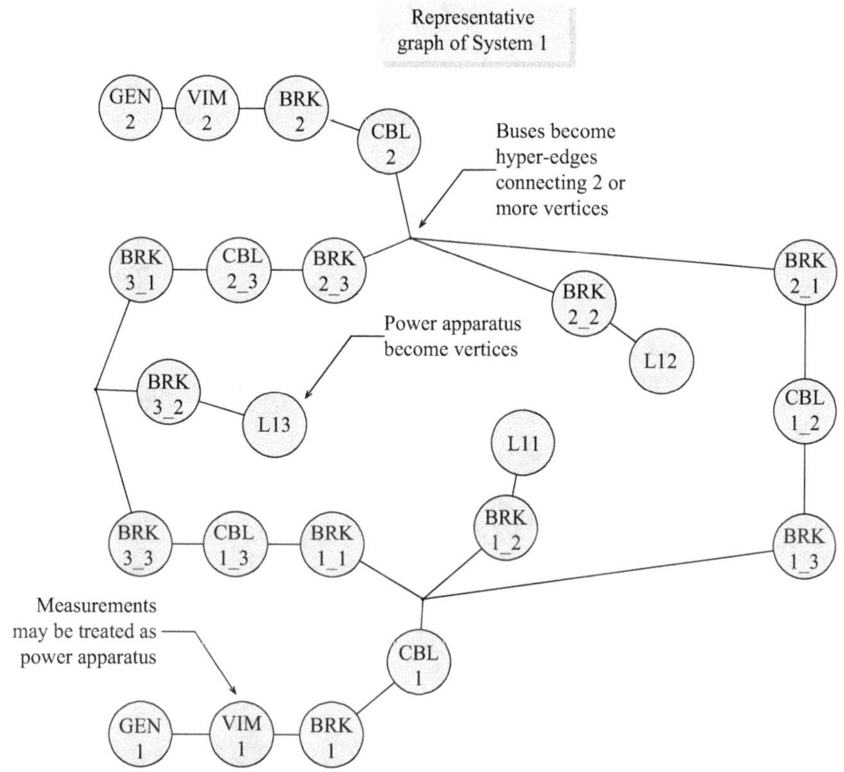

Fig. 7.36 Representative graph of System 1 shown in Fig. 7.35. MTCs (or power apparatus) are mapped to vertices and buses are mapped as hyper-edges

hMetis balance the graph partitions, which affects the locations of the disconnection points. The choice of weights, however, is the responsibility of the user. The approach used here to assign weights is based on the squared of the number of equations of a power apparatus (cubed for time-varying power apparatus). For example, a power apparatus model containing six equations has a vertex weight of 36. Power electronic converters, because they are time-varying components, require more computational effort and their equation counts are cubed instead of squared.[73] For example, the rectifier model shown in Fig. 5.7 has five equations. Therefore, its vertex weight in the graph is $5^3 = 125$. Another approach to assign weights based on arithmetic operation counts is given in Reference 160. Balancing the graph partitions through vertex weights is important for balancing the computational load of each subsystem

[73] Readers are encouraged to test different vertex-weight assignments on their own graphs.

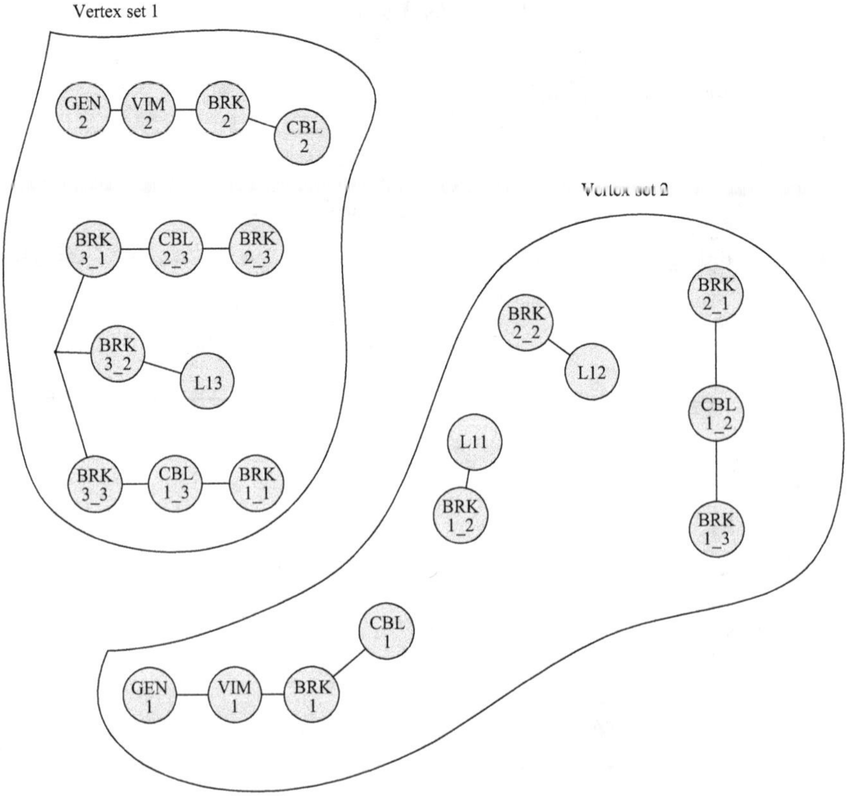

Fig. 7.37 Partitioned graph as one possible output given by hMetis

in a parallel simulation. This is a fundamental issue in parallel computing and one easy to get wrong.

Fig. 7.37 shows one possible vertex set output[74] provided by *hMetis*. This figure is a graphical representation of the text content found in the output text file produced by *hMetis*. Each vertex set represents a subsystem or partition that, ideally, should be computationally balanced.[75] Using the same aforementioned mapping scheme to create the graph, the results of *hMetis* can be mapped *back* to the original power system (i.e., *System* 1) as shown in Fig. 7.38. After *hMetis* suggests the disconnection points (i.e., torn buses) and the power apparatus belonging to each subsystem, the $\mathbf{A}_i$ and $\mathbf{D}_i$ matrices can be formed to solve (7.21) and (7.22).

[74] The output of *hMetis* also depends on its other program settings in addition to vertex weight.

[75] A subsystem (a partition or vertex set) will contain exactly one electrical subsystem and one control subsystem as was illustrated by Fig. 3.3.

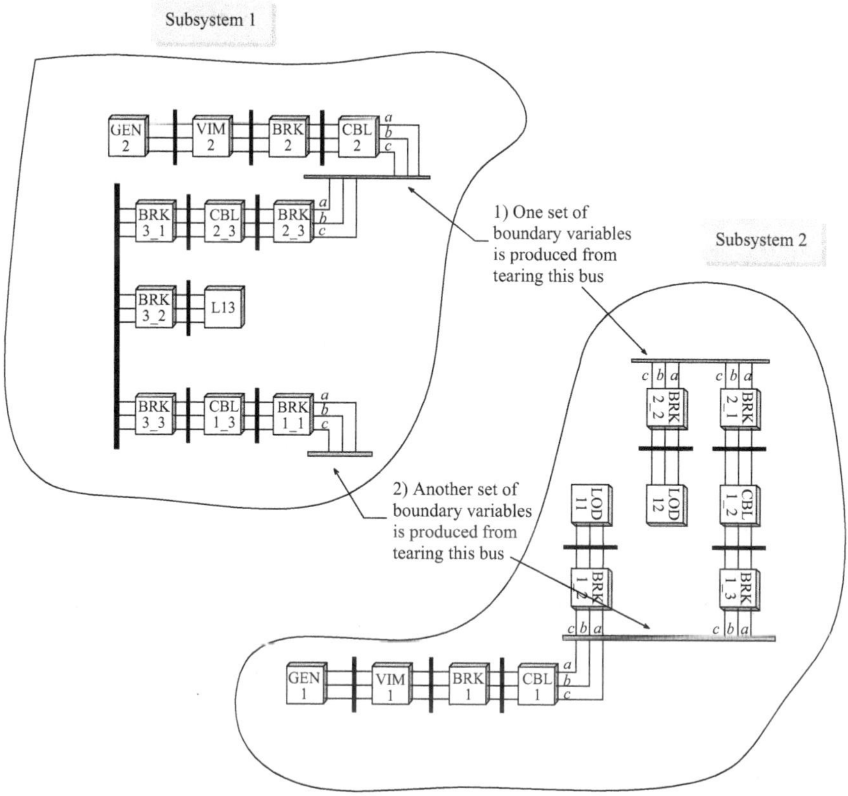

Fig. 7.38 Partitioned power system according to its representative graph produced by hMetis

7.9 Overall difference between mesh and node tearing

Throughout this chapter, it has been highlighted that the mesh tearing partitioning approach produces fewer boundary variables (denoted as r, or denoted by the column dimension of $\mathbf{D}_i$ or by the row dimension of $\mathbf{u}$) than the node tearing approach does. This is an important consideration to bear in mind when deciding to use mesh or nodal formulations.

To generalize this important concept, consider the oversimplified single-phase bus in Fig. 7.39. This bus corresponds to a hypothetical (not shown) circuit that includes a ground. At this bus disconnection point, five partitions ($p = 5$) are formed from the *same* boundary. Shown on the left is the disconnection point as seen in a mesh formulation, where the number of boundary variables $r = 1$ corresponds to the boundary voltage source u_1. On the right is the same disconnection point as seen in a nodal formulation, where the number of boundary variables $r = 4$ corresponds to the boundary current sources u_1, u_2, u_3, and u_4. The value of r in each

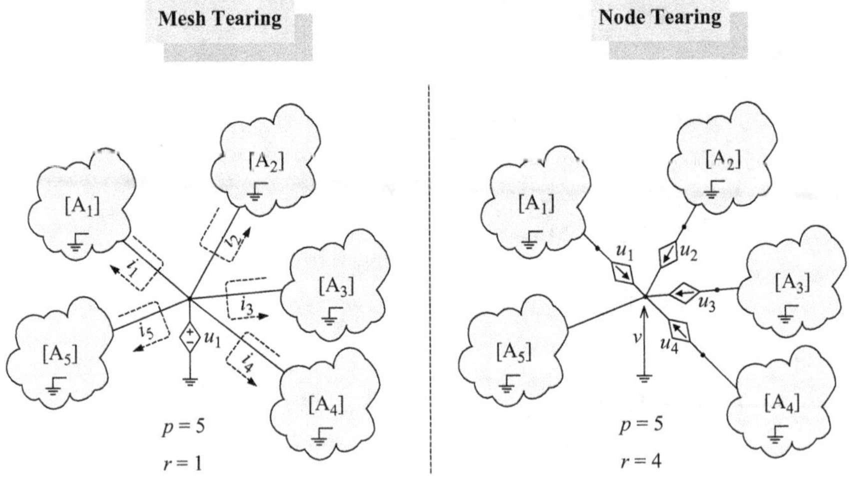

Fig. 7.39 General difference between mesh and node tearing

case is noticeably different. Furthermore, this difference in r is *exacerbated* in larger three-phase networks, such as *System* 4 (Fig. 2.1), as the frame time in each timestep eventually becomes governed by the solution time of $\mathbf{u}$ (which depends on r).

7.10 Summary

This chapter explained how to partition power system models for their simulation in parallel environments. While there are many power system partitioning techniques available (references atop), the ones covered here were diakoptics, node tearing, and mesh tearing techniques. Several graphical illustrations were provided to show the differences in partitioning. Particular details were given to show how to create $p = 2$ and $p > 2$ partitions. The latter was referred to as tearing *bus* disconnection points, which resulted in a different number of boundary variables for mesh and node tearing.

Several tearing examples were provided in this chapter to elucidate node and mesh tearing alike. Each example showed the $\mathbf{A}_i$ and $\mathbf{D}_i$ matrices of each subsystem required to solve the parallel equations summarized in Fig. 7.17. The formation of immittance matrices was the subject of Chapter 6, and was not reiterated in this chapter. Readers should be aware that the methods of Chapter 6 apply to forming immittance matrices at the subsystem level as well; these procedures were not repeated in this chapter.

Formation of the $\mathbf{D}_i$ matrices was demonstrated by graphical examples. It was shown that these matrices consist only of -1 or $+1$ values. The signs of these values relate the orientation of the each boundary source to the variables internal to each subsystem. It was also highlighted in the node tearing examples that all boundary variables must be independent.

It was mentioned that finding the best *number* of power system partitions and the locations of *where* to produce the partitions are not trivial issues. The answers to these issues are challenging to get right, but may be obtained empirically. Knowing where to tear was approximated by asking *hMetis* to make the recommendation. How well the solver performs based on the recommendation provided by *hMetis* clearly depends on how closely and how reliably the representative graph represents the power system model under consideration. This requires the user to specify vertex weights, and also to specify appropriate parameters when calling *hMetis*.[76] The influence of such parameters on the vertex set output are best understood by referring to its user manual available in Reference 192.

A graph of *System* 1 was created to demonstrate how a power system model maps to a graph, and how a partitioned graph maps back to the partitioned power system model. To do so, MTCs were mapped to graph vertices, and buses were mapped to graph edges. It was shown that buses connecting three or more MTCs map to *hyper*-edges, which is the reason why *hMetis* was chosen as a suitable outsource to partition the power system models.

It was shown that *hMetis* outputs vertex sets. These vertex sets *implicitly* revealed the locations of the disconnection points. Expertise and knowledge of the disconnection points is required to form the $\mathbf{A}_i$ and $\mathbf{D}_i$ matrices of each subsystem; but it is also a choice that requires some heuristic to get right. That is, the selection of disconnection points when partitioning large power system models is an on-going power system partitioning issue whose solution (herein) was outsourced to the well-established graph partitioning tool *hMetis*.

Finally, it was highlighted that the number of boundary variables r (i.e., order of $\mathbf{u}$) can adversely affect the performance of parallel simulations. Regarding this consideration, it becomes important to decide whether a solver will implement nodal or mesh formulations early in the software design stage. The answer to this choice is not always apparent, but its impact should be taken into consideration earlier rather than later.

[76] Readers should pay attention to the *UBFactor* parameter. This parameter specifies the permissible unbalance between graph partitions during recursive bisections.

Chapter 8
Multithreading

The previous chapter derived the equations used by the multicore solver to parallelize the simulation of the power system models (*Systems* 1–4 in Figs. 2.4–2.6 and 2.1, respectively). The examples of the previous chapter showed how to form the matrices of these equations and how to identify the disconnection points of the models. This chapter explains how to implement these equations in a solver using *C#* as the programming language. The first half of this chapter introduces the solution procedure; that is, the sequence of steps that the multicore solver takes during each timestep. The second half of this chapter provides a simple *C#* program to show how to parallelize the solution of the equations using threads,[77] and how to synchronize these threads in a "fork/join" algorithm [183]. Each thread will represent a subsystem, which implies that communication between threads will be necessary. Thread synchronization is necessary to exchange data (results) across subsystems at each timestep of the simulation, and it is an important consideration in developing multithreaded solvers.

8.1 Solution procedure

The parallel equations to solve the electrical subsystems were given in (7.21) and (7.22). To guide readers on how to solve these equations at each timestep, the equations are enumerated as substeps *a* through *e* as depicted in Fig. 8.1. This enumeration shows readers the order (or sequence) of operations the multicore solver carries out to solve these equations at each timestep of the simulation.

Referring to the principle of a "fork/join" algorithm, the multicore solver forks-out substeps *a*, *b*, and *c* and executes them in parallel. When the results of substep *c* become available, the solver joins all threads and uses the results of each electrical subsystem in substep *d*. The work executed in substep *d* is the sequential part of the solution (i.e., the "join" part) and a principal reason for diminishing speedups as the number of partitions p increases. As the order of vector $\mathbf{u}$ (the order is denoted as r) increases, so does the time spent in substep *d*.

[77] A *thread* is an independent path of code execution deemed as an asynchronous worker, slave, or agent. Threads naturally execute independently (asynchronously) of one another. To control thread execution, thread synchronization is required.

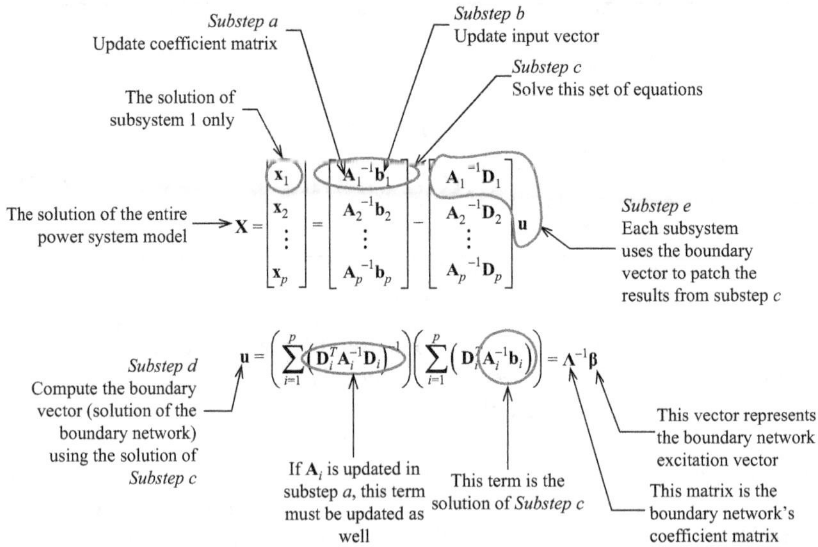

Fig. 8.1 Procedure (substeps) to solve parallel equations at each timestep (same for mesh and node tearing)

The value of r depends on the number of partitions chosen by the user, the weights assigned in the representative graph, and the formulation method (nodal or mesh). Therefore, there is uncertainty in how much time a multicore solver spends in substep d as it is an indirect function of p. If such prediction were possible, a solver could predict a good value for p and guarantee its best performance. As was discussed in the previous chapter, good values of p can be determined empirically by sweeping p until a "rule of thumb" is formed over sufficient engineering experience.

The last routine labeled as substep e patches (corrects, fixes, or compensates) the results obtained earlier in substep c using the solution of the boundary network. In this substep, the contributions of all electrical subsystems propagate across all partitions to finalize the electrical network solution at the given timestep. As the reader may recall, this partition coupling was depicted with the cross-partition arrows shown in Fig. 3.3.

The procedure illustrated by Fig. 8.1 implies (but does not show) when threads are synchronized. A more common way to represent parallel/sequential (or fork/join) work to show thread synchronization is via a *swim lane* diagram [210,212] as shown in Fig. 8.2. This diagram illustrates the substeps performed in parallel and sequentially. Each thread "swims" down its swim lane to perform work in parallel (substeps a, b, c, e, f, h, i) or sequentially (thread 1 executes substeps d and g while threads 2–4 wait). The black circles represent a thread executing a substep in parallel. The white circles indicate that a thread is sleeping (synchronized to wait) and is incurring computer dead time.

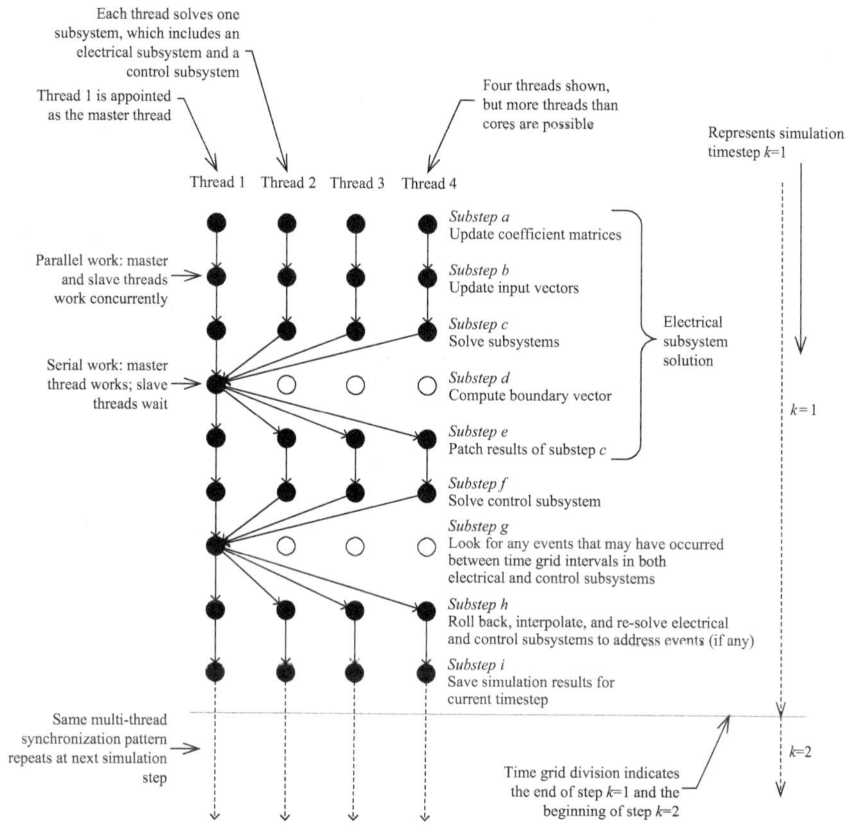

Fig. 8.2 Swim lane diagram identifying solution steps during the time loop

Readers should be aware that threads execute asynchronously (as soon as they can), can execute concurrently (in parallel) or sequentially, and are oblivious to the states of other threads. This asynchrony can easily lead to data corruption, deadlocks, contention for computer resources [22], numerical instability, program instability, and unacceptable results and unexpected program behavior. The asynchronous nature of threads suggests exercising care in the design and testing phases of multithreaded programs. Multithreaded applications are not difficult to write, but it is easy to introduce bugs and not find them until sometime later when it may become extremely difficult to debug. Another difficulty related to multithreaded applications is their maintenance. Making changes to multithreaded applications written too far back in the past is an invitation to bugs. (Proper code documentation reduces the possibility of introducing such bugs, however.)

A multithreaded solver must enforce coordinated work to prevent data corruption [56]. Of the threads spawned to perform multithreaded simulations (e.g., $p = 4$ in Fig. 8.2), one thread is appointed the *master* thread. The master thread is the thread responsible for doing *both* parallel and sequential work. *Slave* threads, on the other hand, only execute parallel work and can labor upon the master thread's command. Each of the substeps shown in Fig. 8.2 are described next.

Substep a: At the beginning of each simulation timestep, each thread updates its subsystem's $\mathbf{A}_i$ matrix if required (e.g., due to switching action). If $\mathbf{A}_i$ is updated, the coefficient matrix $\mathbf{\Lambda}$ in substep *d* must be re-factored as well.

Substep b: Each thread updates its input (excitation) vector $\mathbf{b}_i$, which includes the values of the historical sources and independent voltage/current sources. Historical branch sources are updated according to the discrete branches they belong to, whereas independent sources are updated as a function of time.

Substep c: Each thread produces an interim solution that does not account for the influence of neighbor subsystems.

Substep d: This is the bottleneck of the partitioned solution. This bottleneck is observed in several diakoptics-based [213] algorithms as well, which require thread synchronization and serial computation at each timestep. In fact, many computer scientists and engineers oppose partitioning due to this overhead and advocate increasing the efficiency of unpartitioned solutions instead. While there is reason behind this preference, a motive to partition power systems is to utilize parallel computing resources available in laptop and desktop computers, computer clusters, supercomputers, among other hardware.

Using the results provided by all threads in substep *c*, at substep *d* thread 1 computes the boundary vector **u** while threads 2, 3, and 4 idle. This idle time is counter-productive as threads 2–4 cannot continue until thread 1 releases the solution for **u**. Idle time is exercised by spinning threads shortly before they go into a "sleep" state. "Waking up" threads from a sleep state is expensive—more so when it happens often. This undesirable overhead suggests that substep *d* should constitute lightweight work when compared to substeps *a* or *c* in order to prevent threads from entering sleep states.

Substep e: After thread 1 computes **u**, each thread patches the result from substep *c*. The numerical interpretation of this substep is a solution by superposition—that is, after each thread solves the electrical network in substep *c*, the boundary variables in **u** apply counteracting electrical forces[78] to correct the solution.

Substep f: During this substep, each subsystem computes the solution of its internal control subsystem. The outputs of the control subsystem are used by the electrical subsystem during interpolation and/or during the next timestep. The reader may remember that the relationship between subsystem, electrical subsystem, and control subsystem was represented earlier in the book in Fig. 3.3.

Substep g: Events may occur inside an electrical subsystem *or* inside a control subsystem. This substep searches for possible events that may have occurred between the present and last time-grid division. The approximate time at which

[78] Reverse voltage impressions in mesh formulations and negative current injections in nodal formulations.

the first event occurs is denoted as t_z. If an event occurred between time grid divisions, the solver interpolates all partitions back to t_z [214,215], where a new interim solution is found. If no events were detected, the simulation advances forward as normal. The reader may further remember that this interpolation procedure was explained in considerable detail in section 3.2.

Substep h: At this substep, the solution of all partitions is complete. All instantaneous voltages, currents, and control outputs are saved, and the partitioned solution advances to the next time grid division.

Substep i: At this substep, the simulation data is saved to memory or disk. Saving can be expensive, and it should therefore be avoided or implemented property. If too much data is generated by a solver and all such data is needed for post-simulation analyses, the data can be persisted to disk. Persisting data to disk should not take place at each timestep. It should only take place if data buffers fill up. If data is not persisted to disk, memory depletion occurs. This substep, therefore, represents the procedure of saving data to memory arrays and not to disk.

Expanding on substep *i*, the concept of memory depletion should not be taken lightly. Time domain simulations generate large amounts of data. For example, it is common for users to want to observe voltages and currents everywhere in a model. This "greedy" requirement poses an excessive burden on solvers, and users often disregard it. Assume that a user places measurement blocks at several locations on a model to monitor instantaneous and RMS voltages and currents. Saving the instantaneous and RMS voltages and currents requires saving 12 quantities at each timestep: three instantaneous voltages, three RMS voltages, three instantaneous currents, and three RMS currents. Consider saving 12 values for 100 buses, and further that $t_{stop} = 20$ s. The total memory allocation estimated with (8.1) suggests that considerable amounts of memory may be allocated to store the data users are interested in monitoring.

$$\begin{matrix}\text{Memory}\\\text{Storage}\end{matrix} \rightarrow \underbrace{100}_{\text{buses}} \times \underbrace{12}_{\text{values}} \times \underbrace{400,000}_{\text{num. of steps}} \times \underbrace{8}_{\text{bytes}} = 3.84 \times 10^9 \approx 3.8\ \text{GB} \quad (8.1)$$

To close, this section presented the steps to solve the parallel equations of a partitioned power system model. Most of these steps are parallelizable, but some—like substeps *d* and *g*—are not. Unfortunately, serial (non-parallelizable) steps are required to exchange information across subsystems. Substeps *d* and *g* are two substeps that require synchronization, are the serial part [216] of the partitioning approach presented in this book, and cannot be parallelized.[79]

8.2 Parallel implementation in *C#*

The multicore solver developed for this book was written strictly in *Microsoft C#* 4.0: an objected-oriented, new generation language that is becoming prominent

[79] This statement is open to question because, while the solution to vector **u** *can* indeed be parallelized, it may not be efficient to do so.

in the field of scientific computing [215,217]. While the parallel equations in Fig. 8.2 are language-agnostic, the performance of the solution varies by programming language, hardware, and operating system. Many recommend legacy languages *C*, *Fortran*, and *C++* over *C#* due to their well-established and widely recognized performance. Others recommend *C#* over the legacy languages due to its effectiveness in having less lines of code, more human-readable syntax, having managed memory, the reduced effort to maintain and debug code, and the growing availability of a vast community of users and rich documentation.

The author recognizes that all languages have strengths and weaknesses, and does not advocate any one language over another. It is equally important to realize that it may be required to use third-party numerical libraries to carry out matrix and vector arithmetic. In the case of *C#*, acquiring a commercial-grade numerical library, such as *NMath* [218], enables having the low-level, hardware-optimized performance of legacy languages and reaping the benefits from the growing acceptance of *C#*.

8.2.1 NMath and Intel MKL

Developing code to produce numerical solutions requires objects for matrices and vectors. Unfortunately, *C#* does not include the classes to produce such objects. Instead, developers are expected to either develop or acquire third-party libraries *not* provided by *Microsoft*. Some of these libraries are free and some are not. While both free and paid libraries (should) produce the same numerical results, there can be significant differences in performance, amount and *quality* of online documentation, availability of human support, and in the amount of time required to debug and find answers to common problems.

The numerical library used during the development of the multicore solver for this book was *NMath* 5.2: a commercial-grade high-performance library produced by *CenterSpace Software*, Inc. *NMath* includes structured sparse matrix classes and factorizations, which are very efficient in solving the equations of electrical subsystems. Additionally, there are general matrix decompositions, least squares solutions, random number generators, Fast Fourier Transforms (FFTs), numerical integration and differentiation methods, function minimization, curve fitting, root-finding, linear and nonlinear programming. *NMath* provides object-oriented components for mathematical, engineering, scientific, technological, and financial applications on the *Microsoft .NET* platform. These components can be called from any *.NET* language, including *C#*, *Visual Basic*, and *F#*. For several computations, *NMath* uses the *Intel Math Kernel Library* (MKL), which contains highly optimized versions of the *C* and *FORTRAN* public domain computing packages BLAS (Basic Linear Algebra Subroutines) and LAPACK (Linear Algebra PACKage). This affords *NMath* performance levels comparable to *C*, and often results in performing an order of magnitude faster than non-platform-optimized implementations.

8.2.2 Program example

So far, the book has introduced and demonstrated how to formulate subsystem-level matrices (immittance matrices $\mathbf{A}_i$ and disconnection matrices $\mathbf{D}_i$), and how

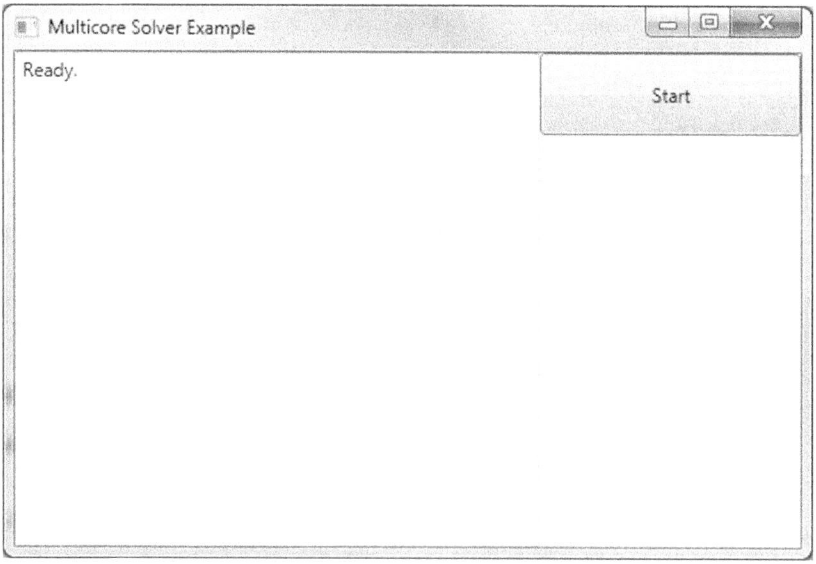

Fig. 8.3 Simplified solver UI developed in WPF

to solve the power system equations in parallel and account for thread synchronization. With this useful background in context, it is possible and pertinent to introduce now the structure of a basic *C#* multicore solver program. (Readers interested in object-oriented program design are referred to References 219 and 220.) The focus of this program is primarily on the time loop, as it is commonly the only part of interest. In passing, the entire multicore solver developed for this book has over 10,000 lines of *C#* code, which, to say the least, is more than the number of pages of this book.

Consistent with the expectation of *Windows* programs, the example program under examination includes a simple front end (user interface, UI) that calls its engine (solver). To learn how the *C#*-based multicore solver and its UI function, consider the over-simplified UI shown in Fig. 8.3 developed using XAML code in *Microsoft*'s *Windows Presentation Foundation* (WPF). This window has a text box to display solver messages and a start button to initiate the simulation. This UI is purposefully minimalistic and is meant to show only thread messages. In practice, UIs are elegant, complex, elaborate, and complete. The XAML code for this UI is shown in Snippet 8.1. Clicking on the *Start* button initiates the multicore solver and displays the messages shown in Fig. 8.4.

Snippet 8.2 shows the code that executes the multithreaded time loop. Lines 10–12 define the class fields. The fields include the number of partitions manually set to $p = 4$. As mentioned in the preceding chapter, a good (correct) value for p is not easy to forecast, but, as an alternative, this value can satisfactorily be set equal to the number of cores c a machine has as an initial guess to get one started. If the simulation performance is poor, the value of p can be adjusted.

Snippet 8.1 WPF XAML code for solver UI

```
<Window x:Class="BookMultithreadedExample.MainWindow"
        xmlns="http://schemas.microsoft.com/winfx/2006/xaml/presentation"
        xmlns:x="http://schemas.microsoft.com/winfx/2006/xaml"
        Title="Multicore Solver Example" Height="350" Width="525">
    <Grid>
        <Grid.ColumnDefinitions>
            <ColumnDefinition Width="2*" />
            <ColumnDefinition Width="*" />
        </Grid.ColumnDefinitions>
        <Grid.RowDefinitions>
            <RowDefinition Height="*" />
            <RowDefinition Height="5*" />
        </Grid.RowDefinitions>
        <TextBox Grid.Column="0"
                 Grid.RowSpan="2"
                 Name="txtMessages"
                 TextWrapping="Wrap"
                 VerticalScrollBarVisibility="Auto">Ready.</TextBox>
        <Button Grid.Column="1" Name="btnStart"
                Click="btnStart_Click">Start</Button>
    </Grid>
</Window>
```

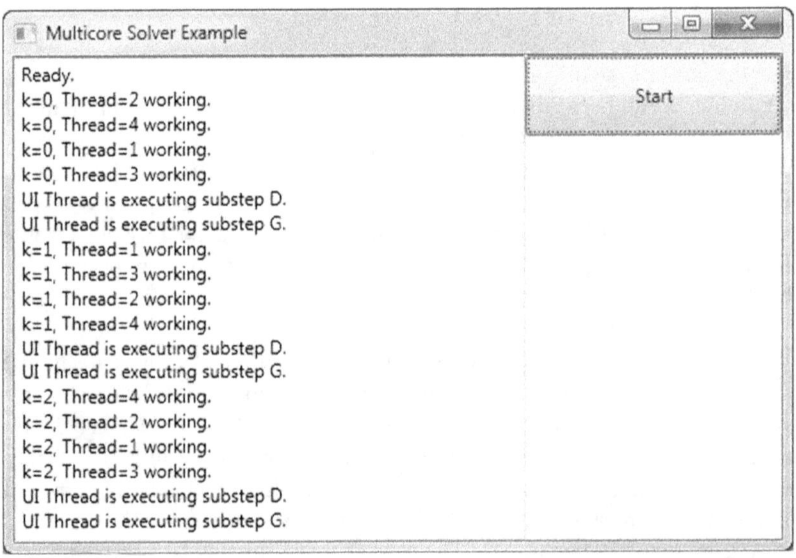

Fig. 8.4 Output messages produced by a multithreaded multicore solver over three timesteps

The *Barrier* objects are used to synchronize the master and slave threads. Referring to Fig. 8.2, a barrier was created for substeps *d* and *g* as these are the only two synchronization steps. In practice, however, there could be more synchronization steps. These two barriers are initialized in the *MainWindow* constructor in

Snippet 8.2 Example code representing multicore solver's engine

```
using System;
using System.Threading;
using System.Windows;

namespace BookMultithreadedExample
{

    public partial class MainWindow : Window
    {
        private const int numPartitions = 4;
        readonly Barrier SubstepDBarrier;
        readonly Barrier SubstepGBarrier;

        public MainWindow()
        {
            InitializeComponent();

            // Name UI thread
            Thread.CurrentThread.Name = "UI Thread";

            // Create barriers
            SubstepDBarrier =
                new Barrier(numPartitions, delegate { SubstepD(); });
            SubstepGBarrier =
                new Barrier(numPartitions, delegate { SubstepG(); });
        }

        private void btnStart_Click(object sender, RoutedEventArgs e)
        {
            for (int i = 0; i < numPartitions; i++)
            {
                // Define subsystem number
                int s = i + 1;

                // Use one thread per subsystem
                ThreadPool.QueueUserWorkItem(delegate
                    {
                        Thread.CurrentThread.Name = String.Format("Thread{0}", s);
                        TimeLoop(s);
                    });
            }
        }

        private void TimeLoop(int s)
        {
            for (int k = 0; k < 3; k++)
            {
                // Report progress
                Dispatcher.Invoke((Action)(() =>
                    {
                        txtMessages.AppendText(String.Format("\nk={0}, Thread={1} working.", k, s));
                    }));

                // SubstepA(); Readers to insert their code
                // SubstepB(); here for substeps a, b, and c.
                // SubstepC();

                // Slaves wait for master
                SubstepDBarrier.SignalAndWait();

                // SubstepE(); Readers to insert their code
                // SubstepF(); here for subtep e and f.

                // Slaves wait for master
                SubstepGBarrier.SignalAndWait();

                // SubstepH(); Readers to insert their code
                // SubstepI(); here for substeps h and i.
```

Snippet 8.2 (Continued)

```
69                 // SubstepJ();
70             }
71         }
72 
73         private void SubstepD()
74         {
75             // Report progress
76             Dispatcher.Invoke((Action)(() =>
txtMessages.AppendText(String.Format("\n{0} is executing substep D.",
Thread.CurrentThread.Name))));
77 
78             // Readers to insert their code for substep d here.
79         }
80 
81         private void SubstepG()
82         {
83             // Report progress
84             Dispatcher.Invoke((Action)(() =>
txtMessages.AppendText(String.Format("\n{0} is executing substep G.",
Thread.CurrentThread.Name))));
85 
86             // Readers to insert their code for substep g here.
87         }
88     }
89 }
```

line 14 (Snippet 8.2). The constructor of each barrier requires specifying both the number of threads or participants (e.g., $p = 4$ in this case) and the method to call while the slaves are idling. These methods are named *SubstepD*() (line 73) and *SubstepG*() (line 81), respectively. That is, at each timestep of the simulation after all slaves (and master) have reached the barrier, method *SubstepD*() (or *SubstepG*()) is called upon while the slave threads wait. This demonstrates the "fork/join" pattern mentioned earlier.

When the user clicks on the *Start* button of the UI (Fig. 8.3), the program begins execution at line 30. This code *fetches* threads from the *Windows* thread pool, assigns a name to them (i.e., "Thread *s*," where *s* is the subsystem number the thread solves), and assigns work to them as well. The work assigned to the threads is the method called *TimeLoop*() as shown on line 44 in Snippet 8.2.

Starting with *Microsoft .NET* 4.0, the *Task Parallel Library* makes it easier to spawn (or re-use) *Windows* threads via *Tasks*. Tasks are lightweight work units that can be scheduled for threads to pick up assignments. In many cases, several tasks are solved by the same thread to avoid thread context-switching overhead. In other cases, however, long-running tasks are assigned to one thread. While using tasks is the new, preferred way to create concurrent work in *Windows*-based programs, performance-profiling tools normally present analyses in terms of threads instead of tasks. This is likely so because threads are *workers* while tasks are *work*. Thus, if one is concerned with thread core-migration, time spent synchronizing (waiting), or with thread performance in general, it is then preferable to visualize performance in terms of threads instead of tasks. If one prefers to define work units and would rather comply to the new, preferred way of doing parallel work, tasks should then be considered instead. Readers should experiment with both ways of creating parallel work.

The time loop starting at line 44 is the time loop executed by *each* thread. Note that each thread solves one subsystem. For purposes of demonstration, in this time loop, the timestep counter only increments from $k = 0$ to $k = 3$. But in realistic scenarios, time loops can increase from $k = 0$ to $k = 20{,}000$ (e.g., when $\Delta t = 50$ μs and $t_{stop} = 1$ s) or more. The progress report call on line 48 shows the present timestep (k) and which thread is reporting the progress. In practice, reporting progress so frequently (at each timestep) is time consuming, inefficient, and bad practice. Instead, progress should be reported less frequently, for instance, every 1 s.

Immediately after reporting progress, each thread executes substeps a, b, and c (shown as commented functions in Fig. 8.2). As soon as each thread (1–4) finishes substep c, each thread notifies its arrival to the first barrier and *waits* at line 59. This *wait* at line 59 exemplifies thread synchronization. The barrier will not execute *SubstepD*() until all four participants (threads) notify the barrier of their arrival. When the barrier receives the four notifications, the master thread proceeds to execute *SubstepD*() on line 73 (while all other slaves wait). As seen from the output messages given in Fig. 8.4, in this example the UI thread (*UIThread)* executed *SubstepD*() due to the dispatcher call. The dispatcher call places work in the UI thread's work queue, which should *never* undergo heavy lifting.

In practice, the master thread executing *SubstepD*() is chosen by the *Windows* thread scheduler based on resource availability. However, referring to the swim lane diagram in Fig. 8.2 it was suggested that thread 1 executes *SubstepD*(). This is not guaranteed unless additional (unnecessary) thread coordination logic is added to ensure thread 1 is the master thread. In practice, which thread number is used as the master thread is unimportant. The purpose of assuming thread 1 was the master thread in Fig. 8.2 was to pedagogically illustrate the concepts and relationship between master and slave threads and their synchronization. However, in practice is better to let *Windows* appoint the master thread to execute the serial substeps.

After *SubstepD*() completes, the parallel loop releases the threads waiting at line 59. Each waiting thread proceeds onto lines 61 and 62 before they reach the next barrier on line 65. When the second barrier is reached, like for the first barrier, all threads await for *SubstepG*() to complete. Execution of *SubstepG*() is also determined by the *Windows* thread scheduler. When *SubstepG*() completes, each waiting thread continues onto lines 67–69. As each individual thread reaches the end of the time loop, each commences the next time loop iteration by incrementing its value of k by one. Notice there is no synchronization to enter the next time loop iteration; the threads enter the next loop iteration independently and do not join again until they reach the first barrier (*SubstepD*()). The fork/join process is repeated until the time loop reaches t_{stop}. It should be stressed that, in this particular example, the program ends after it completes loop iteration $k = 2$.

By now, it should be clear that barriers are "necessary evils." Partitioning a power system does not decouple it. Partitioning a power system changes (does not remove) the numerical coupling such that each subsystem is coupled to the boundary network rather than being coupled to other subsystems. Readers developing a multi-core solver are encouraged to time the average time spent in each substep. Substep timing information is useful to find bottlenecks when using a different number of partitions, and besides it is useful to find unusual timings that may reveal bottlenecks.

Lastly, profiling tools often come handy and are very functional, but they are not always free. If a profiling tool, such as the concurrency visualizer in *Microsoft Visual Studio Ultimate* or *JetBrains dotTrace,* is not available, it should be clear to the reader by now that the compute-intense part of a power system solver is the time loop. To end, readers are encouraged to measure the performance of their solvers using profiling tools. If these tools are not available, readers can profile their solvers (freely) by measuring the time spent in each substep using the *System.Diagnostics.Stopwatch* object available in *.NET*.

8.3 Summary

This chapter presented the sequence of steps to follow when solving the parallel equations introduced in the previous chapter. The swim lane diagram identified the substeps performed in parallel and sequentially, and showed *when*, during a timestep, the threads required synchronization.

The numeric arithmetic internal to a solver involves the use of matrices and vectors. While *C#* does not include libraries with such capabilities, many numerical libraries exist to complement the available classes in the *.NET* framework. *NMath*, for instance, is one numerical library that performs well and was used for this book.

The structure of a simple multicore solver was presented and analyzed using *C#* as the implementation program language. Snippet 8.2 presented this code to show a possible implementation of the theory discussed in this book. The example program structure illustrated was particularly important because it suggested which *.NET* synchronization structure works well with the fork/join algorithm required by the partitioning method (node or mesh tearing). This program also showed code to fetch (re-use) threads from the *Windows* pool.

Finally, the determination to use the *Barrier* class came about after much testing, investigation, and performance evaluation (by the author and by others [221] as well). It is also possible that future releases of *.NET* may include coordination approaches that supersede the *Barrier* class. Readers are encouraged to start from the example program structure provided in this chapter and progressively grow into a more meaningful program by filling-in the commented substep routines with working code.

Chapter 9
Performance analysis

The sequence of steps taken by the multicore solver when solving the parallel equations presented in Chapter 7 were examined in Chapter 8; the solution procedure was carried out using Microsoft *C#* as the programming language. This chapter evaluates runtimes as power systems are partitioned several times. Chapter 9 also evaluates the potential speedups of using node and mesh tearing—and of other metrics as well—of simulating *Systems* 1, 2, 3, and 4 with the multicore solver developed for this book. Such multicore solver, written entirely in *C#*, implements all concepts introduced in earlier chapters. In other words, this multicore solver:

- simulates the notional shipboard power system model (and its variants) presented in Chapter 2,
- implements the time domain concepts presented in Chapter 3,
- discretizes the power apparatus using the root-matching technique presented in Chapter 4,
- uses the power apparatus and control block models introduced in Chapter 5,
- formulates the subsystem-level matrices as presented in Chapter 6,
- produces representative power system graphs and partitions them by calling *hMetis* as explained in Chapter 7, and
- uses the numerical library and multithreaded fork/join program structure suggested in Chapter 8.

The stop time of the simulations was set to $t_{stop} = 1$ s and the timestep to $\Delta t = 46.296$ μs. Regarding this choice of Δt, the power frequency of the three-phase sources was fixed at 60 Hz and the PWM carrier frequencies at 1.8 kHz as suggested by the compatible frequencies in Appendix A.

The hardware specifications and the 10 different software programs used for the development of the multicore solver are listed in Tables 9.1 and 9.2, respectively.

The order in which the software listed in Table 9.2 were used (and their relations to one another) is as follows. After compiling sufficient data[80] for the power system models, the next step is to (incrementally[81]) build the power system models in

[80] Often a time-consuming and laborious task that can require man-months to man-years to finish depending on whether or not the data is available and whether or not their bearers are willing and ready to share it.
[81] By this is meant that it is good practice to create smaller variants before building the final notional power system model.

Table 9.1 Multicore computer used for performance analysis

Computer item	Hardware specification
Brand/Model	Dell/Precision T7500
Memory (RAM)	12 GB
Operating system (OS)	Windows 7 (64-bit) with Service Pack 1
Processor	Intel Xeon E5630, 2.53 GHz, (quad-core)
Number of cores	4
Computer type	desktop

Table 9.2 Manufacturer, name, version, and description of software used for the development of the multicore solver

Software manufacturer	Software name	Version	Description
CenterSpace Software, Inc.	*NMath*	5.2	Numerical library used by multicore solver to invoke high-performance numerical routines not available in *.NET*
Intel	*Math Kernel Library*	10.3	Hardware-optimized math library invoked by *NMath*
Karypis Lab, University of Minnesota	*hMetis*	1.5.3	Graph partitioning software
Microsoft	*.NET*	4.0	Development platform for building *Windows* apps, which supports various programming languages, and includes the common language runtime and an extensive class library
Microsoft	*C#*	4.0	Main development programming language
Microsoft	*Visual Studio*	2010	Integrated Development Environment (IDE)
Microsoft	*Windows*	7	Operating system on multicore computer
The MathWorks, Inc.	*MATLAB*	2012a	General-purpose mathematical environment
The MathWorks, Inc.	*SimPowerSystems*	5.6	Power apparatus model library (block set) to create power system models in *Simulink*
The MathWorks, Inc.	*Simulink*	7.9	Graphical environment integrated with *MATLAB*

MATLAB/Simulink using the *SimPowerSystems* blockset. Once the models are sufficiently parameterized and reach credible behavior, the development of the external multicore solver (an .exe file) can commence.

To develop a multicore solver in *C#* one uses *Visual Studio* as the main integrated development environment (IDE). A robust IDE facilitates program authoring and, most importantly, program *debugging*, where developers invest most of their time. Developing a *C#* program implies referencing *Microsoft .NET* assemblies (.dll files) and consuming the classes available to developers. *Microsoft .NET* is a large commonwealth of fully documented classes that are expected to be utilized in *.NET*-compatible languages such as *C#*. Unfortunately, *.NET* does *not* have classes to represent sparse matrices nor matrix arithmetic.[82] Thus, readers seeking to develop high-performance scientific applications in *C#* should use third-party numerical libraries. The solver developed for this book uses *NMath* (introduced in Chapter 8). Internally, *NMath* calls *Intel*'s *Math Kernel Library*: a low-level hardware-optimized numerical library that exhibits performance comparable to *C* and *Fortran* programs.

Parallelizing power system simulation equations requires *partitioning* the intended power system model *a priori* according to its representative graph. Admittedly, producing graphs to represent power system models is challenging. Forming an *adequate* graph is responsibility of the solver developer; partitioning this graph may be conveniently outsourced to *hMetis*. Graph-partitioning software *hMetis* recommends the best disconnection points based on how well its input graph represents the power system model and the settings specified by the user (e.g., *UBFactor*). On completion of the graph partitioning stage by calling *hMetis*, the multicore solver interprets the disconnection points recommended by *hMetis* in order to define where the subsystem boundaries are.[83]

9.1 Performance metrics

The six performance metrics presented in this chapter to benchmark the multicore solver under examination are as follows:

1. Speedup
2. Runtime
3. Frame time
4. Number of boundary variables
5. Subsystem order
6. Number of non-zeros

Speedup

The first performance metric is speedup. Speedup is defined [222] as the ratio of unpartitioned (t_{unp}) to partitioned (t_{part}) simulation runtime, where runtime was

[82] Large multi-dimensional arrays in *C#* do not perform well in time-critical scientific applications.

[83] *hMetis* outputs one vertex set per partition. Each vertex set corresponds to the power apparatus contained in one partition. The power apparatus contained in one partition implicitly delimits the electrical subsystem's boundary.

defined earlier (see Table 2.6) as the *solution* time of the time loop (not as the total runtime including initialization time). Speedup is computed with (9.1) [210]. For a partitioned simulation to be considered useful, the speedup value should be (much) larger than unity.

$$\text{Speedup} = \frac{t_{unp}}{t_{part}} \tag{9.1}$$

t_{unp} = unpartitioned simulation runtime experienced in *Simulink* (seconds)
t_{part} = partitioned simulation runtime experienced when using the multicore solver developed for this book (seconds)

There is a subtle caveat in (9.1): it has to do with the numerator t_{unp}. Readers should be aware that it is common in the literature to find t_{unp} as the runtime of the unpartitioned version of a custom-made solver. There are two shortcomings in using such value for t_{unp}. The first is that it hinders the ability to relate published results against commercial simulation tools users might have. The second is that, unless developers spend an unbiased amount of time and material resources into speeding-up unpartitioned simulations, the speedup metric in (9.1) will be *effortlessly* high due to a (convenient) large value of t_{unp}.

In this book, t_{unp} is chosen as the time it takes for *Simulink* to complete a simulation run using: its fixed-step discrete solver, the backward Euler integration method, a timestep of $\Delta t = 46.296$ µs, and running under a normal simulation mode. As readers may know, *MATLAB/Simulink* has a well-established and respected (rightfully earned) reputation, exhibits commercial-grade performance, and is a preferred modeling tool in the research community worldwide. Since the author is not involved in the development of *Simulink*, and by assuming that the developers of *Simulink* have invested a considerable amount of resources into optimizing its performance, it may then be more fair to use *Simulink*'s runtime as t_{unp} and the multicore solver's runtime as t_{part}. Stated differently, assuming the developers of *Simulink* strive to minimize t_{unp}, and knowing that the author also strives to minimize the denominator t_{part}, the ratio of these two minimized quantities is intended to be an unbiased metric.

It should be noticed that the results presented in this chapter were averaged over several runs, and that they are specific to the machine operating conditions at the time the simulations took place. Because the machine that conducted the simulations was an ordinary *Windows* desktop machine (see Table 9.1) running simultaneously other background programs, some variance in the simulation runtimes was eliminated by averaging the runtimes of each run to produce a final result.

The specific speedups results reported here are reference values only, and they should not be generalized to any other shipboard power system model. That is, the various results presented in this chapter are not applicable to all cases in general. Remember, there are many other models that run faster in *Simulink* than they do with the multicore solver. In fact, for most daily simulation tasks, the author himself uses *Simulink* as his primary and preferred modeling tool. However, the author uses the multicore solver developed for this book to design, parameterize, simulate, and study large and complex models that are too time consuming to run repeatedly.

Runtime

The second performance metric is runtime. Runtime (as defined here) is the time it takes for the time loop to finish, not the total wait time users experience after clicking *Start*. (This runtime was termed *solution* time in Table 2.6.) The runtime, then, is the time measured from the moment the time loop starts its execution at $t = 0$ s to the time the simulation time reaches $t > t_{stop}$. In contrast to the speedup metric, the runtime metric provides the "wall-clock" time to help determine whether the approach carried out in this book verifies the early proposition that low-order, sparse, and partitioned simulations implemented on a multicore desktop computer significantly reduces the runtime of power system simulations without having to acquire new equipment and power system simulation software. In addition, this metric helps compare the methodology implemented herein with that adopted by the reader.

Frame time

The third performance metric is frame time. The frame time is the *average* time (ms) it takes for the solver to advance one timestep. This value is computed as the ratio dh/dk, where dh is the number of milliseconds (ms) the solver takes to advance dk number of steps. This relation is useful to determine how much time a solver spends in each timestep when compared to a real time solver. For example, in a simulation where $\Delta t = 100$ μs, a real time solver guarantees it takes $<\Delta t$ of wall-clock time to advance each timestep. An offline solver, on the other hand, offers no guarantee and can spend a variable amount of wall-clock time in each timestep. Essentially, the frame time shows how close (or how far) the multicore solver is to matching the performance of a real time solver.

Number of boundary variables

The fourth performance metric is the number of boundary variables. As the number of partitions (denoted p) of a power system is increased, so does the size (order) of the boundary network that forms as a result of introducing disconnection points. As seen earlier, the boundary network order (denoted r) is equal to the column dimension of any $\mathbf{D}_i$ matrix, or to the row dimension of the vector $\mathbf{u}$. To reduce the frame time overhead incurred by parallelizing power system simulations (an important issue), the number of boundary variables r should remain small. This metric, therefore, shows how a partitioned simulation's overhead increases with p.

The percent-ratio of r to the average subsystem order n gives a measure of how large the boundary network is with respect to the average subsystem size (in terms of equation count). A situation where $r > n$ indicates that the power system partitions are fine-grained and will result in the solver spending too much time solving the boundary network. On the other hand, the reverse where $r < n$ indicates that the power system partitions are coarse-grained, and that they will more likely result in noticeable speedups. It is of importance then to underscore that the time spent solving the boundary network is critical, and that it has a significant impact on the overall performance of fork/join algorithms in general. Thus, runtimes and their associated r/n ratios (%) are useful criteria to determine whether simulations should be executed as fine- or coarse-grained.

Subsystem order

The fifth performance metric is the subsystem order. Depending on where in the power system model the disconnection points are located, the subsystems will present a different number of equations (order) to the matrix solver (e.g., to *NMath*). Since the number of equations in each subsystem is (typically) different, the average subsystem order n gives the average equation-count across all subsystems. This metric provides a rough estimate of the computational burden experienced by each thread. As will be shown later, the ratio r/n can help estimate a good value for p.

Number of non-zeros

The sixth performance metric is the number of non-zeros. The number of non-zero entries in an electrical network coefficient matrix is a strong indicator of computational burden. (Examples of matrix structures were shown in Figs. 7.9 and 7.16). Although not covered in this book, the computational efficiency of a *sparse* matrix solver (e.g., Intel's MKL) depends on the number of non-zeros in the resulting coefficient matrix factors. In contrast to the average subsystem order n, which is more amenable to humans, sparse matrix solvers are sensitive to the number of non-zeros rather than to the number of equations (although they are loosely related). In fact, to this end, there are several well-known algorithms (such as Tinney's [46,185]) that reduce the number of non-zeros of the resulting factors by pre-ordering the rows and columns of coefficient matrices. Observing the number of non-zeros in the matrices gives an indication as to how the work of the sparse matrix solvers falls as p rises.

9.2 Benchmark results and analysis

The results of benchmarking *Systems* 1, 2, 3, and 4 are presented in this section. Each result set includes concluding remarks. After presenting and discussing these results, an overall summary of results is presented in section 9.4 following the discussion of *System* 4 below.

9.2.1 System 1

The benchmark results and system order information for *System* 1 using the multicore solver are shown in tabular form in Table 9.3 and graphically in the six charts illustrated in Figs. 9.1 and 9.2. Referring to the speedup chart in Fig. 9.1, the nodal formulation reached its maximum speedup of 5.2 at $p = 2$ while the mesh formulation reached it at $p = 4$. Additionally, inspection of the runtime magnitudes shown in Table 9.3 suggests that the runtime was 1.6 s for both the nodal (at $p = 2$) and mesh cases (at $p = 4$), while in *Simulink* it was 8.3537 s (shown below the table of runtimes). Since the best runtime of 1.6 s was observed for both the nodal and mesh cases, neither method offers any clear advantage over the other when parallelizing *System* 1. However, since these runtimes were averaged over several runs, it should be pointed out that marginal runtime differences do exist between runs, and they can be attributed to background processes discussed earlier. During runtime, it is likely

Table 9.3 Benchmark and subsystem size information for System 1

Speedup		
Num. partitions	**Mesh**	**Nodal**
2	4.9	5.2
4	5.2	3.5
6	2.3	2.3
8	1.9	2.1
10	1.7	1.9
12	1.5	1.7

Number of boundary variables (r)				
Num. partitions	**Mesh**	**Nodal**	**Mesh ratio (%)**	**Nodal ratio (%)**
2	6	6	23	19
4	12	15	86	83
6	8	21	89	162
8	22	30	275	273
10	20	33	286	367
12	24	39	400	488

Runtime (s)		
Num. partitions	**Mesh**	**Nodal**
2	1.7	1.6
4	1.6	2.4
6	3.6	3.6
8	4.4	3.9
10	4.8	4.4
12	5.4	4.9

Simulink runtime: 8.3537 s

Subsystem order (n)			
Num. partitions	**Mesh**	**Nodal**	**Average**
2	26	32	29
4	14	18	16
6	9	13	11
8	8	11	10
10	7	9	8
12	6	8	7

Simulink number of state variables: 44

Frame time (average, ms)		
Num. partitions	**Mesh**	**Nodal**
2	0.07	0.06
4	0.07	0.11
6	0.16	0.16
8	0.20	0.17
10	0.22	0.20
12	0.24	0.22

Simulink frame time: 0.2265 s

Number of non-zeros				
Num. partitions	**Mesh**	**Nodal**	**Average**	**Sparsity (%)**
2	72	83	78	91
4	36	44	40	84
6	24	30	27	78
8	19	24	22	76
10	16	19	18	73
12	12	17	15	70

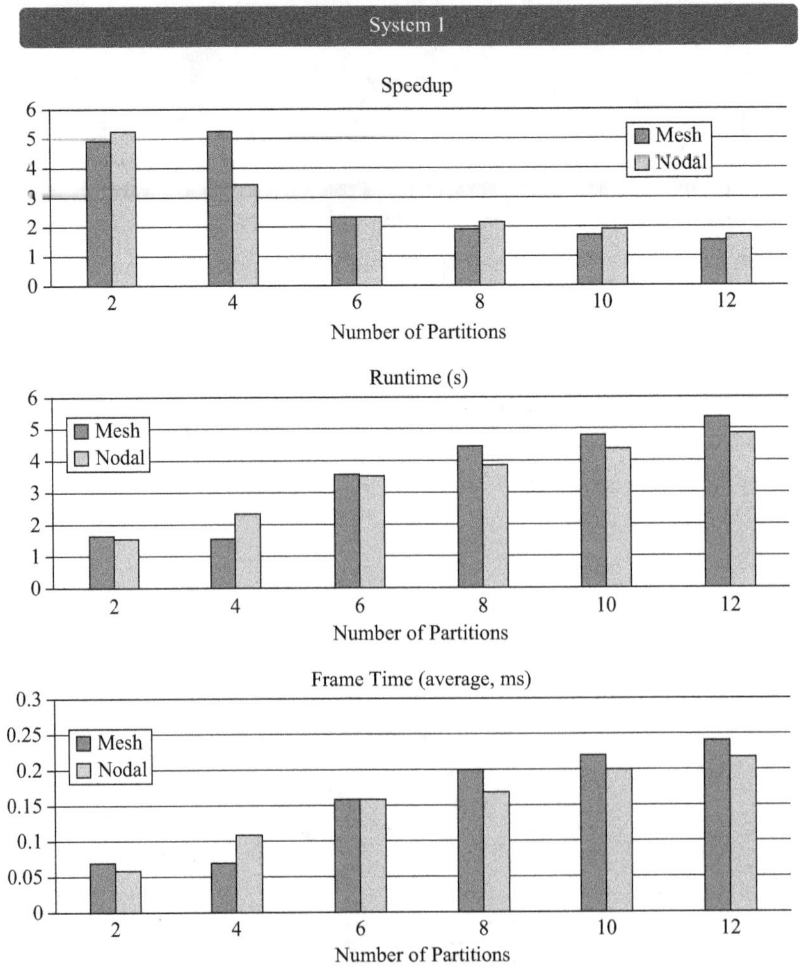

Fig. 9.1 Benchmark results for System 1

that background operating-system events may result in an unfair processor sharing time. For example, inadvertently checking email or allowing antivirus scans during a simulation would adversely affect these results. In addition, *Windows*-based machines are not dedicated machines and determinism should not be expected.

Rather than focusing on the speedup for this smaller power system, it may be more relevant to ask: *was it necessary to partition System* 1*?* The answer is *no*. Looking at the runtime for *Simulink*, *System* 1 executes rather quickly in *Simulink*, and it does not significantly benefit from partitioning. In such cases, the time invested in developing a multicore solver to reduce runtime by a few seconds is not justified. The objective of partitioning this model was to show the poor runtime gains and the

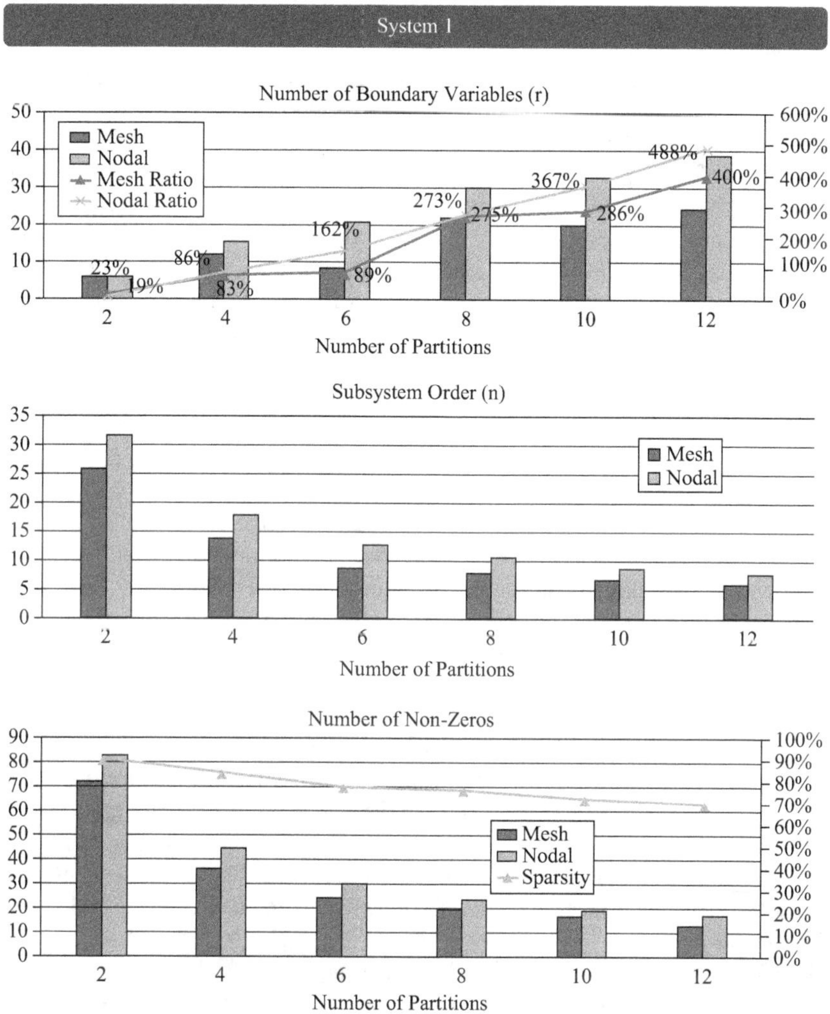

Fig. 9.2 Order information for System 1

speedup *progression* from a simple power system model (like *System* 1 under analysis) to a more complex one (like *System* 4 to be explained later). This progression or tendency should provide readers with a pattern of how speedups may vary with system size and complexity.

With respect to the frame time chart in Fig. 9.1, it can be gathered that the average time spent on each substep in the mesh and nodal formulations is similar. This is an indication that the work in both formulations is nearly the same, and that the choice of formulation method can be subjective in this case. What can also be gathered from the chart is that the frame time varies in similar fashion in both methods as the number of partitions increases through $p = 12$.

Regarding the number of boundary variables shown as columns in the top chart of Fig. 9.2, it is noted that the mesh method keeps a noticeable low r in comparison to the nodal method. The upward sloping lines in the top chart of Fig. 9.2 show r as a percentage of the average subsystem order n for both the mesh and nodal methods. Comparing the r/n percentages, it is interesting to discover that in the nodal case this ratio surpassed 100% when $p = 6$. This implies that for a small model such as *System* 1, more time was spent by solving the boundary network (substep d) than solving the subsystems (substep c). This leads to an important observation: in nodal formulations, as p increases, r more rapidly approaches n than it does in the mesh method. This percentage metric may be a useful indicator to estimate a good value for p; however, engineering experience and judgment cannot be put aside.

The subsystem order chart in the center of Fig. 9.2 and its corresponding values in Table 9.3 show that the average subsystem order n seen by each thread is inversely proportional to p as expected. With respect to the number of state variables achieved in *Simulink* (44, shown at the bottom of the center table included in Table 9.3), the total number of nodes and meshes is still large; however, as the table illustrates, partitioning reduces the work required from each thread.

Referring to the number of non-zeros chart in Fig. 9.2, the columns show how the number of non-zeros in $\mathbf{A}_i$ for each formulation method varies as p increases. As each subsystem matrix ($\mathbf{A}_i$) gets smaller, so does the number of non-zeros. The average of the non-zero counts for both formulations is expressed as average sparsity (%) to show a decaying trend toward the right-hand side of the chart. It is noticed that sparsity reduces at first ($p = 6$) before leveling off. However, the runtimes did not follow this same trend: they kept increasing instead, as Table 9.3 and Fig. 9.1 show.

In small models such as *System* 1, the overhead of partitioning can rapidly dominate simulation runtime even if sparsity is sustained. The partitioning overhead implies thread synchronization time, shared-memory data exchange, and computation of the boundary network. Referring to the high r/n percentages shown in Table 9.3 and in the top chart of Fig. 9.2, it is suggested that the time spent solving the boundary network is quite impactful and cannot be counteracted by sparsity.

It also recognized that although sparsity shows a falling progression (lower chart in Fig. 9.2), not by virtue of having more decoupled equation sets, but by having immittance matrices of reducing order.[84] Another interesting result gathered from the number of non-zeros counts shown is that even for small models as *System* 1, the mesh method produces sparsity comparable to the nodal method.

9.2.2 System 2

The benchmark results and system order information for *System* 2 using the multicore solver are shown in tabular form in Table 9.4 and graphically in the six charts illustrated Figs. 9.3 and 9.4. Referring to the speedup chart in Fig. 9.3, the nodal and

[84] As a simple and extreme example, the maximum sparsity of a non-singular 3×3 system is only 66%.

Table 9.4 Benchmark and subsystem size information for System 2

Speedup		
Num. partitions	Mesh	Nodal
2	7.2	8.3
4	8.7	8.8
6	6.4	6.5
8	5.4	5.9
10	5.2	4.7
12	4.4	3.0

Number of boundary variables (r)				
Num. partitions	Mesh	Nodal	Mesh ratio (%)	Nodal ratio (%)
2	4	6	3	5
4	8	21	13	34
6	14	33	34	79
8	18	36	58	113
10	18	48	72	178
12	24	66	400	275

Runtime (s)		
Num. partitions	Mesh	Nodal
2	12.9	11.2
4	10.7	10.5
6	14.5	14.3
8	17.1	15.7
10	17.8	19.6
12	21.2	30.6

Simulink runtime: 92.7 s

Subsystem order (n)			
Num. partitions	Mesh	Nodal	Average
2	118	114	116
4	60	61	61
6	41	42	42
8	31	32	32
10	25	27	26
12	6	24	15

Simulink number of state variables: 215

Frame time (average, ms)		
Num. partitions	Mesh	Nodal
2	0.56	0.50
4	0.49	0.48
6	0.66	0.65
8	0.78	0.72
10	0.81	0.89
12	0.97	1.18

Simulink frame time: 4.1 ms

Number of non-zeros				
Num. partitions	Mesh	Nodal	Average	Sparsity (%)
2	387	338	363	97
4	184	173	179	95
6	118	117	118	93
8	90	88	89	91
10	70	72	71	89
12	12	61	37	84

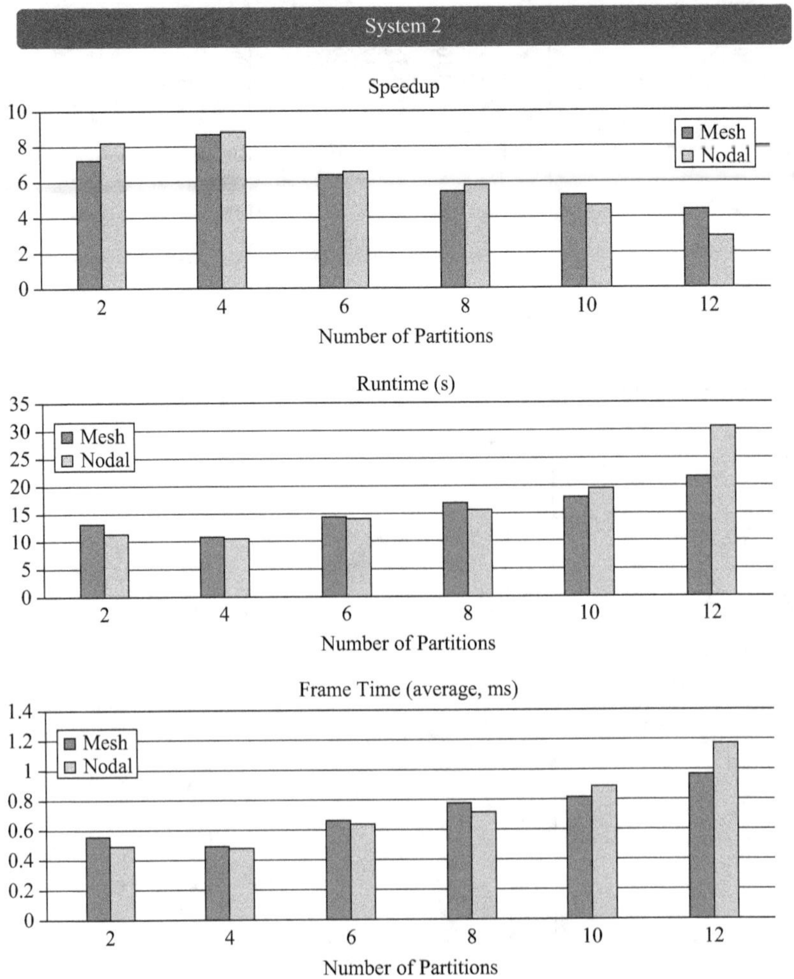

Fig. 9.3 Benchmark results for System 2

mesh formulations show a relatively similar performance for all p. But for both cases, the best performance was when $p = 4$. The speedup decay beyond $p > 4$ is closely similar for both formulations, which indicates that for this system size and complexity neither method exhibits an advantage.

Referring to the runtime values displayed in seconds in Table 9.4, *Simulink* took a little over a minute and a half to complete the simulation. This runtime is also acceptable for a single run, but readers should remember that in practice dozens (even hundreds) of simulations may be conducted while designing or parameterizing a model. Often times models are re-run each time a parameter changes to see if abnormal behavior is introduced as a result of a change. These re-runs and changes,

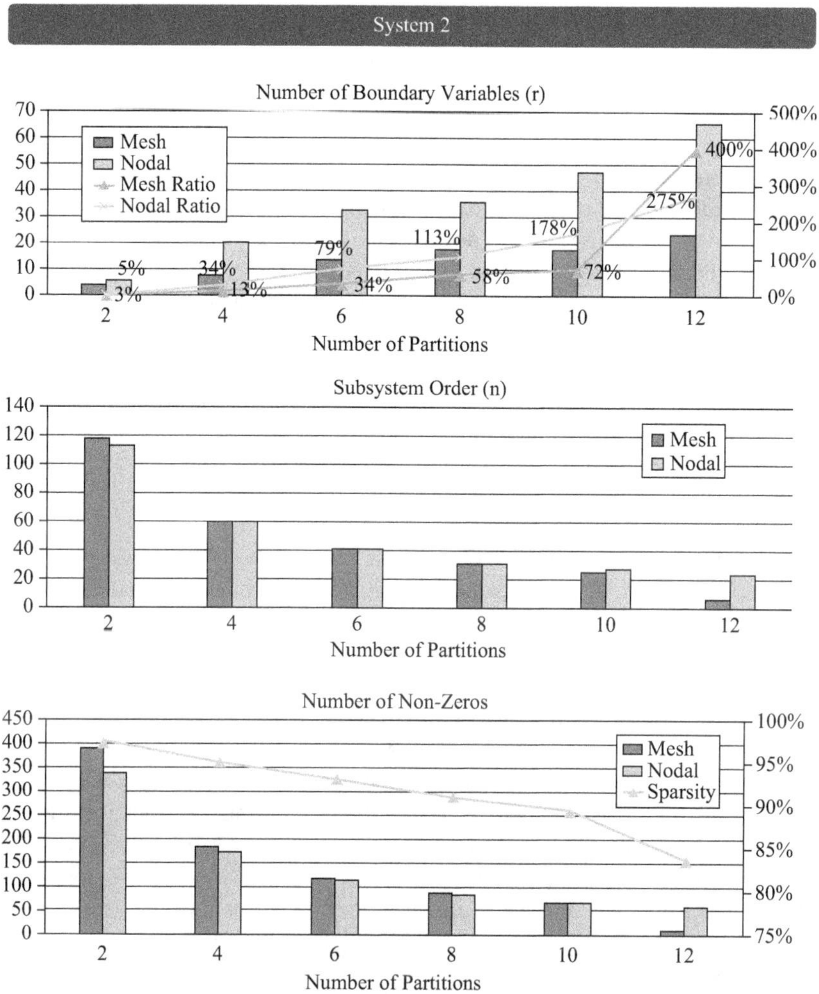

Fig. 9.4 Order information for System 2

in turn, can be (or become) a tedious process due to the wait times. It should be underlined that the worthiness of partitioning depends on the model: at times speedups are negligible; and at other times they are significant. Partitioning is beneficial if it allows researchers to conduct more case studies per day. It is re-emphasized anew that when simulations do *not* have to be run frequently,[85] then waiting a few minutes, or a few hours, for a single run might be (relatively) acceptable and may not (relatively speaking) warrant the development of a multicore solver.

[85] Frequently here means several times per day. Infrequently may mean once per week or longer.

Similar to the speed and runtime charts depicted and analyzed above, the frame time chart in Fig. 9.3 shows that the average time spent on each substep in the mesh and nodal formulations is similar at each p. This similarity is another indication that the both formulation methods appear to perform equally well.

Referring to the number of boundary variables chart in Fig. 9.4, the columns clearly indicate that r grows much more rapidly in the nodal case than it does in the mesh case. Additionally, the r/n ratios (ascending lines) for the mesh case show that the boundary network size remains small when compared to the average subsystem size. In the nodal case, however, this is not so. For example, consider the nodal case for $p = 8$; at this value of p, the boundary network size was 113% of the average subsystem size, which suggests that the boundary network is likely larger than any one electrical subsystem (in terms of equation count). For the mesh case, on the other hand, the boundary network size does not surpass the average electrical subsystem size until $p = 12$. Nevertheless, for a model size such as *System* 2, the rapid growth of r in the nodal case did not appear to be impactful as the nodal speedups are comparable to the mesh ones.

The average[86] number of equations that each formulation presents to the solver are shown in the second chart of Fig. 9.4. This chart shows that the average number of meshes is similar to the average number of nodes, and that both methods produce similar equation counts. This, however, is not true of all system models. It is trivial to reduce mesh counts by, for example, removing line capacitances from all cables. But by engaging in such removals one would unfairly favor mesh tearing over node tearing, which was purposefully not studied here.

The average number of non-zeros is shown at the bottom chart of Fig. 9.4. For $p = 2$, the number of non-zeros for the mesh count exceeds the count for the nodal formulation. This occurs when meshes intersect at shunt impedances common to various power apparatus. For example, if several cables including line capacitances are all interconnected at the same bus, it produces dense regions in the mesh resistance matrix. Dense regions in a coefficient matrix increases the number of non-zeros, which is not desirable. Additional considerations of mesh and nodal formulations are given in Appendix B.

9.2.3 *System 3*

The benchmark results and system order information for *System* 3 using the multicore solver are shown in tabular form in Table 9.5 and graphically in the six charts illustrated in Figs. 9.5 and 9.6. Referring to the speedup chart in Fig. 9.5, the nodal and mesh formulations showed the best performance when $p = 4$. Additionally, two peculiar patterns are noticed. The first is that speedups first increase and then decrease; this occurs for most models that warrant partitioning, which are mostly larger models rather than smaller ones. Second, the maximum speedups occur when $p = c$. Interestingly enough, this occurrence has also been the case from experiences in running parallel simulations on quad-core desktop computers. Observations of such recurrent

[86] The *average* equation count is used because each subsystem does have a different number of equations.

Table 9.5 Benchmark and subsystem size information for System 3

Speedup		
Num. partitions	Mesh	Nodal
2	14.4	20.3
4	25.7	28.9
6	19.9	21.2
8	17.0	16.4
10	15.8	17.7
12	14.1	14.0

Number of boundary variables (r)				
Num. partitions	Mesh	Nodal	Mesh ratio (%)	Nodal ratio (%)
2	6	9	1	2
4	10	33	5	16
6	12	54	8	38
8	14	81	13	73
10	18	84	20	94
12	16	120	22	154

Runtime (s)		
Num. partitions	Mesh	Nodal
2	14.3	10.1
4	8.0	7.1
6	10.3	9.7
8	12.1	12.5
10	13.0	11.6
12	14.6	14.7

Simulink runtime: 205.3 s

Subsystem order (n)			
Num. partitions	Mesh	Nodal	Average
2	434	410	422
4	218	211	215
6	145	144	145
8	109	111	110
10	88	89	89
12	73	78	76

Simulink number of state variables: 873

Frame Time (average, ms)		
Num. partitions	Mesh	Nodal
2	0.57	0.39
4	0.32	0.29
6	0.42	0.41
8	0.51	0.55
10	0.55	0.50
12	0.62	0.64

Simulink frame time: 8.6 ms

Number of non-zeros				
Num. partitions	Mesh	Nodal	Average	Sparsity (%)
2	1504	1280	1392	99
4	737	646	692	98
6	480	434	457	98
8	352	329	341	97
10	278	264	271	97
12	235	223	229	96

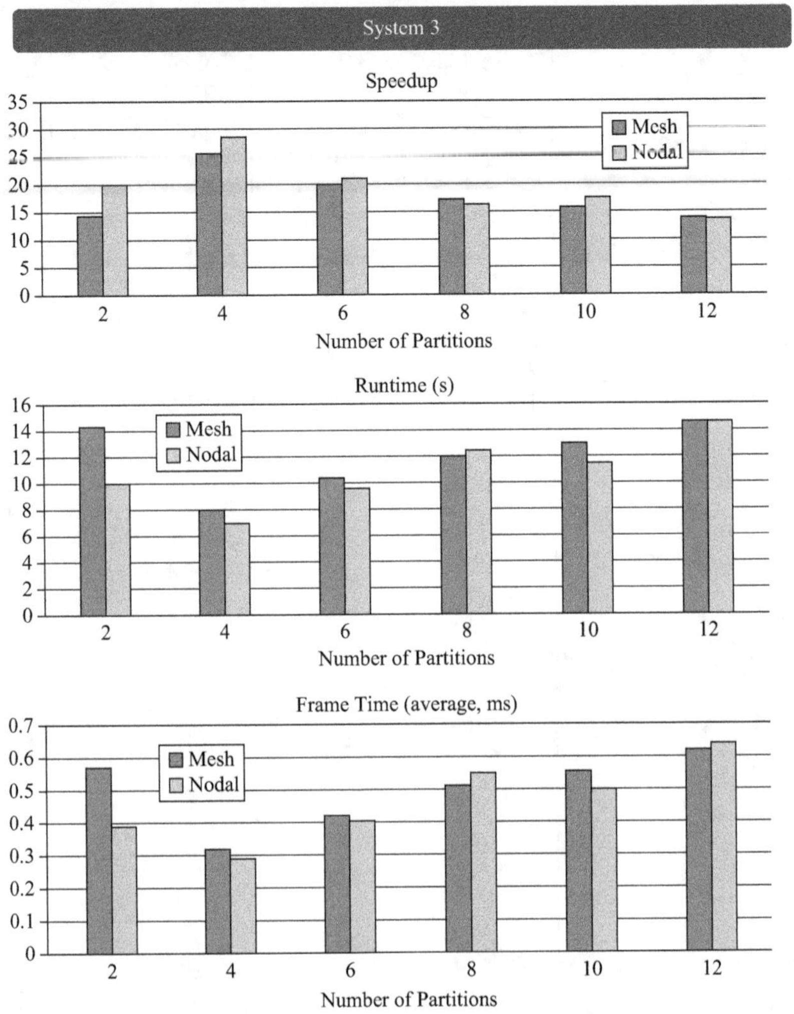

Fig. 9.5 Benchmark results for System 3

patterns constitute simulation engineering experience which can be ported to computer programs to predict reliable good values for p. Comparing the speedups of *System* 3 (first increasing, then decreasing) against those for *System* 1 (only decreasing) suggests a system may not be complex enough for partitioning if its speedup progression only decreases.

Referring to the runtime values shown on the center left-hand side of Table 9.5, *Simulink* took only a few minutes (205.3 s) to complete this simulation. While this runtime is acceptable for one run, the design of a power system model as large as *System* 3 is time consuming due to the number of buses and power apparatus that need to be asserted for credible results (e.g., voltage levels and power flows). In general, it is

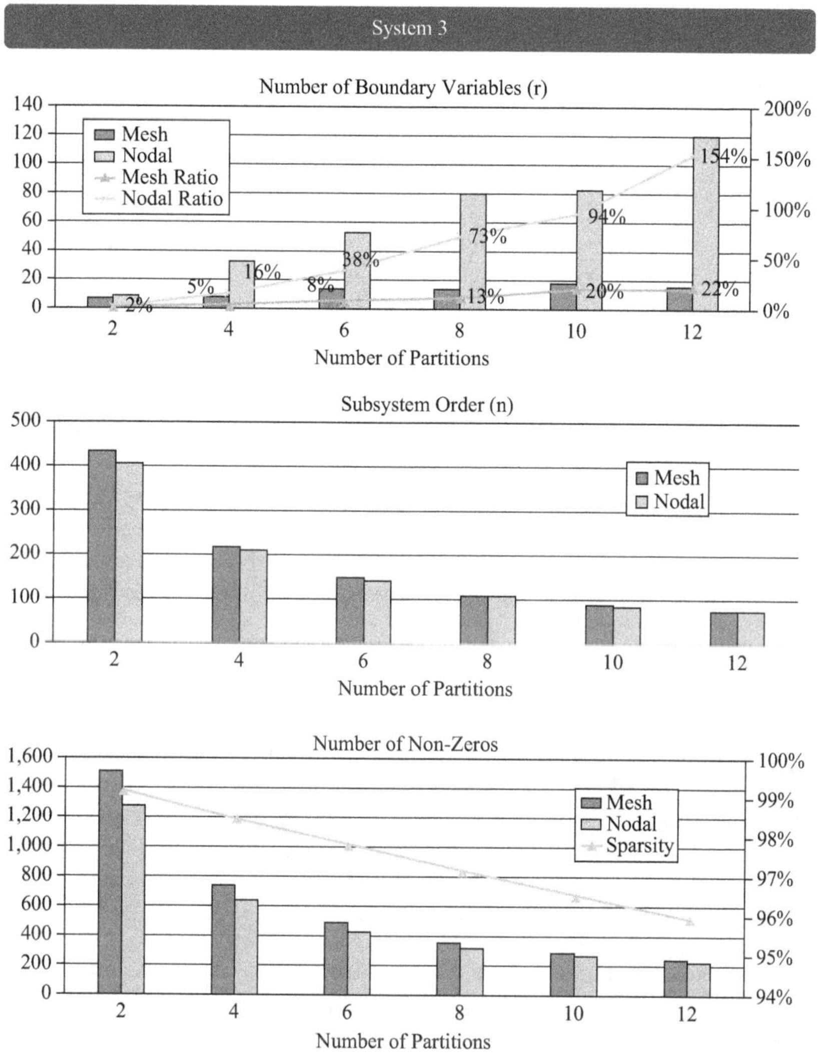

Fig. 9.6 Order information for System 3

agreed that power system models are run many times while measuring instantaneous voltage, current, and average power *throughout* the power system—not only at one power apparatus. Such instantaneous measurements require re-starting the simulation run when network parameter values change and this is a time-consuming process. To counteract such time investment and effort, a multicore solver can be useful in these demanding situations if the multicore solver is readily available.

Similar to the speed and runtime charts, the frame time chart in Fig. 9.5 shows that the average time spent at each substep in the mesh and nodal formulations is in the microsecond range. This frame time shows that multicore simulations of systems such as *System* 3 offer simulation speeds near real-time performance.

Referring to the number of boundary variables in the top chart of Fig. 9.6, a model such as *System* 3 also shows that r grows more rapidly in the nodal formulation than it does in the mesh formulation. However, even when r was comparable (or even larger) than n (shown as % with ascending lines), both formulations showed comparable performance when $p = 4$. For example, at $p = 4$, r was 16% in the nodal formulation and 5% in the mesh formulation. This did not hinder the nodal method from performing closely well, however.

The average subsystem order for the mesh and nodal formulations is similar, which shows that the graph partitioning routine approach produced partitions of similar size for both formulation types. It should also be pointed out that although *hMetis* sees the same representative power system graph in both cases, the number of nodes and meshes is not the same. This can be seen from the slight variations in the subsystem order shown in Table 9.5.

The partition sizes also relate to how well the representative graph represents a power system model and how well *hMetis* can balance the graph partitions according to the constraints specified by the user. This means that for *hMetis* to produce well-balanced partitions, it is the user's responsibility to set appropriate values as edge and vertex weights. Additionally, there are other ways to define a representative power system graph that can lead to different partition sizes. For example, choosing to map electrical nodes as graph vertices and electrical branches as graph edges is a common way to create representative graphs as well [187,193].

If subsystem size is of concern, it should be re-emphasized then that the subsystem order could have been reduced more easily in the mesh formulation than in the nodal formulation. For example, by removing all protective devices shunt impedances used for voltage measurements and converting all three-phase cables to series RL segments can reduce the mesh count by over one hundred equations. This task was not exercised here as it would have unfairly favored the mesh method.

For a system as large as *System* 3, the number of non-zeros and sparsity illustrated in the bottom chart of Fig. 9.6 points out that the mesh method offers comparable matrix structures to the nodal method. Apparently, this is not common knowledge. What is common to postulate is that mesh methods produce dense matrices when compared to nodal methods. This, however, is true when meshes are defined from the links (or chords) of a depth-first-based spanning tree [196, 223]. Therefore, the mesh resistance matrix can be as sparse as the nodal conductance matrix when the internal power apparatus meshes are defined manually (as illustrated in Chapter 5). Additionally, the mesh method offers the flexibility of reducing the overall equation counts by eliminating unnecessary shunt branches from power apparatus models.

9.2.4 System 4

The benchmark results and system order information for *System* 4 using the multicore solver are shown in tabular form in Table 9.6 and graphically in the six charts

Table 9.6 Benchmark and subsystem size information for System 4

Speedup		
Num. partitions	Mesh	Nodal
2	84.8	79.7
4	116.1	95.3
6	98.2	69.3
8	88.1	55.7
10	85.8	45.7
12	80.4	37.5

Number of boundary variables (r)				
Num. partitions	Mesh	Nodal	Mesh ratio (%)	Nodal ratio (%)
2	6	9	1	2
4	10	33	4	14
6	14	57	8	34
8	14	75	11	59
10	14	96	14	92
12	18	111	21	126

Runtime (s)		
Num. partitions	Mesh	Nodal
2	34.5	36.7
4	25.2	30.7
6	29.8	42.2
8	33.2	52.5
10	34.1	64.0
12	36.4	78.1

Simulink runtime: 2,925 s

Subsystem order (n)			
Num. partitions	Mesh	Nodal	Average
2	510	476	493
4	256	244	250
6	171	167	169
8	128	127	128
10	103	104	104
12	86	88	87

Simulink number of state variables: 1,004

Frame time (average, ms)		
Num. partitions	Mesh	Nodal
2	1.4	1.5
4	1.1	1.3
6	1.3	1.9
8	1.4	2.4
10	1.5	2.9
12	1.6	3.6

Simulink runtime: 135.5 ms

Number of non-zeros				
Num. partitions	Mesh	Nodal	Average	Sparsity (%)
2	1723	1504	1614	99
4	854	758	806	99
6	552	509	531	98
8	411	384	398	98
10	328	309	319	97
12	268	259	264	97

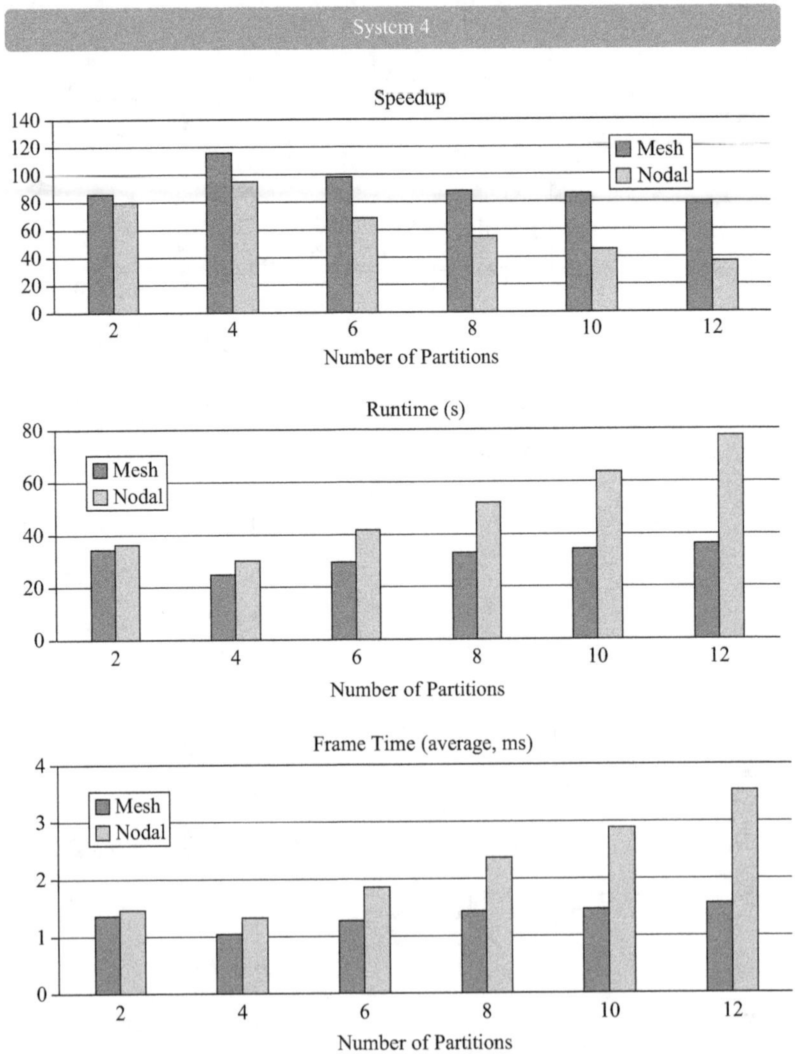

Fig. 9.7 Benchmark results for System 4

illustrated in Figs. 9.7 and 9.8. Referring to the speedup chart in Fig. 9.7, both the mesh and nodal formulations exhibit peak performance at $p = 4$ (consistent with the number of available cores). Interestingly, when using the mesh formulation method, the maximum observed speedup broke the barrier of two-orders-of-magnitude. An important observation from the speedup chart is that the *rate* at which the gains in speed falls as p rises is noticeably different. Beyond (and including) $p = 6$, the nodal speedups decayed much faster than the mesh ones did. This is an interesting result that will be explained later in this section.

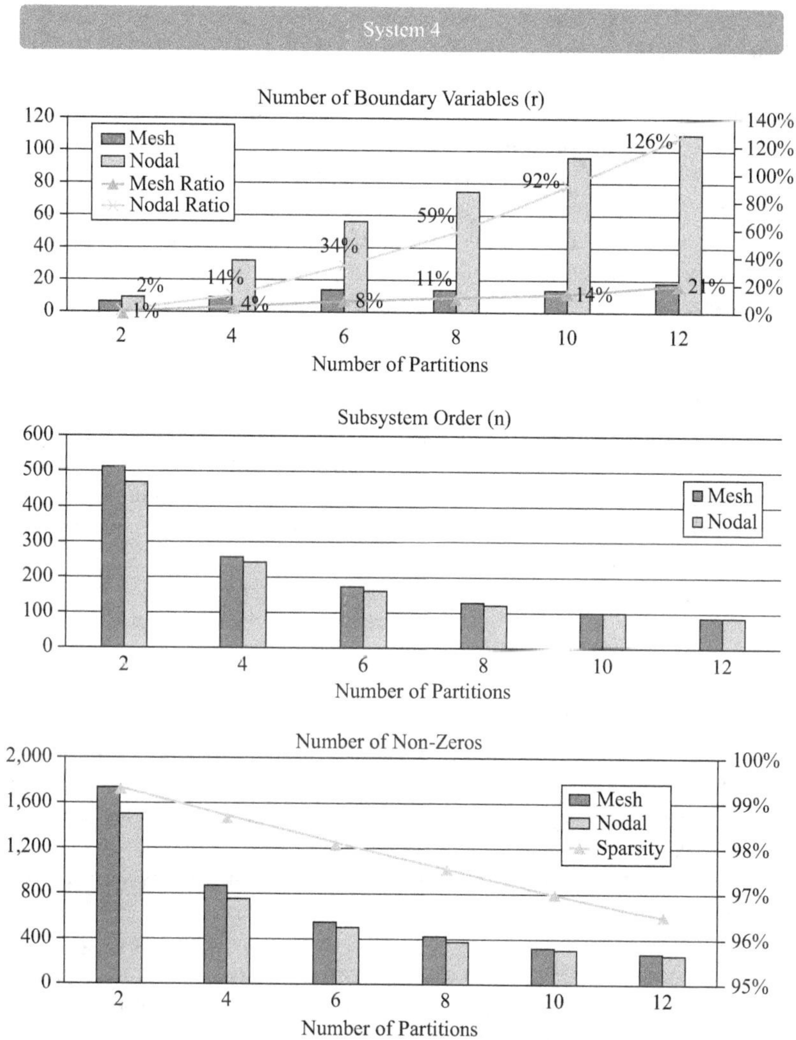

Fig. 9.8 Order information for System 4

Referring to the runtime chart in Fig. 9.7, it is seen that the multicore solver in the nodal formulation method reduced the runtime of the power system simulation to approximately 30 s. This is a significant and a highly desirable result as it allows: (1) designing power systems faster, (2) running large simulation models without incurring undesirable wait times, and (3) potential savings of research resources. It should be mentioned that this runtime reduction did not require the acquisition of any additional hardware. Stated differently, when power system models are large and complex, checking for credible behavior can take significant resources (e.g., billable time, machine hours, among other resources). The simulation runtime produced with the multicore solver took less than a minute, which allows models to be parameterized, tuned, and run multiple times without incurring in the use of additional resources. The runtime reduction of *System* 4 was from 48.75 min (using *Simulink*) to 25.2 s using the multicore solver developed for the book on a desktop multicore computer running *Windows* as the operating system.

Referring to the frame time chart shown at the bottom of Fig. 9.7, the mesh and nodal frame times for $p = 4$ are not much different (see values in the lower-left tabulation in Table 9.6.) However, this difference becomes noticeable when the simulations are run for longer. Another important fact about frame times is their order of magnitude. Both the mesh and nodal methods are very close to breaking the frame time from $O(10^{-3})$ s to $O(10^{-6})$ s. This result is also of importance in the direction of desktop-computer real time simulation, which is not possible today for large models due, in part, to the computational complexity and the high non-determinism of *Windows*-based machines.

Referring to the number of boundary variables shown in the top chart of Fig. 9.8, it is noted once again that, in the nodal formulation, r increases much faster than it does for the mesh formulation. This result is related (in *System* 4) to the decreasing growth rate of the speedup gains mentioned above. Comparing the speedups and number of boundary variables for the nodal and mesh methods, it is suggested[87] that the mesh method may be better suited for parallel environments than the nodal method. This is an interesting result that can often times can be prematurely overlooked during the software design stage.

Comparing the percentage results (r/n) for the mesh and nodal methods in the top chart of Fig. 9.8, this percentage reaches 21% for the mesh method and 126% for the nodal method for $p = 12$. This suggests that mesh formulations in the boundary network remains smaller than the average subsystem size. This result leads to still another observation: it appears that partitioned nodal formulations are limited to coarse-grained scenarios while mesh formulations perform equally well in both the coarse- and fine-grained cases.

Referring to the subsystem order chart shown in the center of Fig. 9.8, it is interesting to note what the order of a partitioned simulation is in comparison to an unpartitioned simulation. The average subsystem order shows that as p increases, the average subsystem order decreases that is, it is inversely proportional to p. Comparing the subsystem order for $p = 4$ against the number of state variables (1,004) reported by *Simulink*, it is detected that the work performed by each thread is less in partitioned simulations.

The number of non-zeros shown in the lower chart of Fig. 9.8 is one way to assess the arithmetic computation burden in a solver. The number of non-zeros matters to the matrix solver because it is related to the number of operations required from the factorization (substep *a*) and solution stages (substep *c*) in each subsystem. However,

[87] This result may be confirmed (or refuted) by repeating the benchmarks on a *Windows* machine with an eight-core processor. At the time of this writing, such processor is not available in a single socket.

the average subsystem order as a metric may be more meaningful and useful to readers, as it is a measure of equation count and is easier to relate to.

Lastly, the sparsity line also represented in the lower chart of Fig. 9.8 shows an inverse relationship with p, that is, sparsity decreases as p increases. Although the trend is downward-sloping, there is little concern (in this case) of sparsity as it is high for all p. What is important to note from the sparsity comparisons is that sparsity was close to 99% in both mesh and nodal formulations.

This finding yields two important results. First, when choosing a network formulation method, the sparsity of the matrix is an important consideration, which should be decided on early on the development process as it contributes to overall work reduction. Second, sparsity in nodal and mesh formulations may be similar, but this similarity depends on which power apparatus are included in the model. For example, buses interconnecting various power apparatus having line-to-line branches affect the density of the mesh resistance matrix as many meshes can be incident to these branches. These shunt branches, however, do not affect the density of the nodal conductance matrix. Similarly, power systems including hundreds of cables with three-phase mutual inductances affect the density of the nodal conductance matrix, but they do not affect the density of the mesh resistance matrix. As a result, the sparsity of subsystem immittance matrices can vary by changing the formulation method.

9.3 Summary of results

The development of a *Windows*-based multicore solver resulted in speeding up the simulation of a notional shipboard power system[88] by two orders of magnitude (mesh case, $p = 4$, speedup: 116.1). In addition, the speedup values reported herein are an intrinsic function of the model size, complexity, topology, power apparatus types, and are not general in scope. For the models examined in this book, the multicore solver results of the runtime and speedup reductions are summarized in Table 9.7 and Fig. 9.9.

Table 9.7 Summary of runtime and speedup results

System number	Total runtime (s)				Speedup		
	Simulink	Mesh	Nodal	Best	Mesh	Nodal	Best
1	8.4	1.6	1.6	1.6	5.2	5.2	5.2
2	92.7	10.7	10.5	10.5	8.7	8.8	8.8
3	205.3	8.0	7.1	7.1	25.7	28.9	28.9
4	2925.0	25.2	30.7	25.2	116.1	95.3	116.1

[88] The models were built in *MATLAB/Simulink* (.mdl files) and then imported into the multicore solver developed for this book.

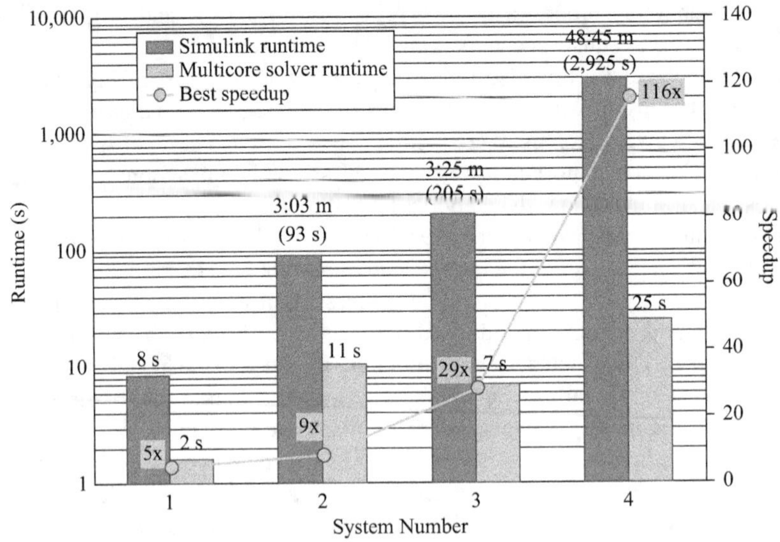

Fig. 9.9 Summary of runtime and speedup

An immediate observation from the above results presented in Table 9.7 is that the larger and more complex the power system model is, the greater is the speedup.[89] Besides, a micro-look at the data included in this table reveals that low-order, sparse, and partitioned simulations properly implemented on a multicore desktop computer running *Windows* as the operating system significantly reduced the runtime of *System* 4 without having to acquire specialized hardware and software.

Furthermore, it does not appear beneficial to parallelize the simulation of smaller, less complex models due to the overheads of multithreaded synchronization. But a more important reason not to parallelize smaller and less complex models is the amount of resources that go into the development of a multicore solver. Such resource allocation is justified if it aims to speed up large models rather than small ones. This fact is readily observable by perusing how the speedup follows the unpartitioned runtime in Fig. 9.9.

On the basis of (9.1), it should be re-emphasized that speedups depend heavily on where the numerator is taken from. The result will be biased if a development team does not spend equal resources to minimize the numerator. This numerator can be high if there is little, or none, resource allocation to it which can result in high speedups. The approach to compute speedup in this book used the runtime of a well-established, commercial simulator in a transparent effort to remove possible prejudice or distortion of the estimates of the runtimes and speedups.

[89] Informal conversations with software manufacturers tend to support this observation.

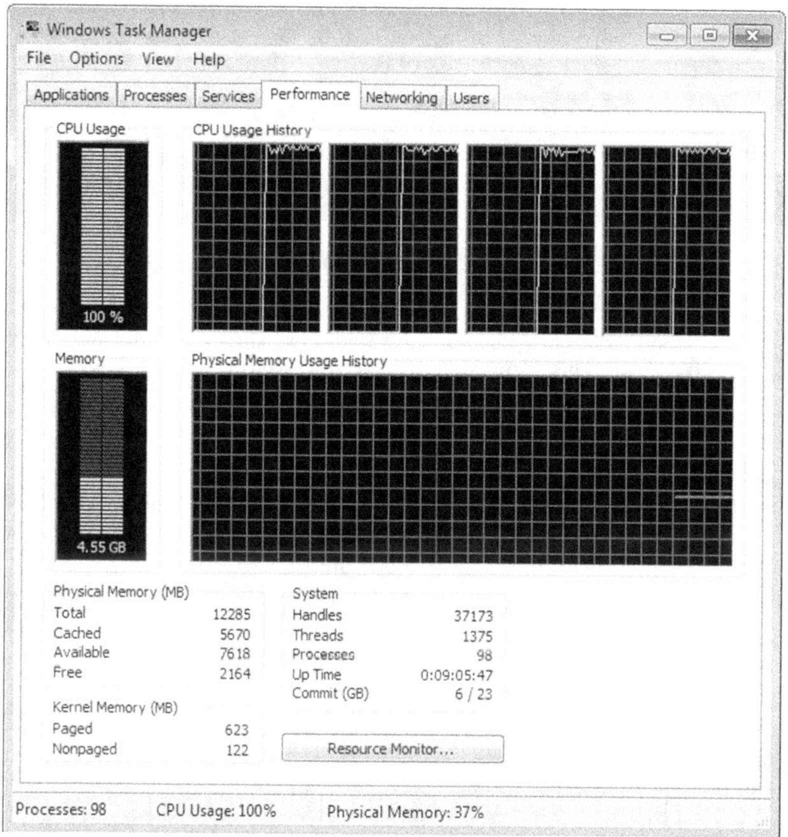

Fig. 9.10 CPU usage in a multicore simulation (four physical cores)

On the other hand, a screenshot of *Microsoft Windows'* task manager showing CPU[90] usage immediately after starting the $p = 4$ case for *System* 4 is shown in Fig. 9.10.[91] This CPU corresponds to the multicore computer specified in Table 9.1. As noticed, all cores are properly utilized. Readers can expect this type of CPU usage level during parallel simulations, which adequately exploits the often-untapped multicore technology available on desktop computers today.

The threads that produce the CPU and core usage shown in Fig. 9.10 are configured to execute with *Normal* priority.[92] Although the CPU usage is 100%, the *core* usages show some "head room," which indicates the application is likely to be responsive to user input. Elevating thread priority may increase simulation performance, but it may jeopardize UI responsiveness which is contrary to usability rules [224].

[90] Central processing unit.

[91] The CPU usage is similar in both mesh and nodal formulations.

[92] Process and thread priorities are configurable in *.NET*.

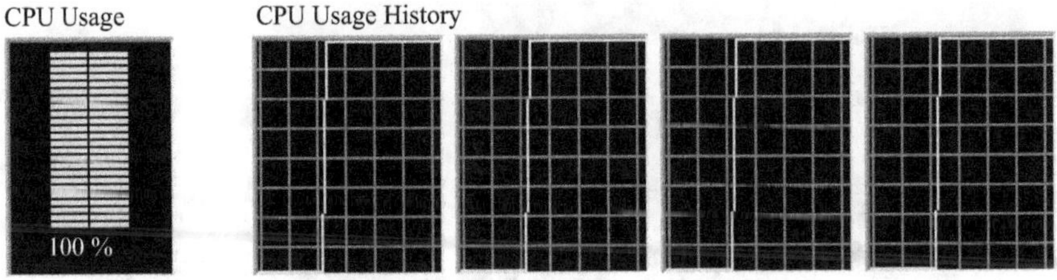

Fig. 9.11 CPU usage in a multicore simulation with elevated thread priorities

Fig. 9.11 shows the CPU usage of the multicore solver running with elevated thread priorities. Both the CPU and its cores are fully utilized (100%), which indicates the application is making full use of the available multicore (quad-core) processor. However, this type of performance may cause the application to be unresponsive to user keyboard and mouse input.

In this regard, programmers have a choice. If the target computer runs without intervention, thread and process priorities may be elevated to higher-than-normal values. On the other hand, if the target computer is used for daily computing and simultaneously for simulation,[93] the simulation threads and process should retain their default, normal priorities.

Lastly, the typical CPU usage of an application that may not be multicore ready is shown in Fig. 9.12. This quad-core performance is common of many applications that use mainly one thread for compute-intense work—that is, the performance of an application that does not parallelize its work. It should also be noted that not all applications require parallelization. Many applications execute in acceptable time without the need to parallelize its algorithms.

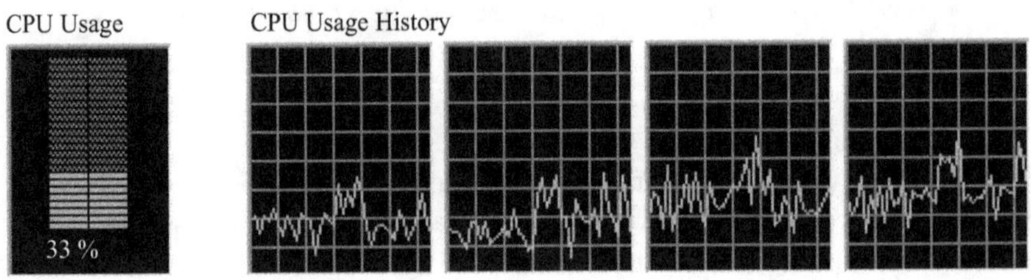

Fig. 9.12 Typical CPU usage of non-parallel programs

Although Fig. 9.12 shows the performance of a *single*-threaded application, all cores appear to be doing work. This type of core usage is a result of *Windows* moving different program thread across the cores based on core availability. Moving

[93] Long-running power system simulations are often run in the background while daily computing tasks occur in the foreground (e.g., word processing).

threads across the cores (cross-core migration) allows threads to get attention from a core when hardware resources become available.

9.4 Summary

This chapter presented basic performance metrics to assess the effectiveness of the multicore solver developed for this book. Perhaps the most interesting metric and finding of this book has to do with runtimes and speedups. As was discussed throughout the chapter, speedup refers to the gains in executing a custom multicore solver versus a commercial one. It should be highlighted, however, that the multicore solver does not constitute a replacement for a proven valuable tool such as *MATLAB/Simulink*. Instead, the custom multicore solver serves as an alternative way to produce results in demanding long-running power system simulation scenarios. For all other simulation scenarios, the author himself relies on *MATLAB/Simulink*.

Several factors were observed that limited speedups. Among these are: other *Windows* processes running simultaneously, computational imbalance, computation of boundary variables, matrix non-zero counts, programming efficiency,[94] processing power, available memory cache, false sharing,[95] thread affinity, priority of UI thread over masters/slave threads, matrix re-factorization algorithms, to name a few. Despite somc of these speed-hindering factors, the efficiency of the partitioning approach was demonstrated by reducing runtime for larger models than for smaller models.

Computation of the boundary network is one of the three bottlenecks in diakoptics-based approaches, and it becomes dominant as p and r increase (i.e., in fine-grained simulations). The other two bottlenecks are substeps a and c, but these are common to unpartitioned simulations as well. *Substep d* only exists in partitioned simulations and requires thread synchronization. Thread synchronization incurs thread-coordination delays and wake-up times that are detrimental to parallel simulation performance.

Lastly, an important observation is that the simulation runtime reduced for certain values of p. For most models, $p = 4$ was found to be satisfactory; but for small models, it can be shown that $p = 1$ is can be the best choice. There appears to be limited engineering experience in predicting the correct number of partitions (and also where to tear) to produce the fastest simulations. However, there is a 'folk theorem' (with little proof) that p should match the number of available computer cores c. This run-of-the-mill view of matching the number of partitions to the number of cores is "simplistic in its analysis" [85]. In *praxis*, however, for the development of the multicore solver for this book (see Table 9.1), this view appeared to signal a good place to start at.

[94] This subtle fact can significantly impact speedup in unpartitioned simulation scenarios as well.

[95] *False sharing* constitutes unnecessary data-migration across core cache-lines, and it can lessen simulation performance.

To close this chapter, the benchmark results were shown for *both* nodal and mesh formulations. A comparison and contrast of both methods is rare in current literature. Perhaps, this is because most programs only implement nodal formulations rather than both formulations. Although it is generally accepted that nodal formulations are the best choice, there is little evidence in parallel environments to support this belief. As a whole, this chapter showed that mesh formulations appear to be a stronger candidate for parallel scenarios than nodal formulations. This finding was demonstrated by the superior speedup and achieved by the lesser rate at which r grew in relation to n.

Chapter 10
Overall summary and conclusions

This book demonstrated a methodology to reduce the runtime of power system transient simulations by parallelizing the solution on a multicore desktop computer running *Windows* as the operating system. It was said at the beginning that electromagnetic transient simulation was notoriously slow, that it limited the number of case studies per day, and that it consumed significant research resources. It was also said at the outset that to counteract the problem of slow and time-consuming simulations in research environments, faster runs were needed. This book directed research and parallel programming efforts toward the solution of this problem by properly exploiting the rapid advancement in multicore technology.

This book introduced an approach to partition and parallelize the simulation of power systems on multicore computers. Before the proliferation of multicore computers, parallel simulation first saw its application on distributed computers [179]. These simulations efforts motivated many of the well-known partitioning methods in use today. Today, parallel simulation does not require a distributed computer. Parallel simulations can be carried out on a personal desktop or laptop computer. This low-cost hardware availability is prompting experts in the field to re-visit partitioning methods suitable for such multicore shared-memory machines.

Parallel simulation is not suitable for all simulation scenarios. However, when it is, it can reduce runtime by one or two orders of magnitude on select cases. This book introduced several power system models to demonstrate how runtime reduction is possible. Although the models (*Systems* 1, 2, 3, and 4) were rooted on the same notional shipboard power system, the sizes and complexities of each, as highlighted, were noticeably different. Noticeably different complexities gave noticeably different results, and they showed that the potential speedup is directly proportional to the model size at hand. That is, the larger the model, the larger the speedup benefit. Although this result is not general, it is true and valid for the specific power system models considered in this book.

Model *size* is an important metric to quantify. Size is required to compare other published models [115,225–227] readers may have against the size of the notional shipboard power system presented in Chapter 2. Chapter 2 suggested a few ways to quantify model size. *System* 4, the largest of all models treated in the book, presented a considerable large number of state variables and switches, which affects its offline runtime and makes it challenging to simulate in real time. There are also models more complex than *System* 4. Researchers of electric ship technology routinely work with such models and routinely experience even lengthier runtimes that the ones reported in this book.

System 4 was chosen for consideration because it presented sufficient computational complexity, completeness, and relevancy to demonstrate the research issue advanced at the beginning of the book. *System* 4 was also helpful to estimate the runtime and potential speedup—among other useful metrics included in the book—of parallelizing its simulation on a desktop *Windows*-based multicore computer. Although *System* 4 has necessary complexity to demonstrate the benefits of parallelization, it also lacks sufficient complexity such as machines and non-linear controls. Although these complexities were not treated herein, their inclusion does not change the power system parallelization methodology presented by this book.

There was motive behind selecting *Systems* 1–3 and the notional Navy shipboard power system model (*System* 4) to illustrate partitioning. These systems, similar to terrestrial microgrids [1,228], have particular traits that make their simulation challenging. In the case of shipboards, the essential goal behind their simulation is to assess continuous mobility, and power (including thermal management) for combat systems despite major disruptions involving cascading failures [225,229,230]. In addition, power electronics converters (motor drives and power supplies) are highly non-linear, inject significant harmonics, and challenge traditional simulation techniques. Examples of the challenging aspects of these types of microgrids are short cable lengths, which prevent using natural propagation delays to produce partitioning. Additionally, the inclusion of cable capacitance (six per cable) on over one hundred cables rapidly increases the state-variable count. Other examples that make the simulation of these four systems challenging is the requirement of using small timesteps and the inclusion of hundreds of switches. Small timesteps increase accuracy, but reduce simulation performance. Further, the inclusion of hundreds of switches requires frequent matrix re-factorizations and increases the number of interpolations required at each timestep as discussed in Chapters 3 and 4.

Time domain electromagnetic transient simulation is the most comprehensive simulation type in power engineering, but it also is the most time consuming. Whether or not to conduct this simulation type depends on the set of credible contingencies defined by the user. These contingencies (or run sets) should be defined *a priori* to determining the simulation type. In many cases there are alternatives to transient simulation, where it may suffice to conduct time-domain load flow [140] or transient-stability simulations [32] rather than high-fidelity electromagnetic transient simulations. Time-domain load flow simulations execute very fast when compared to electromagnetic transient simulation, but there is a pronounced loss of fidelity as waveform-detail is not available. Transient-stability simulations also execute much faster than electromagnetic transient simulation, but they are meant to study mechanical transients rather than electrical ones [231].

Nonetheless, time domain electromagnetic transient simulations have their own place. If the pre-defined credible contingencies of interest call for high fidelity, electromagnetic transient simulation, then the concepts introduced in Chapter 3 are important to master. They defined and clarified important aspects of power system solvers in general. In particular, Fig. 3.3 showed a possible paradigm of solving electrical and control network as subsystems inside a power system partition. This type of illustration is a guideline on how software may be designed. For example, classes in object-oriented programming can be designed to follow this diagram as the class relations (i.e., one-to-one and one-to-many) are readily apparent.

Common uses of electromagnetic simulation [32,58] include testing for insulation stress levels, the study of ferroresonance, inrush, transient recovery voltage, islanding, arcing faults [113,202], harmonic content, power converter performance [232], motor drive design [57], among other. Before choosing the simulation type, however, users should spend the necessary time to define the credible contingencies of interest. This choice leads to the appropriate simulation *type*.

As introduced in Chapter 4, time domain simulation required discretizing computers models before their simulation. In the power system models shown, this was accomplished by discretizing electrical branches containing inductors and capacitors. The discretization methods covered were backward Euler, trapezoidal rule, tunable integration, and root-matching. These approaches all have strengths and weaknesses, which leaves the choice of the discretization method to the reader. It was also said that it is common to use the trapezoidal rule for sinusoidal networks, but not for switching networks. For networks with switching elements, backward Euler is a more conventional approach due to its numerical stability. The technique adopted in this book, however, was root matching. Root matching provides both high accuracy and numerical stability [81,82] while reducing the dependence on small Δt values to obtain accurate results.

The choice of discretization method also affects the performance, accuracy, and numerical stability of the control network solution. Instabilities may occur when interfacing [233] control networks with electrical networks as reported in References 80,83, and 147. Chapter 4 also showed how to discretize the control network using tunable integration as it sufficed for the modeling scenarios considered here. During the implementation of a solver, readers should remember that, while the trapezoidal rule provides high accuracy, when the slope of a state variable goes to zero over two consecutive timesteps numerical chatter appears. This is also true of control networks and requires careful attention. There are, however, well known methods to eliminate this chatter [42,61,70,80].

Navy shipboards [29,30,234] have hundreds of power apparatus onboard, but only a few power apparatus types were considered in this book. The power apparatus models in Chapter 5 were simple "place holder" models. Their use was valid for the main purpose of this book, which, was to demonstrate a methodology to partition and parallelize the simulation of power systems on a multicore computer. The power apparatus models were presented as enclosed by gray boxes. This subtle enclosure promotes software modularity. Enclosing power apparatus isolates them from other power apparatus and permits treating them as "miniature networks." This feature thus facilitates their analyses, troubleshooting in code, and the solution and viewing (charting) of their internal states.

It was also shown that solving the electrical side of a partitioned power system required formulating subsystem-level matrices. In nodal analysis, the nodal conductance matrix is commonly formed using the branch-stamping method [49]. This approach works well, but is not easily transferable to mesh analysis. Chapter 6 presented a tensor-based formulation approach that allowed formulating both the nodal and mesh immittance matrix with a single algorithm. By first block-diagonalizing the "miniature networks" (the gray-enclosed power apparatus models) presented in Chapter 5 and then using an interconnection tensor, the subsystem-level immittance matrices were formed by matrix multiplication.

If readers are interested only in the nodal method to formulate power system equations, branch stamping is the recommended approach. If readers are interested in comparing the performance of nodal and mesh formulations over a wide range of networks, the tensor approach of Chapter 6 is recommended. The reasons to compare nodal and mesh analysis become important in parallel scenarios, where, as seen, the size of the boundary network in each case grew at different rates. Large boundary networks are detrimental to parallel simulation as they increase the solution time of the boundary network (e.g., substep *d*), or the serial part of the "fork/join" algorithms.

Although power system partitioning is well understood [186], its issues remain challenging. Two partitioning issues not addressed frequently in power engineering literature are the answers to *where to partition* and *how many partitions to create*. The answers to these questions are challenging to get right other than through empirical methods. This book outsourced the question of *where to partition* to the graph partitioning tool *hMetis*. Outsourcing this question, however, requires that the input graph represent well the power system model, that the vertex weights be set commensurately to power apparatus complexity, and that the program parameters of *hMetis* be given proper consideration.

An approach to adequately represent a power system model as a graph was given in Chapter 7. This approach mapped power apparatus to graph vertices and buses to graph edges. The vertex weights were functions of the equation count of each power apparatus. The mapping technique produced a coarsened graph that forced *hMetis* to tear only at power apparatus terminals. Forcing the tearing at power apparatus terminals is a deliberate choice to simplify graph partitioning, and follows the principles of multi-terminal component theory, software modularity, and the use of miniature networks. Although this approach worked well to partition *Systems* 1, 2, 3, and 4, there are other approaches to represent electrical networks using graphs as well [95,97,150,160].

The answer to *how many partitions to create* was shown to be related to the ratio of boundary network and the average subsystem order (r/n ratio) [22]. Although a closed-form expression to prove this observation may not exist, the simulation results provide insights that speedup may be related to this ratio. For instance, the different model sizes returned their highest speedups at different ratios of boundary network to average subsystem order sizes. For each power system model, the maximum speedup also varied as a function of the formulation type. Although the speedup results (and intuition) suggest that the number of partitions p should equal the number of cores c, the value of p also appears to be related to the r/n ratio.

In the power system models and quad-core computer used in this book, the best number of partitions *did* equal the number of cores. However, this result is not general and cannot be assumed true for many-core machines. A reason that hinders extrapolating this result to many-core machines is the uncertainty that thread synchronizations brings about as more threads are created. It is also not clear how well an application scales to more cores as there is a critical amount of time that can be spent on the serial part of a solution before the serial overhead outweighs the parallel advantages. Nonetheless, the best performance of a multicore solver can be sweeped on many-core desktop computers becoming available.

As the number of partitions increased, partitions became less coarse-grained and more fine-grained. Fine-grained simulations did not perform as well as coarse-grained ones did. Readers are encouraged to experiment with a varying number of partitions until one's right "recipe" for success develops. Closed-form expressions to predict the best number of partitions are highly desirable and have been done before [50,160], but they require knowledge of the exact number of floating-point operations inside a solver. This type of analysis is prone to encountering, eventually, a "black box" if a solver depends on libraries developed by third parties. Compiled assemblies [195] limit the ability to dissect the internal methods and count the number of floating-point operations.

The multicore solver developed for this book was developed in *C#* using *NMath* as a third-party numerical library. The choice of *C#* was purely subjective, and it does not represent an endorsement of the product to the detriment of others. The choice of *C#* also facilitates the integration of the solver and its user interface. In recent years, the integration of *C#* and *Windows Presentation Foundation* [235] has motivated the scientific computing community to consider *C#* as an underlying language for user interfaces [190]. Chapter 8 presented a simple *C#* program to demonstrate this proclivity. A simple user interface was implemented using the XAML language [217,236] where a button click called the *C#* solver behind it. The call from XAML to *C#* shows the integration between a user interface and the solver behind it. An alternate approach to having a user interface tightly coupled with the solver is to develop two separate executables and have them communicate via memory pipelines [212]. This approach favors performance, but increases development time, debugging complexity, and requires lower-level programming constructs.

The program structure presented in Chapter 8 also demonstrated the implementation of a parallel time loop. The time loop fetched multiple threads from the *Windows* thread pool and used them in a coordinated fashion to execute parallel (concurrent) operations [216]. The parallel operations implemented the fork/join pattern introduced with the swim lane diagram. The program structure showed serial sections in code that forced slave threads to sleep while the master thread executed the serial part [213]. This serial part is the major bottleneck of the partitioning algorithm presented herein, and it is common to power system partitioning methods using the fork/join paradigm.

It was highlighted throughout the book that the purpose of this work was to develop an alternative simulation approach to simulate power system transients for readers interested in the development of a *Windows*-based multicore solver for their own use. As seen, this inherent purpose was accomplished by Chapters 4–8. In this sense, readers are strongly encouraged to continue where this book stops by starting from the program structure provided in Chapter 8, and by expanding it into a more meaningful and useful compute-intense program. Although this program example appears simple, it is useful and important as it presents the structure to parallelize the simulation of power system models on multicore machines running *Windows*. The program also indicates (via commented code) where readers should insert code. With regard to thread coordination (or synchronization), the *Barrier* structure was tested against alternatives and was found to be suitable for the simulation scenarios under analysis. However, it also likely that this structure will be improved (or superseded) eventually by lighter-weight alternatives in future releases of *.NET.*

Chapter 9 analyzed the performance of the partitioning method presented in this book. The performance was measured principally on the basis of the runtime and speedup, but other metrics were presented as well. The definition of speedup, however, needed a transparent caveat which had to do with its numerator. By using the runtime of custom solvers as the numerator, it was underscored that developers must have spent an unbiased amount of time and other resources to optimize unpartitioned simulations before optimizing partitioned ones. To address this caveat, the approach taken in this book was to use the unpartitioned runtime of a commercial simulator to reduce biasing from the speedup comparison.

The overall analysis of the performance metrics was presented through six tables and six charts. The performance of parallelizing *System* 1 was poor: there were insufficient gains on the basis of the utilized resources that went into developing the multicore solver. The objective of partitioning *System* 1, however, was not to show runtime gains, but to show the speedup *progression* from a simple power system model (*System* 1) to a more complex one (*System* 4). *System* 2 also represented a small system, but it added complexity to the simulation by including nine three-phase rectifiers and nine three-phase inverters. The progression of the speedups observed between *Systems* 1 through 4 provides a pattern of how speedups may follow system size and complexity.

The parallelization of *Systems* 2, 3, and 4 showed two important findings. First, speedups first increased and then decreased. It was noted that this behavior occurs for systems that may warrant partitioning (i.e., for larger models, not for smaller ones such as *System* 1). Second, the maximum speedups occurred when the number of partitions p equaled the number of cores c, or when $p = c$. This has also been the case over years of experience in running parallel simulations on quad-core computers. One test this "equality rule" may confront in future research is whether the same experience will scale with the many-core shift; that is, on forthcoming computers with many more cores than four.

The simulation of *System* 4 stood out the performance of node and mesh tearing. In mesh tearing, for instance, the maximum speedup broke the barrier of two-orders-of-magnitude. This runtime reduction from 48.75 min to 25.2 s verified the working hypothesis advanced at the outset of this work. In addition, this finding suggests that it is possible to run several case studies per day which ameliorates the use of scarce research resources.

Another important finding was that in the node tearing, the boundary network size increased a lot faster than it did in mesh tearing. This finding suggests that nodal and mesh formulations do *not* perform the same in parallel scenarios. Also, comparing the speedups and number of boundary variables for the nodal and mesh methods, it was found that mesh tearing may be better suited for parallel environments than node tearing. This result may be easily overlooked prematurely during software design [191]. Another observation from the partitioned results, obtained empirically from sweeping the number of partitions, is that node tearing appears to perform better in coarse-grained scenarios while mesh tearing performs equally well in both the coarse- and fine-grained cases.

To close, the runtimes and speedups conclusions drawn from the evidence (performance metrics) have important relevance to power systems engineering as verified in the book. Readers should remember that the runtimes and speedups reported herein are *specific* to the multicore solver developed, and the shipboard power system models under analyses. The speedups are an intrinsic function of programming efficiency, numerical library, model size, complexity, topology, power apparatus types, and cannot be generalized. Therefore, different readers implementing the methodologies outlined in this book will likely obtain different performance results.

Appendix A

Compatible frequencies with Δt

The listings in Table A.1 below show frequencies (Hz) compatible with $\Delta t =$ 46.296 μs calculated in section 3.4. The timestep Δt is said to be "compatible" with a signal if it fits the signal's period an *integer*-number of times.

The first line on Table A.1 reads as follows: "A timestep of $\Delta t = 46.296$ μs fits exactly 4 times in a period of 185.2 μs. The period of 185.2 μs corresponds to a signal of frequency 5,400 Hz. Therefore, $\Delta t = 46.296$ μs *is compatible* with 5,400 Hz." The compatibility ensures PWM carrier frequencies (assumed constant) are sampled exactly at their peaks and zero-crossings, which avoids aliasing. (Signals with frequencies greater than 5.4 kHz are not considered compatible with $\Delta t = 46.296$ μs.)

It can be noticed that $\Delta t = 46.296$ μs fits the period of a 60-Hz signal exactly 360 times. This exact (compatible) fit avoids the 2nd-harmonic problem in RMS measurements mentioned in section 4.3.3. It is also assumed that the signal frequency (60 Hz) is constant.

Table A.2 shows frequencies compatible with the more widely used $\Delta t = 50$ μs. Referring to the first row, it is implied that carrier signals with frequencies >5 kHz require a smaller Δt (e.g., 46.296 μs as shown in the preceding table) to avoid aliasing. Similarly, referring to the last row, $\Delta t = 50$ μs is not naturally compatible with a 60-Hz signal, and it can therefore exhibit the 2nd-harmonic problem in RMS measurements. (Signals with frequencies greater than 5 kHz are not considered compatible with $\Delta t = 50$ μs.)

The *MATLAB* code in Snippet A.1 of this appendix plots in nine diagrams, as depicted in Fig. A.1 and Fig. A.2, overlays of a carrier (triangular) signal and a reference (60 Hz, sine) signal. Each plot shows a carrier signal of different frequency, which corresponds to the first nine rows of Table A.1 ($\Delta t = 46.296$ μs) and Table A.2 ($\Delta t = 50$ μs), respectively.

Table A.1 Frequencies compatible with Δt = 46.296 μs

dt (s)	Fits	Period (s)	Frequency (Hz)
46.296E-6	4	185.2E-6	5400
46.296E-6	8	370.4E-6	2700
46.296E-6	12	555.6E-6	1800
46.296E-6	16	740.7E-6	1350
46.296E-6	20	925.9E-6	1080
46.296E-6	24	1.1E-3	900
46.296E-6	28	1.3E-3	771.4285714
46.296E-6	32	1.5E-3	675
46.296E-6	36	1.7E-3	600
46.296E-6	40	1.9E-3	540
46.296E-6	44	2.0E-3	490.9090909
46.296E-6	48	2.2E-3	450
46.296E-6	52	2.4E-3	415.3846154
46.296E-6	56	2.6E-3	385.7142857
46.296E-6	60	2.8E-3	360
46.296E-6	64	3.0E-3	337.5
46.296E-6	68	3.1E-3	317.6470588
46.296E-6	72	3.3E-3	300
46.296E-6	76	3.5E-3	284.2105263
46.296E-6	80	3.7E-3	270
46.296E-6	84	3.9E-3	257.1428571
46.296E-6	88	4.1E-3	245.4545455
46.296E-6	92	4.3E-3	234.7826087
46.296E-6	96	4.4E-3	225
46.296E-6	100	4.6E-3	216
46.296E-6	104	4.8E-3	207.6923077
46.296E-6	108	5.0E-3	200
46.296E-6	112	5.2E-3	192.8571429
46.296E-6	116	5.4E-3	186.2068966
46.296E-6	120	5.6E-3	180
46.296E-6	124	5.7E-3	174.1935484
46.296E-6	128	5.9E-3	168.75
46.296E-6	132	6.1E-3	163.6363636
46.296E-6	136	6.3E-3	158.8235294
46.296E-6	140	6.5E-3	154.2857143
46.296E-6	144	6.7E-3	150
46.296E-6	148	6.9E-3	145.9459459
46.296E-6	152	7.0E-3	142.1052632
46.296E-6	156	7.2E-3	138.4615385
46.296E-6	160	7.4E-3	135
46.296E-6	164	7.6E-3	131.7073171
46.296E-6	168	7.8E-3	128.5714286
46.296E-6	172	8.0E-3	125.5813953
46.296E-6	176	8.1E-3	122.7272727
46.296E-6	180	8.3E-3	120
46.296E-6	184	8.5E-3	117.3913043
46.296E-6	188	8.7E-3	114.893617
46.296E-6	192	8.9E-3	112.5
46.296E-6	196	9.1E-3	110.2040816
46.296E-6	200	9.3E-3	108
46.296E-6	204	9.4E-3	105.8823529
46.296E-6	208	9.6E-3	103.8461538
46.296E-6	212	9.8E-3	101.8867925
46.296E-6	216	10.0E-3	100
46.296E-6	220	10.2E-3	98.18181818
46.296E-6	224	10.4E-3	96.42857143
46.296E-6	228	10.6E-3	94.73684211
46.296E-6	232	10.7E-3	93.10344828
46.296E-6	236	10.9E-3	91.52542373
46.296E-6	240	11.1E-3	90
46.296E-6	244	11.3E-3	88.52459016
46.296E-6	248	11.5E-3	87.09677419
46.296E-6	252	11.7E-3	85.71428571
46.296E-6	256	11.9E-3	84.375
46.296E-6	260	12.0E-3	83.07692308
46.296E-6	264	12.2E-3	81.81818182
46.296E-6	268	12.4E-3	80.59701493
46.296E-6	272	12.6E-3	79.41176471
46.296E-6	276	12.8E-3	78.26086957
46.296E-6	280	13.0E-3	77.14285714
46.296E-6	284	13.1E-3	76.05633803
46.296E-6	288	13.3E-3	75
46.296E-6	292	13.5E-3	73.97260274
46.296E-6	296	13.7E-3	72.97297297
46.296E-6	300	13.9E-3	72
46.296E-6	304	14.1E-3	71.05263158
46.296E-6	308	14.3E-3	70.12987013
46.296E-6	312	14.4E-3	69.23076923
46.296E-6	316	14.6E-3	68.35443038
46.296E-6	320	14.8E-3	67.5
46.296E-6	324	15.0E-3	66.66666667
46.296E-6	328	15.2E-3	65.85365854
46.296E-6	332	15.4E-3	65.06024096
46.296E-6	336	15.6E-3	64.28571429
46.296E-6	340	15.7E-3	63.52941176
46.296E-6	344	15.9E-3	62.79069767
46.296E-6	348	16.1E-3	62.06896552
46.296E-6	352	16.3E-3	61.36363636
46.296E-6	356	16.5E-3	60.6741573
46.296E-6	360	16.7E-3	60

Table A.2 Frequencies compatible with $\Delta t = 50\ \mu s$

dt (s)	Fits	Period (s)	Frequency (Hz)
50.0E-6	4	200.0E-6	5000
50.0E-6	8	400.0E-6	2500
50.0E-6	12	600.0E-6	1666.666667
50.0E-6	16	800.0E-6	1250
50.0E-6	20	1.0E-3	1000
50.0E-6	24	1.2E-3	833.3333333
50.0E-6	28	1.4E-3	714.2857143
50.0E-6	32	1.6E-3	625
50.0E-6	36	1.8E-3	555.5555556
50.0E-6	40	2.0E-3	500
50.0E-6	44	2.2E-3	454.5454545
50.0E-6	48	2.4E-3	416.6666667
50.0E-6	52	2.6E-3	384.6153846
50.0E-6	56	2.8E-3	357.1428571
50.0E-6	60	3.0E-3	333.3333333
50.0E-6	64	3.2E-3	312.5
50.0E-6	68	3.4E-3	294.1176471
50.0E-6	72	3.6E-3	277.7777778
50.0E-6	76	3.8E-3	263.1578947
50.0E-6	80	4.0E-3	250
50.0E-6	84	4.2E-3	238.0952381
50.0E-6	88	4.4E-3	227.2727273
50.0E-6	92	4.6E-3	217.3913043
50.0E-6	96	4.8E-3	208.3333333
50.0E-6	100	5.0E-3	200
50.0E-6	104	5.2E-3	192.3076923
50.0E-6	108	5.4E-3	185.1851852
50.0E-6	112	5.6E-3	178.5714286
50.0E-6	116	5.8E-3	172.4137931
50.0E-6	120	6.0E-3	166.6666667
50.0E-6	124	6.2E-3	161.2903226
50.0E-6	128	6.4E-3	156.25
50.0E-6	132	6.6E-3	151.5151515
50.0E-6	136	6.8E-3	147.0588235
50.0E-6	140	7.0E-3	142.8571429
50.0E-6	144	7.2E-3	138.8888889
50.0E-6	148	7.4E-3	135.1351351
50.0E-6	152	7.6E-3	131.5789474
50.0E-6	156	7.8E-3	128.2051282
50.0E-6	160	8.0E-3	125
50.0E-6	164	8.2E-3	121.9512195
50.0E-6	168	8.4E-3	119.047619
50.0E-6	172	8.6E-3	116.2790698
50.0E-6	176	8.8E-3	113.6363636
50.0E-6	180	9.0E-3	111.1111111

dt (s)	Fits	Period (s)	Frequency (Hz)
50.0E-6	184	9.2E-3	108.6956522
50.0E-6	188	9.4E-3	106.3829787
50.0E-6	192	9.6E-3	104.1666667
50.0E-6	196	9.8E-3	102.0408163
50.0E-6	200	10.0E-3	100
50.0E-6	204	10.2E-3	98.03921569
50.0E-6	208	10.4E-3	96.15384615
50.0E-6	212	10.6E-3	94.33962264
50.0E-6	216	10.8E-3	92.59259259
50.0E-6	220	11.0E-3	90.90909091
50.0E-6	224	11.2E-3	89.28571429
50.0E-6	228	11.4E-3	87.71929825
50.0E-6	232	11.6E-3	86.20689655
50.0E-6	236	11.8E-3	84.74576271
50.0E-6	240	12.0E-3	83.33333333
50.0E-6	244	12.2E-3	81.96721311
50.0E-6	248	12.4E-3	80.64516129
50.0E-6	252	12.6E-3	79.36507937
50.0E-6	256	12.8E-3	78.125
50.0E-6	260	13.0E-3	76.92307692
50.0E-6	264	13.2E-3	75.75757576
50.0E-6	268	13.4E-3	74.62686567
50.0E-6	272	13.6E-3	73.52941176
50.0E-6	276	13.8E-3	72.46376812
50.0E-6	280	14.0E-3	71.42857143
50.0E-6	284	14.2E-3	70.42253521
50.0E-6	288	14.4E-3	69.44444444
50.0E-6	292	14.6E-3	68.49315068
50.0E-6	296	14.8E-3	67.56756757
50.0E-6	300	15.0E-3	66.66666667
50.0E-6	304	15.2E-3	65.78947368
50.0E-6	308	15.4E-3	64.93506494
50.0E-6	312	15.6E-3	64.1025641
50.0E-6	316	15.8E-3	63.29113924
50.0E-6	320	16.0E-3	62.5
50.0E-6	324	16.2E-3	61.72839506
50.0E-6	328	16.4E-3	60.97560976
50.0E-6	332	16.6E-3	60.24096386
50.0E-6	336	16.8E-3	59.52380952
50.0E-6	340	17.0E-3	58.82352941
50.0E-6	344	17.2E-3	58.13953488
50.0E-6	348	17.4E-3	57.47126437
50.0E-6	352	17.6E-3	56.81818182
50.0E-6	356	17.8E-3	56.17977528
50.0E-6	360	18.0E-3	55.55555556

Snippet A.1 MATLAB code to display various carrier frequencies

```
% Center figure window on screen
figure
set(gcf,'Position',[0, 0, 1024, 768]); % set desired size 1st
scrsz = get(0,'ScreenSize');     % return screen size
currPos = get(gcf,'Position');   % get current position
xr = scrsz(3) - currPos(3);      % get initial dx
xp = round(xr/2);                % now round and half it
yr = scrsz(4) - currPos(4);      % get dy
yp = round(yr/2);                % round it off
set(gcf,'Position',[xp yp currPos(3) currPos(4)]);

% Clear workspace and close all windows
clear all
bdclose
clc
ctr=1;

% Simulation time step
dt = 46.296e-6;
dt = 50e-6;
Tstop = (1/60)/4;
timeArray = 0:dt:(Tstop/dt);

% Sampling frequency
fs = 1 / dt;

% Sine wave frequency
f60 = 60;        % constant

disp('-------------------------')
fprintf('\tdt = %g usecs\n', dt*1e6);
disp('-------------------------')
for N=1:9
    % Set carrier signal frequency to 1/(4N) of sampling freq.
    fc = fs/(4*N);

   % Carrier signal period
    Tc = 1/fc;

    % Compute the nr. of carrier signal periods
    % per simulation time step
    samplesPerPeriod = Tc/dt;

    % Display console output
    fprintf('%g)\tfc = %2.2f Hz\tfits %g steps/cycle\n', ctr, fc, 4*N);
    ctr=ctr+1;

    % Plot carrier signal ("-" sign adds 180 phase)
    subplot(3,3,N)
    plot(timeArray, -2/pi*asin(sin(2*pi*fc*(timeArray))),'r.-', 'MarkerSize',5);

    % Plot sine signal
    hold on;
    subplot(3,3,N)
    plot(timeArray, sin(2*pi*f60*(timeArray)),'b--');

    % Add a top-title to each plot
    title({['f_{carr} = ',num2str(fc), ' Hz; T_{carr} = ', num2str(1e6*Tc), ' us']; ['(dt = ', num2str(1e6*dt), ' us fits ', num2str(samplesPerPeriod), 'x in Tcarr)']})

    % Add x-axis label to bottom plots only
    if (N>=7)
        xlabel('Time (secs)')
    end
    grid on
    xlim([0 Tstop])
end
subplot(3,3,1), legend('Carr.', 'Ref.')
```

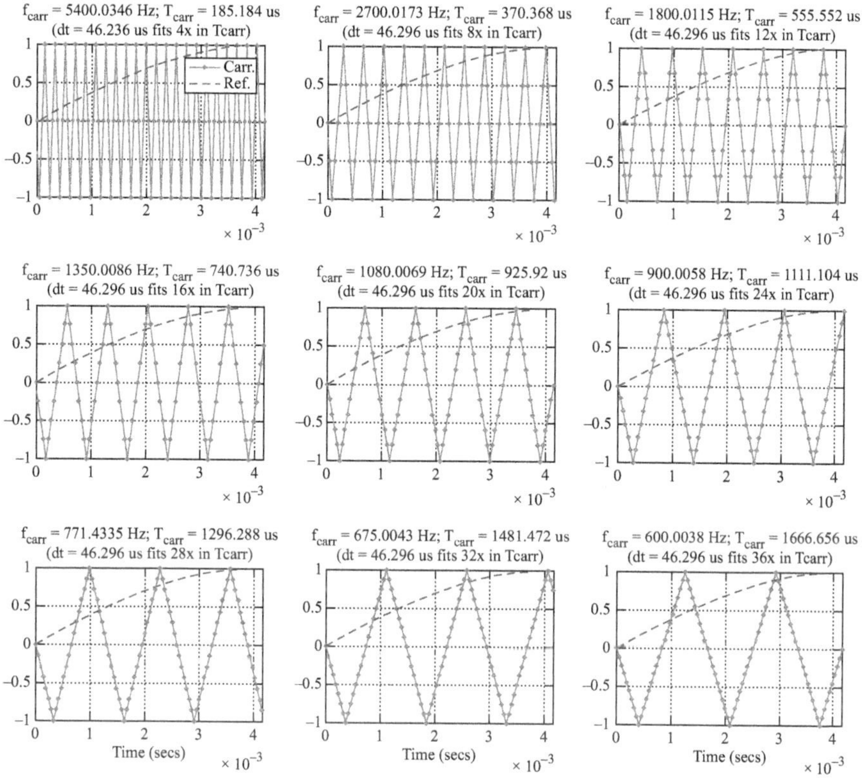

Fig. A.1 Carrier frequencies compatible with $\Delta t = 46.296\ \mu s$

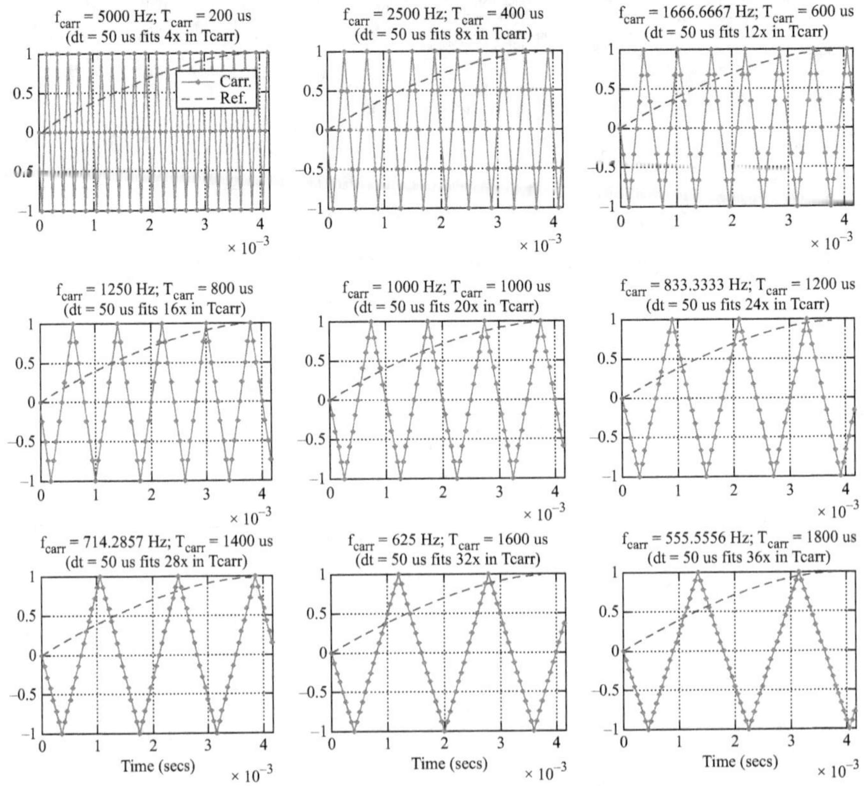

Fig. A.2 Carrier frequencies compatible with $\Delta t = 50\ \mu s$

Appendix B
Considerations of mesh and nodal analysis

The parallel equations of the multicore solver were presented using both mesh and nodal formulations. The reason for including both methods was to compare their performance in a parallel scenario. It was determined that the main difference between these two formulation methods in a parallel environment is the *rate* at which the boundary network grows with increasing number of partitions. This is an important consideration to produce fine-grained parallel simulations. In addition to this outcome, and for completeness, some contrast considerations and practical experiences in working with both methods are summarized in this appendix. Before the summary, a clarification between *mesh* and *loop* analysis is provided.

B.1 Mesh vs. loop analysis

It is common to interchange the terms *mesh* analysis and *loop* analysis when referring to the method of writing voltage equations in electrical networks. While both formulations methods are based on Kirchhoff's voltage law (KVL), the difference between them is in the *approach* each adopts to define the contour (circular) branch paths.

In mesh analysis, one writes KVL equations when there is visual access to a network problem and when the network's planes can be visually identified. As a result of its use, and perhaps also as a result of the way it is taught, it is commonly accepted that mesh analysis is *only* suitable for planar networks of small size and limited to academic examples. This is a misconception: there exist algorithms that can identify planes in large networks and, as a result, mesh analysis can be applicable. Additionally, this book showed that by using a tensor, the internal meshes of power apparatus can be interconnected to form a large mesh network.

As far as how meshes are defined, the mesh analyst normally defines all meshes in a network to circulate in the same direction (e.g., clockwise), defines mesh paths of small lengths, aims to minimize the number of branches common to two or more meshes, and ensures that meshes do not enclose other meshes. This approach requires visual inspection of a network and results in a sparse immittance (mesh impedance) coefficient matrix.

Loop analysis, on the other hand, resorts to graph theory to find contour paths for which independent KVL equations can be written. This makes loop analysis blindfolded (i.e., visual inspection of a network is not required) and (apparently) a more-flexible routine than mesh analysis. It also removes any restrictions on how users connect branches inside or outside power apparatus. When using loop analysis, large non-planar electrical networks are first represented as *directed* graphs (digraph), which consist of vertices and *directed* edges. The correspondence between an electrical network and a graph is—typically—that the electrical network nodes map to graph vertices and branches map to directed edges. This mapping is common when representing electrical circuits as graphs.

The *approach* to define the circular branch paths in loop analysis uses a spanning-tree search algorithm wherein the digraph is decomposed into two sets of edges: one set of connected twigs forming a spanning tree (many spanning trees are possible from the same digraph), and another set of floating links (or chords). The union of twigs and links makes up the original graph.

By adding a link, one at a time, to the spanning tree, exactly one closed (circular) path is formed. This closed path, made up of several twigs and one link, is mapped to the electrical network as a single KVL equation. After defining the KVL equation, the link that was added to the spanning tree is discarded and the process is repeated for the next link. After all links have been used, the loop equations for the entire electrical network are obtained. Since a link can only be used once in this approach, twigs are re-used several times, which results in loops having many intersecting branches and a dense immittance (loop impedance) coefficient matrix.

To summarize the contrast between mesh analysis and loop analysis, it can be said that mesh analysis and loop analysis are:

- *the same* as both methods use KVL equations to describe electrical networks, but
- *different* in the approach implemented to identify the KVL equations, and
- *different* in the resulting sparsity of the coefficient matrix (mesh analysis yields sparser immittance matrices).

B.2 Mesh/loop analysis vs. nodal analysis

There are several reasons why nodal analysis has been, and still is, the preferred formulation approach in power system and circuit solvers. Some reasons include guaranteed sparsity, the *ease* by which the nodal conductance (or admittance) matrix is formed, modeling open terminals in power apparatus is trivial, and because node voltages are readily obtained from the solution.

While the aforementioned reasons are well known, there are also good reasons to consider mesh analysis for a power system and circuit solvers. A comparative table illustrating some pertinent items (by no means exhaustive) is presented here. Some of these differences were mentioned throughout the book. Nonetheless, the remaining differences, characteristics, and questions listed in Table B.1 are each briefly discussed next.

B.2.1 Appearance of graph hyper-branches

It is common to represent large electrical network problems as graphs. When doing so, the nodal conductance matrix of the electrical network is mapped to the graph

Table B.1 Considerations of mesh, loop, and nodal analysis

	Description and/or question	Mesh/loop analysis	Nodal analysis
1	Appearance of graph hyber-branches	Anywhere	At mutual inductances only
2	Can model 0 Ω branches?	Yes	No
3	Can model open circuits?	No	Yes
4	Computation of line voltages	Shunt branches are needed	Presents no problem
5	Diakoptics tearing type	Longitudinal	Traversal
6	Equation count	Smaller	Larger
7	Formation of the network matrix	Harder	Easier
8	Is positive-definiteness possible?	Yes	Yes
9	Kirchhoff law	Voltage	Current
10	Requirement of *datum* node	No	Yes
11	Signs of off-diagonals	Positive and negative	Negative
12	Sparsity	High/medium	High

by mapping the matrix's diagonal and off-diagonals to graph vertices and edges, respectively.[96] In nodal analysis, when there is no mutual inductance in a network, branches appear as graph edges connected across two vertices. When there are mutual inductance segments in an electrical network, branches appear as edges connected across *more* than two vertices, which are known as hyper-edges in graph theory.

In mesh analysis, hyper-edges appear when more than two mesh currents intersect the same branch. This is common at buses with shunt branches, and it is a reason it is believed that the mesh resistance matrix is denser than the nodal conductance matrix. Although this reduces sparsity, it was shown that mesh resistance matrix matrices of >99% sparsity were still possible.

B.2.2 Can model 0 Ω branches?

Modeling 0 Ω branches in mesh analysis is accomplished by *excluding* resistances at the desired location. This is not possible when using nodal analysis unless 1) two adjacent nodes are collapsed to force an order reduction or 2) a 0-V voltage sources is introduced in the network. The latter approach requires resorting to *modified* nodal analysis to handle voltage sources, which may alter the (desirable) symmetric positive-definiteness property of the nodal matrix.

B.2.3 Can model open circuits?

Modeling open circuits in nodal analysis branches is trivial. This is accomplished either by not having a branch between two nodes or by using current sources of 0 amp.

[96] The terms *nodes* and *branches* are typically used in both circuit theory and graph theory; the terms *vertices* and *edges* are normally restricted to graph theory.

In mesh analysis this is not possible because meshes are defined as *closed* paths of finite impedance; thus, a finite impedance must exist along the mesh. (Inserting a 1-MΩ resistance in a mesh's path is not considered an open circuit.) In order to obtain infinite impedance in a closed path to model an open circuit, the mesh must be removed from the equation set. This process was demonstrated by marking meshes for removal in the flow chart shown in Fig. 6.5.

B.2.4 Computation of line voltages

Computation of instantaneous line voltages using nodal analysis is found as the voltage difference across two nodes. In mesh analysis, the solution vector returns mesh currents from which branches voltage and currents can be obtained. If line-to-line branches exist, the line voltages *can* be computed. If such branches, wherein voltage measurement is desired, do not exist, the line voltages must be obtained in either of two ways: first, by adding 1 MΩ branches across the nodes of interest; and, second, by searching for a closed path across the nodes of interest and calculate the net voltage drop across these nodes. The latter approach constitutes programming overhead, particularly so when the network is large. This is a limitation of using mesh (or loop) analysis when compared to nodal analysis.

B.2.5 Diakoptics tearing type

There are two types of diakoptics-based tearing: *traversal* and *longitudinal* tearing [95]. Traversal tearing is used when systems are formulated in node voltages as variables. Traversal tearing tears two radially attached networks by removing tie-lines, solving each subsystem's node voltages, and injecting the tie-line currents back into each subsystem. Longitudinal tearing, on the other hand, is used when systems are formulated using mesh (or loop) currents as variables. Longitudinal tearing tears two networks attached adjacently by replacing shunt branches with short circuits, solving each subsystem's mesh currents, and impressing the tie-line voltages back into each subsystem. These differences can be seen graphically by referring to the node and mesh tearing illustrations shown in Chapter 7.

B.2.6 Equation count

Experience teaches that there are typically less meshes than nodes[97] in power systems. This physical property implies that the mesh resistance matrix is of lesser order than the nodal conductance matrix is. Combining the high sparsity of the mesh resistance matrix with its low order makes mesh analysis an excellent alternative to simulation methods based on nodal analysis. This consideration is important when using full-matrix solvers. In practical contexts, however, more important than matrix order is the number of non-zeros in the matrix LU or Cholesky factors.

[97] The numerical relation is strongly dependent on the number of shunt branches included in the network.

B.2.7 Formation of the network matrix

Formation of the mesh (or loop) matrices is typically believed to require graph theory which is not trivial—a reason why nodal analysis is preferred in practice. The nodal conductance matrix can be formed using a netlist;[98] where the branch parameters and node numbers are known, the nodal conductance matrix can be formed with ease. This is still another reason why nodal analysis is preferred in practice. However, this book demonstrated that the mesh resistance matrix can be obtained, contrary to common belief, without graph theory, although some additional programming is required. Formation of (both) the nodal and mesh matrices using a tensor approach was the focus of Chapter 6.

B.2.8 Is positive-definiteness possible?

Fast algorithms to solve systems of equations in the form of $\mathbf{A} \cdot \mathbf{x} = \mathbf{b}$ use Cholesky decompositions instead of LU factorizations. The advantage of a Cholesky decomposition is the efficient forward-backward substitution when using triangular factors $\mathbf{A} = \mathbf{L}\mathbf{L}^{\mathrm{T}}$ which, combined with sparse storage techniques [194], render very efficient computer implementations. To perform Cholesky factorizations, however, the coefficient matrix $\mathbf{A}$ must be symmetric positive-definite. Whether a matrix is symmetric positive-definite depends on the structure of $\mathbf{A}$. For example, in conventional nodal analysis where voltage sources or dependent sources are not permitted, $\mathbf{A}$ is a symmetric positive-definite matrix. However, in modified nodal analysis, the inclusion of dependent sources removes symmetric positive-definiteness properties in favor of added flexibility by using unknown source values as part of the solution vector.

In mesh (or loop) analysis, only voltage sources are permitted. Using current sources or dependent sources in mesh analysis also jeopardizes the symmetric positive-definiteness property. Because it is desirable to use Cholesky factorizations (or decompositions) in the solution of a system of equations such as $\mathbf{A} \cdot \mathbf{x} = \mathbf{b}$, the transformer model in section 5.5 eliminated dependent node and mesh variables instead of using dependent sources to model the voltage and current dependencies. In both, mesh (or loop) analysis and nodal analysis, symmetric positive-definiteness is possible as long as care is exercised to prevent situations such as zero-diagonals and loss of matrix symmetry.

B.2.9 Kirchhoff law

Formation of the mesh resistance matrix or the loop resistance matrix is based on KVL, whereas formation of the nodal conductance matrix is based on Kirchhoff's current law (KCL). The KCL approach, aside from being more intuitive and being used in commercial power system simulators, has the advantage that the nodal conductance matrix can be formed by stamping branches one at a time. Stamping branches in mesh analysis is also possible as long as the meshes numberings are known.

[98] A text file listing all branches and nodes in an electrical circuit.

B.2.10 Requirement of a datum node

Nodal analysis is a special case of cutset analysis. Cutset analysis states that the net flow through a cutset is zero. This resembles a KCL situation at nodes in electrical networks. In this regard, cutsets *can* be identified at electrical network nodes. To write cutset (or KCL) equations at a network node, a *datum* node must exist, however. *Datum* nodes, commonly referred to as *reference* or *ground* nodes, are required when using nodal analysis. In mesh analysis, *datum* nodes are not required which makes mesh analysis inherently suitable for ungrounded networks. If nodal analysis is used in ungrounded networks, either high-resistance branches can be added to the ground plane or an alternate node can be defined as the *datum* node.

B.2.11 Signs of off-diagonals

When using a mesh resistance matrix, there is flexibility in how the meshes are defined. If desired, mesh current directions can be reversed to avoid negative off-diagonals. In nodal analysis, off-diagonals are mostly negative; this outcome cannot be controlled. The only case where off-diagonals in nodal analysis can be positive is in networks with mutual inductances.

B.2.12 Sparsity

The sparsity of a matrix is the percentage of entries that have a value of zero. Sparsity is a highly desirable characteristic of matrices in general, as it reduces their memory storage requirement and the number of arithmetic operations required to perform common operations on them (e.g., factorizations, additions, multiplications, forward-backward substitutions in the sparse factors, to name a few). When using mesh analysis, sparsity is controlled by how many branches are common to two or more meshes. If meshes are defined to be short in length and to minimize the number of intersecting branches, high sparsity is possible. When using nodal analysis, sparsity is determined by the degree of each node (i.e., by how branches are incident at each node). Since node degrees are typically low (2 or 3), this naturally results in sparse matrices.

To end, it is re-emphasized that the nodal matrix is a *special* case of the possible *cutset* analysis and that the mesh matrix is a special case of *tieset* analysis [149]. Both the nodal and mesh matrices are believed to return the sparsest network matrix of their genre. Sparser matrices than the nodal and mesh matrices *may* be possible [194]; however, the deep search required to find them may not be worthwhile in practice due to the associated programming time, computational overhead, and development of good heuristics required to obtain a marginal gain in sparsity. Stated differently, the nodal and mesh matrices are often >99% sparse, which is a sufficient numerical indicator in practice.

References

[1] F. M. Uriarte, R. E. Hebner, A. Kwasinski, A. L. Gattozzi, *et al.*, "Technical Cross–fertilization between Terrestrial Microgrids and Ship Power Systems," submitted to *IEEE Trans. Smart Grid.*

[2] K. L. Butler-Purry, N. D. R. Sarma, C. Whitcomb, H. D. Carmo, *et al.*, "Shipboard Systems Deploy Automated Protection," *IEEE Computer Applications in Power*, Apr. 1998, vol. 11, pp. 31–36.

[3] H. Zhang, K. L. Butler-Purry, and N. D. R. Sarma, "Simulation of Ungrounded Shipboard Power Sytems in PSpice," *IEEE Midwest Symposium on Circuits and Systems*, Notre Dame, IN, 1998.

[4] A. T. Adediran, H. Xiao, and K. L. Butler-Purry, "The Modeling and Performance Testing of a Shipboard Power System," *33rd Annual Frontiers of Power Conference*, Oklahoma State University, Oct. 30–31, 2000.

[5] A. T. Adediran, H. Xiao, and K. L. Butler-Purry, "Fault Studies of an U.S. Naval Shipboard Power System," *North American Power Symposium (NAPS)*, University of Waterloo, Canada, Oct. 23–24, 2000.

[6] A. Adediran, H. Xiao, and K. L. Butler-Purry, "The Modeling and Simulation of a Shipboard Power System in ATP," *International Conference on Power System Transients (IPST)*, New Orleans, USA, 2003.

[7] K. L. Butler-Purry and N. D. R. Sarma, "Visualization for Shipboard Power Systems," *IEEE Hawaii International Conference on System Sciences*, Big Island, Hawaii, Jan. 6–9, 2003, pp. 648–656.

[8] M. M. Medina, L. Qi, and K. L. Butler-Purry, "A Three Phase Load Flow Algorithm for Shipboard Power Systems (SPS)," *IEEE Transmission and Distribution Conference and Exposition*, 2003, vol. 1, pp. 227–233.

[9] L. Qi and K. L. Butler-Purry, "Reformulated Model Based Modeling and Simulation of Ungrounded Stiffly Connected Power Systems," *IEEE Power Engineering Society General Meeting*, 2003, pp. 725–730.

[10] K. L. Butler-Purry, "An ONR Young Investigator Project—Predictive Reconfiguration of Shipboard Power Systems," *Power Engineering Society General Meeting*, 2004, p. 975.

[11] K. Miu, V. Ajjarapu, K. Butler-Purry, D. Niebur, *et al.*, "Testing of Shipboard Power Systems: A Case for Remote Testing and Measurement," *IEEE Electric Ship Technologies Symposium*, Philadelphia, PA, 2005.

[12] S. K. Srivastava and K. L. Butler-Purry, "A Pre-hit Probabilistic Reconfiguration Methodology for Shipboard Power Systems," *IEEE Electric Ship Technologies Symposium*, Philadelphia, PA, 2005.

[13] F. M. Uriarte and K. L. Butler-Purry, "Real-Time Simulation of a Small-Scale Distribution Feeder Using Simulink and a Single PC," *The North American Power Symposium (NAPS)*, Iowa State University. Oct. 24–25, 2005, pp. 213–218.

[14] K. L. Butler-Purry and N. D. R. Sarma, "Geographical Information Systems for Automation of Shipboard Power Systems," *Naval Engineers Journal*, 2006, vol. 118, pp. 63–75.

[15] F. M. Uriarte and K. L. Butler-Purry, "Real-Time Simulation of a Small Power System with xPC," *IEEE Transmission and Distribution Conference and Expo*, Dallas, TX, 2006.

[16] F. M. Uriarte and K. L. Butler-Purry, "Real-Time Simulation of a Small Power System Using a PC," *IEEE Transmission and Distribution Conference (poster)*, Dallas, TX, May 21–24, 2006, pp. 87–88.

[17] F. M. Uriarte and K. L. Butler-Purry, "Real-Time Simulation Using PC-based Kernels," *Power Systems Conference & Expo (PSCE'06)*, Atlanta, GA, Oct. 29–Nov. 1, 2006, pp. 1991–1995.

[18] F. M. Uriarte and K. L. Butler-Purry, "Diakoptics in Shipboard Power System Simulation," *North American Power Symposium (NAPS)*, Southern Illinois University Carbondale, Sep. 17–19, 2006, pp. 201–210.

[19] K. L. Butler-Purry, G. R. Damle, N. D. R. Sarma, F. Uriarte, *et al.*, "Test Bed for Studying Real-Time Simulation and Control for Shipboard Power Systems," *Electric Ship Technologies Symposium (ESTS 2007)*, Arlington, VA, May 21–23, 2007, pp. 434–437.

[20] F. M. Uriarte and K. L. Butler-Purry, "A Partitioning Approach for the Parallel Simulation of Ungrounded Shipboard Power Systems using Kron's Diakoptics and Loop Analysis," *Summer Computer Simulation Conference 2007 (SCSC'07)*, San Diego, CA, Jul. 16–17, 2007.

[21] X. Feng, T. Zourntos, K. Butler-Purry, and S. Mashayekh, "Dynamic Load Management for NG IPS Ships," *Power Engineering Society General Meeting*, Minneapollis, MN, 2010.

[22] F. M. Uriarte and K. L. Butler-Purry, "Multicore Simulation of an AC-radial Shipboard Power System," *Power Engineering Society General Meeting*, Minneapolis, MN, Jul. 25–29, 2010, pp. 1–8.

[23] F. M. Uriarte and K. L. Butler-Purry, "A Partitioning Approach for the Parallel Simulation of Ungrounded Shipboard Power Systems using Kron's Diakoptics and Loop Analysis," *Summer Computer Simulation Conference*, San Diego, CA, Jul. 15–18, 2007.

[24] H. Zhang, K. L. Butler, N. D. R. Sarma, H. DoCarmo, *et al.*, "Analysis of Tools for Simulation of Shipboard Electric Power Systems," *Elsevier Science – Electric Power Systems Research*, Jun. 2001, vol. 58, pp. 111–122.

[25] X. Feng, K. Butler-Purry, and T. Zourntos, "Multi-agent System-based Real-time Load Management for All-electric Ship Power Systems in DC Zone Level," *IEEE Transaction Power Systems*, 2013.

[26] IEEE Std 45, *Recommended Practice for Electrical Installations on Shipboard*, p. i, 1998. doi: 10.1109/IEEESTD.1998.91149.

[27] DoD Military Specification MIL-C-24643A (1994), *Cables and Cords, Electric, Low Smoke, For Shipboard Use, General Specification for* [S/S BY MIL-DTL-24643B].

[28] DoD Military Handbook MIL-HDBK-299 (SH) (1989), *Cable Comparison Handbook - Data Pertaining to Electric Shipboard Cable*, Department of Defense, Washington, DC.

[29] IEEE Std 1709-2010, *Recommended Practice for 1 kV to 35 kV Medium-Voltage DC Power Systems on Ships*, pp. 1, 54, Nov. 2 2010. doi: 10.1109/IEEESTD.2010.5623440.

[30] "Naval Ships' Technical Manual (Ch. 320)—Electrical Power Distribution Systems (rev. 2)," edn., 1998 [Online]. Available: http://www.hnsa.org/doc/nstm/ch320.pdf.

[31] The MathWorks, Inc. (2010). *Simulink 7 User's Guide* [Online]. Available: http://www.mathworks.com/help/toolbox/simulink/.

[32] IEEE Std 399-1997, *IEEE Recommended Practice for Industrial and Commercial Power Systems Analysis (Brown Book)*, pp. 1, 488, Aug. 31, 1998. doi: 10.1109/IEEESTD.1998.88568.

[33] N. H. Doerry, "Next Generation Integrated Power Systems for the Future Fleet," *Corbin A. McNeill Symposium*, Annapolis, MD, 2009.

[34] S. B. V. Broekhoven, N. Judson, S. V. T. Nguyen, and W. D. Ross, "Microgrid Study: Energy Security for DoD Installations," Available [online] http://serdp-estcp.org/2012.

[35] F. Katiraei and J. R. Aguero, "Solar PV Integration Challenges," *Power & Energy Magazine*, May/Jun. 2011, pp. 62–71.

[36] A. Kwasinski, A. Toliyat, and F. M. Uriarte, "Effects of High Penetration Levels of Residential Photovoltaic Generation: Observations from Field Data," *International Conference on Renewable Energy Research and Applications (ICRERA)*, Nagasaki, Japan, Nov. 11–14, 2012.

[37] National Fire Protection Association, *NFPA 70—National Electric Code*, 2005.

[38] Manitoba HVDC Research Centre, Inc., *EMTDC User's Guide—A Comprehensive Resource for EMTDC*. Manitoba, Canada: Manitoba HVDC Research Centre, 2005.

[39] E. Broughton, B. Langland, E. Solodovnick, and G. Croft, *Virtual Test Bed User's Manual*. Columbia, SC: University of South Carolina, 2003.

[40] J. R. Marti, L. Linares, J. A. Hollman, and F. A. Moreira, "OVNI: Integrated Software/Hardware Solution for Real-Time Simulation of Large Power Systems," *Power Systems Computation Conference (PSCC'02)*, Sevilla, Spain, 2002, pp. 1–7.

[41] P. M. Lee, S. Ito, T. Hashimoto, J. Sato, *et al.*, "A Parallel and Accelerated Circuit Simulator with Precise Accuracy," 2002, pp. 213–218.

[42] R. M. Kielkowski, *Inside SPICE*, 2nd edn. New York: McGraw-Hill, 1998.

[43] C. Dufour, J. Mahseredjian, J. Bélanger, and J. L. Naredo, "An Advanced Real-time Electro-magnetic Simulator for Power Systems with a Simultaneous State-space Nodal Solver," *IEEE/PES Transmission and Distribution Conference and Exposition: Latin America*, São Paulo, Brazil, 2010.

[44] I. Dudurytch and V. Gudym, "Mesh-Nodal Network Analysis," *IEEE Transaction Power Systems*, Nov. 1999, vol. 14, pp. 1375–1381.

[45] J. Arrillaga and N. R. Watson, *Computer Modeling of Electrical Power Systems*, 2nd edn. Christchurch, New Zealand: John Wiley & Sons, LTD, 2001.

[46] M. Crow, *Computational Methods for Electric Power Systems*. Rolla, MO: CRC Press, 2003.

[47] H. W. Dommel, *Electromagnetic Transients Program Theory Book (EMTP Theory Book)*. Portland: Bonneville Power Administration, 1986.

[48] P. C. Krause, O. Wasynczuk, and S. D. Sudhoff, *Analysis of Electric Machinery and Drive Systems*, 2nd edn. Piscatawy: IEEE Press, 2002.

[49] T. L. Pillage, R. A. Rohrer, and C. Visweswariah, *Electronic Circuit & System Simulation Methods*. NY: McGraw-Hill, Inc., 1995.

[50] K. Strunz, *Numerical Methods for Real Time Simulation of Electromagnetics in AC/DC Network Systems*. Düsseldorf: VDI Verlag, 2002.

[51] N. Watson and J. Arrillaga, *Power Systems Electromagnetic Transients Simulation*. London: IEE, 2003.

[52] P. W. Sauer and M. A. Pai, *Power System Dynamics and Stability*. Upper Saddle River, NJ: Prentice Hall, 1998.

[53] F. A. Moreira, J. R. Marti, and L. Linares, "Electromagnetic Transients Simulation with Different Time Steps: The Latency Approach," *International Conference on Power System Transients (IPST)*, New Orleans, LA, 2003.

[54] P. Kuffel, K. Kent, and G. Irwin, "The Implementation and Effectiveness of Linear Interpolation within Digital Simulation," *Electrical Power & Energy Systems*, 1997, vol. 19, pp. 221–227.

[55] P. Kuffel, K. Kent, and G. Irwin, "The Implementation and Effectiveness of Linear Interpolation within Digital Simulation," *International Conference on Power System Transients*, Lisbon, 1995, pp. 499–504.

[56] K. Strunz, "Flexible Numerical Integration for Efficient Representation of Switching in Real Time Electromagnetic Transients Simulation," *IEEE Transaction Power Delivery*, Jul. 2004, vol. 19, pp. 1276–1283.

[57] A. M. Gole, A. Keri, C. Nwankpa, E. W. Gunther, *et al.*, "Guidelines for Modeling Power Electronics in Electric Power Engineering Applications," *IEEE Transaction Power Delivery*, Jan. 1997, vol. 12, pp. 505–514.

[58] A. M. Gole, "Electromagnetic Transient Simulation of Power Electronic Equipment in Power Systems: Challenges and Solutions," *Power Engineering Society General Meeting*, Montreal, Quebec, Oct. 2006, pp. 1301–1306.

[59] A. E. A. Araujo, H. W. Dommel, and J. R. Marti, "Converter Simulations with the EMTP: Simultaneous Solution and Backtracking Technique," *Joint International Power Conference (IEEE - NTUA) - Athens Power Tech*, Athens, Greece, 1993, pp. 941–945.

[60] A. E. A. Araujo, H. W. Dommel, and J. R. Marti, "Simultaneous Solution of Power and Control Systems Equations," *IEEE Transaction on Power Systems*, 1993, vol. 8, pp. 1483–1489.

[61] J. R. Marti and J. Lin, "Suppression of Numerical Oscillations in the EMTP," *IEEE Transaction Power Systems*, 1989, vol. 4, pp. 739–747.

[62] J. Langston, S. Suryanarayanan, M. Steurer, M. Andrus, *et al.*, "Experiences with the Simulation of a Notional All-electric Ship Integrated Power System on a Large-scale High-speed Electromagnetic Transients Simulator," *Power Engineering Society General Meeting*, Jun. 18–22, 2006.

[63] Y. Gong, L. Chen, Y. Chen, and Y. Xu, "A Parallel Based Real-time Electromagnetic Transient Simulator for IPS," *Electric Ship Technologies Symposium*, Alexandria, VA, Apr. 10–13, 2011, pp. 96–101.

[64] R. Kuffel, J. Giesbrecht, T. Maguire, R. P. Wierckx, *et al.*, "RTDS—A Fully Digital Power System Simulator Operating in Real Time," *IEEE WESCANEX Comm. Power, and Computing*, 1995, pp. 300–305.

[65] L. R. Linares and J. R. Martí, "A Resynchronization Algorithm for Topological Changes in Real Time Fast Transients Simulation," *Power Systems Computation Conference (PSCC'02)*, Sevilla, Spain, Jun. 24–28, 2002.

[66] M. Zou, J. Mahseredjian, G. Joos, B. Delourme, *et al.*, "Interpolation and Reinitialization in Time-domain Simulation of Power Electronic Circuits," *Electric Power Systems Research*, May 2006, vol. 76, pp. 688–694.

[67] M. O. Faruque, V. Dinavahi, and W. Xu, "Algorithms for the Accounting of Multiple Switching Events in Digital Simulation of Power-electronic Systems," *IEEE Transaction Power Delivery*, 2005, vol. 20, pp. 1157–1167.

[68] A. E. A. Araujo, "Numerical Instabilities in Power System Transient Simulations," *Electrical and Computer Engineering*, vol. PhD, edn. Vancouver: The University of British Columbia, 1993, p. 143.

[69] G. D. Irwin, D. A. Woodford, and A. M. Gole, "Precision Simulation of PWM Controllers," *International Power System Transients Conference (IPST'01)*, Brazil, 2001, pp. 161–165.

[70] T. Funaki, T. Takazaqa, T. Tada, A. Kurita, *et al.*, "A Study on the Usage of CDA in EMTP Simulations," *International Conference on Power Systems Transients*, New Orleans, 2003, pp. 1–6.

[71] J. Mahseredjian, V. Dinavahi, and J. A. Martinez, "Simulation Tools for Electromagnetic Transients in Power Systems: Overview and Challenges," *IEEE Transaction Power Delivery*, Jul. 2009, vol. 24, pp. 1657–1669.

[72] D. Schuller, *C# Game Programming: For Serious Game Creation*. Boston, MA: Course Technology, a part of Cengage Learning, 2011.

[73] R. Hebner, J. Herbst, and A. Gattozzi, "Large Scale Simulations of a Ship Power System with Energy Storage and Multiple Directed Energy Loads," *Grand Challenges in Modeling & Simulation (GCMS 2010)*, Ottawa, Canada, Jul. 11–14, 2010, pp. 430–435.

[74] G. Chen and X. Zhou, "Asynchronous Parallel Electromagnetic Transient Simulation of Large Scale Power System," *International Journal of Emerging Power Systems*, 2005, vol. 2, pp. 1–13.

[75] R. Crosbie, J. Zenor, R. Bednar, D. Word, *et al.*, "Using Attached Processors to Achieve High-speed Real-time Simulation," *2nd International Conference on Advances in System Simulation*, 2010, pp. 140–143.

[76] R. Crosbie, J. Zenor, R. Bednar, D. Word, *et al.*, "High-speed, Scalable, Real-time Simulation Using DSP Arrays," *Parallel and Distributed Simulation,*

2004. PADS 2004. 18th Workshop on 16–19 May 2004, Page(s):52–59, 2004, p. 52.

[77] A. Greenwood, *Electrical Transients in Power Systems*, 2nd edn. New York, NY: John Wiley & Sons, 1991.

[78] A. Hambley, *Electronics*, 2nd edn. Upper Saddle River, NJ: Prentice Hall, 2000.

[79] H. W. Dommel, "Digital Computer Solution of Eelectromagnetic Transients in Single- and Multiphase Networks," *IEEE Transaction Power Apparatus and Systems*, Apr. 1969, vol. PAS-88, pp. 388–399.

[80] J. A. Lima, "Numerical Instability due to EMTP-TACS Interrelation," *EMTP Newsletter*, 1985, vol. 5, pp. 21–33.

[81] N. R. Watson and G. D. Irwin, "Electromagnetic Transient Simulation of Power Systems Using Root-matching Techniques," *IEE Proceedings: Generation, Transmission and Distribution*, 1998, vol. 145, pp. 481–486.

[82] J. M. Smith, *Mathematical Modeling and Digital Simulation for Engineers and Scientists*, 2nd edn. Washington, DC: John Wiley, 1987.

[83] W. Gao, E. Solodovnik, R. Dougal, G. Cokkinides, *et al.*, "Elimination of Numerical Oscillations in Power System Dynamic Simulation," *18th IEEE Applied Power Electronics Conference and Exposition*, Miami, FL, Feb. 2003, pp. 790–794.

[84] C.-W. Ho, "The Modified Nodal Approach to Network Analysis," *IEEE International Symposium on Circuits and Systems*, 1974, pp. 505–509.

[85] H. Chung-Wen, A. Ruehli, and P. Brennan, "The Modified Nodal Approach to Network Analysis," *IEEE Transaction Circuits and Systems*, 1975, vol. 22, pp. 504–509.

[86] C.-W. Ho, A. E. Ruehli, and P. A. Brennan, "The Modified Nodal Approach to Network Analysis," *IEEE Transaction Circuits and Systems*, Jun. 1975, vol. CAS-22, pp. 504–509.

[87] J. Vlach, "Tableau and Modified Nodal Formulations," *The Circuits and Filters Handbook*. vol. 2nd edn., W.-K. Chen, Ed., CRC Press, 2003, Chapter 22, p. 663.

[88] F. M. Uriarte and R. Hebner, "Assessing Confidence in Parallel Simulation Results," in *Electric Ship Technologies Symposium*, Arlington, VA, Apr. 22–23, 2013.

[89] F. M. Uriarte and C. Dufour, "Multicore Methods to Accelerate Ship Power System Simulations," in *Electric Ship Technologies Symposium*, Arlington, VA, Apr. 22–23, 2013.

[90] P. Pejovic and D. Maksimovic, "A Method for Fast Time-domain Simulation of Networks with Switches," *IEEE Transaction Power Electronics*, 1994, vol. 9, pp. 449–456.

[91] H. Macbahi, A. Ba-Razzouk, and A. Chériti, "Decoupled Parallel Simulation of Power Electronics Systems Using Matlab-Simulink," *International Conference on Parallel Computing in Electrical Engineering (PARELEC'00)*, 2000, pp. 232–236.

[92] W. H. Liao, S. C. Wang, and Y. H. Liu, "Generalized Simulation Model for a Switched-Mode Power Supply Design Course Using MATLAB/SIMULINK," *IEEE Transaction Education*, 2012, vol. 55, pp. 36–47.

[93] N. Mohan, T. M. Undeland, and W. P. Robbins, *Power Electronics*, 3rd edn. New York, NY: John Wiley & Sons, 2003.

[94] M. Rashid, *Power Electronics Circuits, Devices, and Applications*, 2nd edn. Pensacola, FL: Pearson Education, 1993.

[95] F. E. Rogers, *Topology and Matrices in the solution of Networks*. London: Iliffe Books, Ltd., 1965.

[96] K. S. Chao, "State-Variable Techniques," *The Circuits and Filters Handbook*. 2nd edn., W.-K. Chen, Ed., CRC Press, 2003, Chapter 26, p. 799.

[97] A. L. Shenkman, *Transient Analysis of Electric Power Circuits*. Holon, Israel: Springer, 2005.

[98] K. Ogata, *Modern Control Engineering*, 2nd edn. Englewood Cliffs, NJ: Prentice-Hall, 1996.

[99] Z. A. Yamayee and J. L. Bala, *Electromechanical Energy Devices and Power Systems*. New York, NY: Wiley, 1994.

[100] M. Armstrong, J. R. Marti, L. R. Linares, and P. Kundur, "Multilevel MATE for Efficient Simultaneous Solution of Control Systems and Nonlinearities in the OVNI Simulator," *IEEE Transaction Power Systems*, Aug. 2006, vol. 21, pp. 1250–1259.

[101] Southwire Company, *Power Cable Manual*, 4th edn., Carrollton, Georgia: Southwire Company, 2005.

[102] IEEE Std 1580-2010 *IEEE Recommended Practice for Marine Cable for Use on Shipboard and Fixed or Floating Facilities*, pp. 0_1, 0_2, 2002. doi: 10.1109/IEEESTD.2002.93624.

[103] M. Mazzola, A. Card, S. Grzybowski, M. Islam, *et al.*, "Impact of Dielectric Requirements on Design of Marine Cabling," *2012 ESRDC 10th Anniversary Meeting*, Austin, TX, Jun. 4–6, 2012.

[104] Naval Sea Systems Command NAVSEA SE000-00-EI M-100 (1983), *Electronics Installation and Maintenance Book: Article 3-3.2—Misconceptions of a Shipboard Ungrounded System*.

[105] J. I. Ykema, "Protective Devices in Navy Shipboard Electrical Power Systems," *Naval Engineers Journal*, 1988, vol. 100, pp. 166–179.

[106] Eaton Corporation plc. (2010). *Circuit Breakers Naval Shipboard Use (PG01218003E)* [Online]. Available: http://www.eaton.com/ecm/groups/public/@pub/@electrical/documents/content/pg01218003e.pdf.

[107] J. L. Blackburn, *Protective Relaying: Principles and Applications*. New York, NY: M. Dekker, 1987.

[108] S. H. Horowitz and A. G. Phadke, *Power System Relaying*, 2nd edn. Blacksburg, VA: John Wiley & Sons, Inc., 1996.

[109] R. D. Garzon, *High Voltage Circuit Breakers: Design and Applications*, 2nd edn. New York, NY: Marcel Dekker, 2002.

[110] V. V. Terzija, M. Popov, V. Stanojevic, and Z. Radojevic, "EMTP simulation and spectral domain features of a long arc in free air," *18th Int'l Conf. Electric Distribution*, Jun. 2005, pp. 1–4.

[111] A. Parizad, H. R. Baghaee, A. Tavakoli, and S. Jamali, "Optimization of Arc Models Parameters Using Genetic Algorithm," *International Conference on Electrical Power & Energy Conversion Systems*, Ottawa, Canada, Nov. 10–12, 2009, pp. 1–7.

[112] J. Andrea, P. Schweitzer, and E. Tisserand, "A New DC and AC Arc Fault Electrical Model," *56th IEEE Holm Conference on Electrical Contacts*, Charleston, SC, Oct. 4–7, 2010, pp. 1–6.

[113] V. V. Terzija and H. J. Koglin, "On the Modeling of Long Arc in Still Air and Arc Resistance Calculation," *IEEE Transaction Power Delivery*, Jul. 2004, vol. 19, pp. 1012–1017.

[114] A. T. Adediran, "Final Report: Modeling of Components of a Surface Combatant Ship with the ATP Software," Power Systems Automation Laboratory, Texas A&M University, College Station, Texas, Dec. 31, 2003.

[115] ESRDC Electrical Integrated Product Team, "Modeling of Shipboard Power Systems," *10th Anniversay Electric Ship Research and Development Consortium Meeting*, Austin, Texas, Jun. 4–6, 2012.

[116] J. Langston, M. Steurer, J. Crider, S. Sudhoff, *et al.*, "Waveform-Level Time-Domain Simulation Comparison Study of Three Shipboard Power System Architectures," *Grand Challenges in Modeling and Simulation*, Genoa, Italy, 2012.

[117] H. Ali, R. Dougal, A. Ouroua, R. Hebner, *et al.*, "Cross-Platform Validation of Notional Baseline Architecture Models of Naval Electric Ship Power Systems," *Electric Ship Technologies Symposium*, Alexandria, VA, Apr. 10–13, 2011, pp. 78–83.

[118] M. Steurer, S. Woodruff, R. Wen, H. Li, *et al.*, "Accuracy and Speed of Time Domain Network Solvers for Power Systems Electronics Applications," *10th European Conference on Power Electronics and Applications (EPE2003)*, Toulouse, France, Sep. 2–4, 2003.

[119] J. R. Marti and T. O. Myers, "Phase-domain Induction Motor Model for Power System Simulators," *IEEE Wescanex Communications, Power, and Computing Conference*, 1995, pp. 276–282.

[120] B. K. Bose, *Modern Power Electronics and AC Drives*. Upper Saddle River, NJ: Prentice Hall, 2002.

[121] R. Krishnan, *Electric Motor Drives Modelling, Analysis, and Control*. Blacksburg, VA: Prentice Hall, 2001.

[122] P. M. Anderson and A. A. Fouad, *Power System Control and Stability*, 5th edn. New York, NY: IEEE Press, 1994.

[123] G. Kron, *Tensors for Circuits (formerly entitled A Short Course in Tensor Analysis for Electrical Engineers)*, 2nd edn. Schenectady, NY: Dover Publications, Inc., 1959.

[124] G. Kron, "Tensorial Analysis of Integrated Transmission Systems," *AIEE*, 1952, vol. 71, pp. 814–822.

[125] G. Kron, *Tensor Analysis of Networks*. Schenectady, NY: John Wiley & Sons, Inc., 1939.

[126] M. A. Pai, D. P. S. Gupta, and K. R. Padiyar, *Small Signal Analysis of Power Systems*. Bangladore, India: Alpha Science International Ltd., 2004.

[127] W. A. Lewis, "A Basic Analysis of Synchronous Machines, Part I," *Transactions AIEE*, 1958, vol. 77, pp. 436–55.

[128] J. S. Mayer and O. Wasynczuk, "An Efficient Method of Simulating Stiffly Connected Power Systems with Stator and Network Transients Included," *Power Systems, IEEE Transaction*, 1991, vol. 6, pp. 922–929.

[129] S. A. Nasar and I. Boldea, *Electric Machines: Dynamics and Control*. Boca Raton, FL: CRC Press, 1992.

[130] T. J. McCoy, "Dynamic Simulation of Shipboard Electric Power Systems," Master's of Science Thesis, Dept. of Ocean Engineering, Massachusetts Institute of Technology, Cambridge, MA, 1993.

[131] P. Kundur, N. J. Balu, and M. G. Lauby, *Power System Stability and Control*. New York, NY: McGraw-Hill, 1994.

[132] P. M. Anderson, *Analysis of Faulted Power Systems*, 5th edn. New York, NY: IEEE Press, 1995.

[133] K.-W. Louie, "Phase-domain Synchronous Generator Model for Transients Simulation," University of British Columbia, M.S. Thesis, 1995.

[134] R. M. Hamouda, M. A. Badr, and A. I. Alolah, "Effect of Torsional Dynamics on Salient Pole Synchronous Motor-driven Compressors," *Energy Conversion, IEEE Transaction on*, 1996, vol. 11, pp. 531–538.

[135] J. Machowski, J. W. Bialek, and J. R. Bumby, *Power System Dynamics and Stability*. Chichester, NY: John Wiley, 1997.

[136] J. R. Marti, "A Phase-domain Synchronous Generator Model Including Saturation Effects," *IEEE Transaction Power Systems*, 1997, vol. 12, pp. 222–229.

[137] J. G. Ciezki and R. W. Ashton, "The Resolution of Algebraic Loops in the Simulation of Finite-Inertia Power Systems," *IEEE International Symposium on Circuits and Systems*, 1998, pp. 342–345.

[138] C.-M. Ong, *Dynamic Simulation of Electric Machinery Using MATLAB/SIMULINK*. Upper Saddle River, NJ: Prentice Hall PTR, 1998.

[139] X. Cao, A. Kurita, H. Mitsuma, Y. Tada, *et al.*, "Improvements of Numerical Stability of Electromagnetic Transient Simulation by Use of Phase-domain Synchronous Machine Models," *Electrical Engineering in Japan*, 1999, vol. 128, pp. 53–62.

[140] H. W. Beaty, *Handbook of Electric Power Calculations*, 3rd edn. New York, NY: McGraw-Hill, 2001.

[141] W. Gao, "New Methodology for Power System Modeling and Its Application in Machine Modeling and Simulation," PhD dissertation, Georgia Institute of Technology, 2002.

[142] E. Solodovnik, "Synchronous Machine with Two Damper Windings: Phase-domain Model," *VTB 2003 Documentaton (PDF in model's help)*, Columbia, SC: University of South Carolina, 2003.

[143] Z. Wu, "3 Phase Synchronous Machine," *VTB 2003 Documentaton (PDF in model's help)*, Columbia, SC: University of South Carolina, 2003.

[144] E. Solodovnik and R. A. Dougal, "Symbolically Assisted Method for Phase-domain Modelling of a Synchronous Machine," *15th IASTED International Conference Modelling and Simulation*, 2004, pp. 113–118.

[145] I. Boldea, *The Electric Generators Handbook. Synchronous Generators*. Boca Raton, FL: CRC/Taylor & Francis, 2006.

[146] IEEE Committee Report, "Computer Representation of Excitation Systems," *IEEE Transaction Power Apparatus and Systems*, 1967, vol. PAS-87, pp. 1460–1464.

[147] X. Cao, A. Kurita, Y. Tada, and H. Mitsuma, "Suppression of Numerical Oscillation Caused by the EMTP-TACS Interface Using Filter Interposition" *IEEE Transaction Power Delivery*, 1996, vol. 11, pp. 2049–2055.

[148] H. R. Martens and D. R. Allen, *Introduction to System Theory*. New York, NY: Charles E. Merrill Publishing Company, 1969.

[149] F. M. Uriarte, "A Tensor Approach to the Mesh Resistance Matrix," *IEEE Transaction Power Systems*, Nov. 2011, vol. 26, pp. 1989–1997.

[150] K. Thulasiraman, "Graph Theory," *The Circuits and Filters Handbook*. 2nd edn., W.-K. Chen, Ed., CRC Press, 2003, Chapter 7.

[151] I. Vago, *Graph Theory: Application to the Calculation of Electrical Networks*. New York, NY: Elsevier Science Pub. Co., 1985.

[152] J. Schutt-Aine, "Latency Insertion Method (LIM) for the Fast Transient Simulation of Large Networks," *IEEE Transaction Circuits and Systems I: Fundamental Theory and Applications*, 2001, vol. 48, pp. 81–89.

[153] S. Esmaeili and S. M. Kouhsari, "A Distributed Simulation Based Approach for Detailed and Decentralized Power System Transient Stability Analysis," *Electric Power Systems Research*, 2007, vol. 77, pp. 673–684.

[154] J. A. Hollman and J. R. Marti, "Real Time Network Simulation with PC-cluster," *IEEE Transaction Power Systems*, 2003, vol. 18, pp. 563–569.

[155] S. Jiwu, X. Wei, and Z. Weimin, "A Parallel Transient Stability Simulation for Power Systems," *IEEE Transaction Power Systems*, 2005, vol. 20, p. 1709.

[156] J. R. Marti and L. R. Linares, "Real-Time EMTP-based Transients Simulation," *IEEE Transaction Power Systems*, 1993, vol. PWRS-9, pp. 1309–1317.

[157] T. Noda and S. Sasaki, "Algorithms for Distributed Computation of Electromagnetic Transients toward PC Cluster Based Real-time Simulations," *International Conference on Power System Transients*, New Orleans, LA, 2003.

[158] Y. Xie, G. Seenumani, J. Sun, Y. Liu, *et al.*, "A PC-cluster Based Real-time Simulator for All-electric Ship Integrated Power Systems Analysis and Optimization," *Electric Ship Technologies Symposium (ESTS)*, Arlington, VA, May 22–23, 2007.

[159] K. Strunz and E. Carlson, "Nested Fast and Simultaneous Solution for Time-domain Simulation of Integrative Power-electric and Electronic Systems," *IEEE Transaction Power Delivery*, 2007, vol. 22, p. 277.

[160] P. Zhang, J. R. Marti, and H. W. Dommel, "Network Partitioning for Real-time Power System Simulation," *International Conference on Power System Transients*, Montreal, Canada, 2005, pp. 1–6.

[161] J. R. Marti, L. R. Linares, J. Calviño, H. W. Dommel, *et al.*, "OVNI: An Object Approach to Real-time Power System Simulators," *International Conference on Power System Technology (Powercon'98)*, Beijing, China, 1998.

[162] A. Brameller, M. N. John, and M. R. Scott, *Practical Diakoptics for Electrical Networks*. London: Chapman & Hall, 1969.

[163] H. H. Happ, *Gabriel Kron and Systems Theory*. Schenectady, NY: Union College Press, 1973.

[164] H. H. Happ, "Diakoptics: The Solution of System Problems by Tearing," *Proceedings of the IEEE*, 1974, vol. 62, pp. 930–940.

[165] T. Watanabe, Y. Tanji, H. Kubota, and H. Asai, "Fast Transient Simulation of Power Distribution Networks Containing Dispersion Based on Parallel-distributed Leapfrog Algorithm," *IEICE Transaction Fundamentals*, 2007, vol. E90, pp. 388–397.

[166] K. K. C. Yu and N. R. Watson, "A Comparison of Transient Simulation with EMTDC and State Space Diakoptical Segregation Methodology," *International Conference on Power System Transients*, Montreal, Canada, Jun. 19–23, 2005.

[167] L. Bergeron, *Water Hammer in Hydraulics and Wave Surges in Electricity*. New York: John Wiley & Sons, 1961.

[168] P. T. Norton, P. Deverill, P. Casson, M. Wood, *et al.*, "The Reduction of Simulation Software Execution Time for Models of Integrated Electric Propulsion Systems through Partitioning and Distribution," *Electric Ship Technologies Symposium (ESTS)*, Arlington, VA, May 22–23, 2007, pp. 53–59.

[169] Y. Zhang, R. Dougal, B. Langland, J. Shi, *et al.*, "Method for Partitioning Large System Models When Using Latency Insertion Method to Speed Network Solution," *Grand Challenges in Modeling and Simulation*, Istanbul, Turkey, 2009.

[170] C. Dufour, J.-N. Paquin, V. Lapointe, J. Bélanger, *et al.*, "PC-cluster-based Real-time Simulation of an 8-synchronous Machine Network with HVDC Link Using RT-LAB and Test Drive," *International Conference on Power Systems Transients*, 2007.

[171] M. Kleinberg, K. Miu, and C. Nwankpa, "A Study of Distribution Power Flow Analysis Using Physically Distributed Processors," *PSCE*, Atlanta, GA, Oct. 2006. Available [online] http://www.ieee.org/portal/cms_docs_pes/pes/subpages/meetings-folder/PSCE/PSCE06/panel20/Panel-20-4_A_Study_of_Distribution_Power_Flow_Analysis.pdf.

[172] M. Kleinberg, K. Miu, and C. Nwankpa, "Distributed Multi-phase Distribution Power Flow: Modeling, Solution Algorithm, and Simulation Results," *Transactions of the Society for Modeling & Simulation International*, 2008, vol. 84, pp. 403–412.

[173] Q. Huang, J. Wu, J. L. Bastos, and N. N. Schulz, "Distributed Simulation Applied to Shipboard Power Systems," *Electric Ship Technologies Symposium, 2007. ESTS '07. IEEE*, 2007, pp. 498–503.

[174] A. Benigni, P. Bientinesi, and A. Monti, "Benchmarking Different Direct Solution Methods for Large Power System Simulation," *Grand Challenges in Modeling and Simulation*, Ottawa, Canada, 2010.

[175] F. M. Uriarte, R. E. Hebner, and A. L. Gattozzi, "Accelerating the Simulation of Shipboard Power Systems," *Grand Challenges in Modeling & Simulation*, The Hague, Netherlands, Jun. 27–30, 2011.

[176] F. M. Uriarte and R. Hebner, "Development of a Multicore Power System Simulator for Ship Systems," *Electric Ship Technologies Symposium*, Alexandria, VA, Apr. 10–13, 2011, pp. 106–110.

[177] F. M. Uriarte, "Multicore Simulation of an Ungrounded Power System," *IET Electrical Systems in Transportation*, Mar. 2011, vol. 1, pp. 31–40.

[178] F. M. Uriarte, "A Partitioning Approach for Parallel Simulation of AC-radial Shipboard Power Systems," *Electrical and Computer Eng.* vol. PhD, College Station: Texas A&M University, 2010, p. 287.

[179] IEEE Power Systems Engineering Committee, "Parallel Processing in Power Systems Computation," *IEEE Transaction Power Systems*, 1992, vol. 7, pp. 629–38.

[180] A. Kalantari and S. M. Kouhsari, "An Exact Piecewise Method for Fault Studies in Interconnected Networks," *International Journal of Electrical Power & Energy Systems*, 2008, vol. 30, pp. 216–225.

[181] Z. Quming, S. Kai, K. Mohanram, and D. C. Sorensen, "Large Power Grid Analysis Using Domain Decomposition," *Design, Automation and Test in Europe, 2006. DATE '06. Proceedings*, 2006, pp. 1–6.

[182] C. Yue, X. Zhou, and R. Li, "Node-splitting Approach Used for Network Partition and Parallel Processing in Electromagnetic Transient Smiulation," *International Conference on Power System Technology*, Singapore, 2004.

[183] K. W. Chan, R. C. Dai, and C. H. Cheung, "A Coarse Grain Parallel Solution Method for Solving Large Sets of Power System Network Equations," *International Conference on Power System Technology (PowerCon '02)*, 2002, pp. 2640–2644.

[184] G. Kron, *Diakoptics: The Piecewise Solution of Large-Scale Systems*. London: MacDonald & Co., 1963.

[185] W. F. Tinney and J. W. Walker, "Direct Solutions of Sparse Network Equations by Optimally Ordered Triangular Factorization," *Proceedings of the IEEE*, 1967, vol. 55, pp. 1801–1809.

[186] A. Klos, "What Is Diakoptics?," *International Journal of Electrical Power & Energy Systems*, 1982, vol. 4, pp. 192–195.

[187] I. S. Duff, A. M. Erisman, and J. K. Reid, *Direct Methods for Sparse Matrices*. Oxford: Oxford University Press, 1986.

[188] H. V. Henderson and S. R. Searle, "On Deriving the Inverse of a Sum of Matrices," *SIAM Review*, 1981, vol. 23, pp. 53–60.

[189] S. Toub. (2010). *Patterns of Parallel Programming—Understanding and Applying Parallel Patterns with the .NET Framework 4 and Visual C#* [Online]. Available: http://www.microsoft.com/en-us/download/details.aspx?id=19222.

[190] G. C. Hillar, *Professional Parallel Programming with C#*, 1st edn. Indianapolis, IN: Wiley Pub., Inc., 2010.

[191] F. M. Uriarte, "On Kron's diakoptics," *Electric Power System Research*, Jul. 2012, vol. 88, pp. 146–150.

[192] G. Karypis and V. Kumar. (1998). *hMETIS: A Hypergraph Partitioning Package Version 1.5.3.* Minneapolis: Department of Computer Science & Engineering, University of Minnesota [Online]. Available: http://glaros.dtc.umn.edu/gkhome/metis/hmetis/download.

[193] T. A. Davis, *Direct Methods for Sparse Linear Systems*. Philadelphia: SIAM, 2006.

[194] IEEE Std 315-1975 (Reaffirmed 1993), *Graphic Symbols for Electrical and Electronics Diagrams*.

[195] J. Richter, *CLR via C#*, 3rd edn. Redmond, WA: Microsoft Press, 2010.

[196] L. Chua and C. Li-Kuan, "On Optimally Sparse Cycle and Coboundary Basis for a Linear Graph," *IEEE Transaction Circuit Theory*, Sep. 1973, vol. CT-20, pp. 495–503.

[197] R. Hebner, J. Herbst, and A. Gattozzi, "Intelligent Microgrid Demonstrator," *ASNE Electric Machines Technology Symposium*, Philadelphia, PA, May 19–20, 2010.

[198] A. L. Gattozzi, F. M. Uriarte, J. Herbst, and R. E. Hebner, "Analytical Description of a Series Fault on a DC Bus," *2012 IEEE Innovative Smart Grid Technologies Conference*, Washington, DC, Jan. 16–19, 2012.

[199] J. D. Herbst, Angelo L. Gattozzi, A. Ouroua, and F. M. Uriarte, "Flexible Test Bed for MVDC and HFAC Electric Ship Power System Architectures for Navy Ships," *Electric Ship Technologies Symposium*, Alexandria, VA, Apr. 10–13, 2011.

[200] A. G. J. Herbst, F. Uriarte, M. Steurer, C. Edrington, *et al.*, "The Role of Component and Subsystem Testing in Early Stage Design," *2012 ESRDC 10th Anniversary Meeting*, Austin, TX, Jun. 4–6, 2012.

[201] F. M. Uriarte, A. L. Gattozzi, H. Estes, T. Hotz, *et al.*, "Development of a Series Fault Model for DC Microgrids," *2012 IEEE Innovative Smart Grid Technologies Conference*, Washington, DC, Jan. 16–19, 2012.

[202] F. M. Uriarte, A. L. Gattozzi, J. Herbst, H. Estes, *et al.*, "A DC Arc Model for Series Faults in Low Voltage Microgrids," *IEEE Transaction Smart Grid*, Dec. 2012, vol. 3, pp. 2063–2070.

[203] J. Langston, K. Schoder, I. Leonard, and M. Steurer, "Considerations for Verification and Validation of Electromagnetic Transient Simulation Models of Shipboard Power Systems," *2012 Summer Simulation Multiconference*, Genoa, Italy, Jul. 8–11, 2012.

[204] J. Langston, K. Schoder, M. Steurer, O. Faruque, *et al.*, "Power Hardware-in-the-Loop Testing of a 500 kW Photovoltaic Array Inverter," *Submitted to IEEE IECON Conference*, Montreal, Canada, Oct. 25–28, 2012.

[205] S. D. Sudhoff, S. Pekarek, B. Kuhn, S. Glover, *et al.*, "Naval Combat Survivability Testbeds for Investigation of Issues in Shipboard Power Electronics Based Power and Propulsion Systems," *Power Engineering Society General Meeting*, 2002, p. 347.

[206] H. Ding, A. A. Elkeib, and R. Smith, "Optimal Clustering of Power Networks Using Genetic Algorithms," *Electric Power Systems Research*, Sep. 1994, vol. 30, pp. 209–214.

[207] G. Karypis, R. Aggarwal, V. Kumar, and S. Shekhar, "Multilevel Hypergraph Partitioning: Applications in VLSI Domain," *Design and Automation Conference*, Minneapolis, 1997, pp. 526–529.

[208] B. Y. Wu and K.-M. Chao, *Spanning Trees and Optimization Problems*. Boca Raton, FL: Chapman & Hall/CRC, 2004.

[209] S. Pemmaraju and S. Skiena, *Computational Discrete Mathematics: Combinatorics and Graph Theory with Mathematica.* Cambridge, MA: Cambridge University Press, 2003.

[210] B. Wagner, *More Effective C#: 50 Specific Ways to Improve Your C#*. Upper Saddle River, NJ: Addison-Wesley, 2009.

[211] B. W. Kernighan and S. Lin, "An Efficient Heuristic Procedure for Partitioning Graphs," *Bell Laboratories Record*, 1970, vol. 49, pp. 291–307.

[212] J. Albahari and B. Albahari, *C# 3.0 in a Nutshell*, 5th edn. Cambridge: O'Reilly, 2007.

[213] G. M. Amdahl, "Validity of the Single Processor Approach to Achieving Large-Scale Computing Capabilities," *AFIPS*, Washington, DC, 1967, pp. 483–485.

[214] W. D. Passos, *Numerical Methods, Algorithms, and Tools in C#*. Boca Raton, FL: CRC Press, 2010.

[215] J. Xu, *Practical Numerical Methods with C#*. Phoenix, AR: UniCAD, 2008.

[216] J. Duffy, *Concurrent Programming on Windows*. Upper Saddle River, NJ: Addison-Wesley, 2009.

[217] G. M. Hall, *Pro WPF and Silverlight MVVM: Effective Application Development with Model-View-ViewModel*. New York, NY: Apress: Distributed to the book trade worldwide by Springer Science+Business Media, 2010.

[218] CenterSpace Software. (2012). *NMath User Guide* [Online]. Available: http://www.centerspace.net/resources/documentation/.

[219] T. Noda, "Object Oriented Design of a Transient Analysis Program," *International Conference on Power Systems Transients (IPST'07)*, Lyon, France, 2007.

[220] B. Hakavik and A. T. Holen, "Power System Modelling and Sparse Matrix Operations Using Object-Oriented Programming," *IEEE Transaction Power Systems*, 1994, vol. 9, pp. 1045–1051.

[221] E. Omara. (2010). *Performance Characteristics of New Synchronization Primitives in the .NET Framework 4* [Online]. Available: http://download.microsoft.com/download/B/C/F/BCFD4868-1354-45E3-B71B-B851CD78733D/PerformanceCharacteristicsOfSyncPrimitives.pdf.

[222] A. Y. Zomaya, *Parallel and Distributed Computing Handbook*. New York: McGraw-Hill, 1996.

[223] A. G. Exposito, A. Abur, and E. R. Ramos, "On the Use of Loop Equations in Power System Analysis," *IEEE International Symposium on Circuits and Systems*, Seattle, WA, 1995, pp. 1504–1507.

[224] K. Wang, "Piecewise Method for Large-Scale Electrical Networks," *Circuits and Systems, IEEE Transaction [Legacy, pre-1988]*, 1973, vol. 20, pp. 255–258.

[225] R. E. Hebner, S. Dale, R. Dougal, S. Sudhoff, *et al.*, "The U.S. ESRDC Advances Power System Research for Shipboard Systems," *43rd International Universities Power Engineering Conference*, Padova, Italy, Sep. 1–4, 2008.

[226] J. H. Beno, R. E. Hebner, and A. Ouroua, "High-frequency Power Generation and Distribution in Multi-megawatt Power Systems," *Electric Ship Technologies Symposium*, Alexandria, VA, Apr. 10–13, 2011.

[227] R. E. Hebner, J. H. Beno, and A. Ouroua, "Dynamic Simulations of a Large high-frequency Power System," *Grand Challenges in Modeling & Simulation (GCMS 2011)*, The Hague, Netherlands, Jul. 27–30, 2011.

[228] S. Chowdhury, S. P. Chowdhury, and P. Crossley, *Microgrids and Active Distribution Networks*. IET Renewable Energy Series, 2009.

[229] E. L. Zivi, "Integrated Shipboard Power and Automation Control Challenge Problem," *Power Engineering Society General Meeting*, 2002, vol. 1, pp. 325–330.

[230] E. L. Zivi and T. J. McCoy, "Control of a Shipboard Integrated Power System," *33rd Annual Conference on Information Sciences and Systems*, 1999. [Online] http://www.usna.edu/EPNES/Zivi_McCoy_CISS99.pdf.

[231] The MathWorks, Inc. (2010). *SimPowerSystems 5 User's Guide* [Online]. Available: http://www.mathworks.com/help/toolbox/physmod/powersys/.

[232] A. Kwasinski and C. N. Onwuchekwa, "Dynamic Behavior and Stabilization of DC Micro-grids with Instantaneous Constant-Power Loads," *IEEE Transaction Power Electronics*, 2011, vol. 3, pp. 822–834.

[233] J. Ledin, *Simulation Engineering: [Build Better Embedded Systems Faster]*. Lawrence, Kan.: CMP Books, 2001.

[234] IEEE Std 45-1998, *Recommended Practice for Electric Installation on Shipboard*, p. i, 1998. doi: 10.1109/IEEESTD.1998.91149.

[235] M. MacDonald, *Pro WPF in C# 2010: Windows Presentation Foundation in .NET 4*. New York, NY: Apress: Distributed to the book trade worldwide by Springer-Verlag, 2010.

[236] M. Dalal and A. Ghoda, *XAML Developer Reference*, Sebastopol, CA: Microsoft Press, 2011.

Index

www.ingramcontent.com/pod-product-compliance
Lightning Source LLC
LaVergne TN
LVHW011915200926
848484LV00002B/632

* 9 7 8 1 8 4 9 1 9 5 7 2 0 *